THE OXFORD ENGINEERING SCIENCE SERIES

GENERAL EDITORS

A. ACRIVOS, J.M. BRADY, F.W. CRAWFORD,
A.L. CULLEN, L.C. WOODS, C.P. WROTH

THE OXFORD ENGINEERING SCIENCE SERIES

Image Restoration and Reconstruction

R.H.T. BATES

*Department of Electrical and Electronic Engineering,
University of Canterbury, New Zealand*

and

M.J. McDONNELL

*Division of Information Technology,
DSIR, New Zealand*

CLARENDON PRESS · OXFORD
1986

Oxford University Press, Walton Street, Oxford OX2 6DP
Oxford New York Toronto
Delhi Bombay Calcutta Madras Karachi
Kuala Lumpur Singapore Hong Kong Tokyo
Nairobi Dar es Salaam Cape Town
Melbourne Auckland
and associated companies in
Beirut Berlin Ibadan Nicosia

Oxford is a trade mark of Oxford University Press

Published in the United States
by Oxford University Press, New York

British Library Cataloguing in Publication Data

Bates, R.H.T.
Image restoration and reconstruction. – (The
Oxford engineering science series; 16)
1. Image processing
I. Title II. McDonnell, M.J.
621.38'0414   TA1632
ISBN 0-19-856176-8

Library of Congress Cataloging in Publication Data

Bates, R.H.T.
Image restoration and reconstruction.
(The Oxford engineering science series; 16)
Bibliography: p.
Includes index.
1. Image processing. I. McDonnell, M.J. II. Title.
III. Series.
TA1632.B36   1986   621.36'7   85-13711
ISBN 0-19-856176-8

Set and Printed in Northern Ireland by
The Universities Press (Belfast) Ltd.

PREFACE

Image processing is a new science, even though its literature is already huge. It is still too amorphous to permit a definitive treatise to be attempted. What we have produced is a compendium, touching all aspects of the subject with which we are familiar and dwelling upon those we have found to be most useful in scientific contexts and in practical engineering applications. While the majority of the subject matter can be considered soundly established, some of it is tentative. We make no apology for this because of the amount of recently developed material treated here. Advances in engineering and science often come from chance associations of ideas. We have therefore thought it proper to mention our insights, convictions, prejudices (call them what you will!) wherever it seems appropriate. Accordingly, the solid results are interspersed with asides and exhortations which we hope may inspire people to have a crack at the many challenging problems that abound in this science.

Here in New Zealand, pleasantly perched on the very edge of the world, we have found ourselves drawn towards six particular areas of image processing. These are various manifestations of holography, aspects of imaging of importance in medical contexts, astronomical data reduction, restoration of blurred photographs, the very new technique which has come to be called speckle imaging, and the massaging and presentation of satellite imagery. There is, consequently, repeated reference throughout this book to applications in these areas. Furthermore, because we are laying claim to have written a compendium, we have tried to make at least brief mention of all of the scientific and technical fields in which useful advances have stemmed from applying image processing principles and techniques.

While this book is certainly classifiable as a research monograph, it can also be used as a graduate text. It has become customary to intersperse such volumes with 'homework problems'. We are none too sympathetic to this practice in general (except for elementary treatments of basic subjects) and we think it is completely inappropriate for providing the kind of insight which is the essence of image processing. The latter is something of an art as well as a science, relying on intuition based on objectively assessable logical deduction. We hold that the only effective way for a student to acquire what is important of any approach to image processing is actually to implement it, more or less immediately after studying its theoretical background and implications.

Accordingly, we have appended to each chapter in this monograph an extended worked example, relating to particular features of the material covered in the chapter. A suitable student exercise would be either to generate an example of the same kind as one of those presented herein or, more demandingly, to devise a similar example based on an alternative feature of the particular chapter being studied. The software needed for the exercise could be acquired from one of the sources quoted in this book, or be generated in the instructor's laboratory (most probably by senior research students or post-doctoral associates), or even be written by the graduate students themselves.

The material covered by virtually every other graduate-level text on image processing is restricted to what we deal with in a single chapter or, in some instances, in one section of a chapter. This is another sense in which this book can be regarded as a compendium. The subject matter of each of our chapters is summarized in introductory comments which quote the relevant literature, including the available graduate texts. We feel that a graduate student need not have to look any further than these introductory comments to find sources for 'further reading'.

Much of this book is a distillate of the labours of students, colleagues (abroad, as well as in New Zealand at the University of Canterbury and at the Physics and Engineering Laboratory of the Department of Scientific and Industrial Research—PEL/DSIR) and visitors, all of whom we hope will feel adequately acknowledged by their listings in the References. Two who have contributed signally and whom we quote in several of the papers listed in the References, are George L. Berzins (Los Alamos Scientific Laboratory) and R.M. (Bob) Hodgson (Canterbury). John H. Andreae's (Canterbury) detailed criticisms of an early draft of the book helped to correct some of our expositional deficiencies. We are grateful for the unwavering support of Mike A. Collins and Peter J. Ellis of PEL/DSIR. Without Alec L. Cullen's (University College London) initial encouragement and subsequent exhortations, this book would not have been completed. Particular thanks must go to the 1984 postgraduate 'information processing' class at the University of Canterbury whose course assignment was to generate the figures for the worked examples for Chapters IV, V, and VI.

Canterbury,   R.H.T.B.
New Zealand,   M.J.McD.
November 1984

For

D^2EG^2JP

and

a man of the skies: Edward Roch McDonnell

CONTENTS

I

SETTING THE SCENE

The purpose of this book is to present the theoretical bases of useful restoration and reconstruction techniques, and to discuss their implementation in the framework of a general image processing system. Everything included here relates directly to experimental methods and/or data reduction procedures that actually work. Specimen processed images, illustrating salient aspects of the text, are included in extended worked examples, one of which serves as a conclusion to each chapter.

The views expressed have been strongly influenced or biased – as is inevitable – by the authors' collective experience in image processing at the University of Canterbury and at the Physics and Engineering Laboratory (PEL), as well as by visits to large and small laboratories around the world. The development of the image processing laboratories at Canterbury and more recently at PEL over the past fifteen years or so has been detached, for largely geographical reasons, from the mainstream of image processing research in the USA. This has led to a somewhat independent approach to a variety of image processing problems. Whenever appropriate, we explain the differences between our approaches and those of other groups and, with the benefit of hindsight, review the ways in which our ideas have evolved.

As far as possible, a narrative style is adopted. This means that in any one section there is minimal reliance on material appearing in later sections. Naturally, there are repeated references to previous sections. Before getting down to the real business of this book, the usual preliminaries must, of course, be disposed of. This is done in the first (i.e. §1) of the sections into which the book is divided.

1 Necessary preliminaries

This book is written in sections which run consecutively from §1 to §54. The sections are arranged in groups covering more or less specific topics. These groups are identified in the Table of Contents by the titles of the chapters. Various comments are inserted at the start of each chapter, indicating the overall drift of the subject matter, listing the major applications and summarizing the worked example with which chosen aspects of the text are illustrated. These introductory comments also contain references relevant to the chapter in general and each individual section in particular. References are avoided within the sections themselves, which we believe contributes to the readability of the book.

This chapter provides an overview of image science and introduces ideas which are needed in the body of the work. In §2 image science is defined and the scope of the book stated. In §3 concepts are collected which are required for categorizing blurred images (which are defined in §4) and point spread functions (which are defined in §4 and discussed further in §5).

References are quoted by inserting a name followed by a comma and (a letter and) a number inside square brackets. Multiple references under the same name are indicated by (letters and) numbers separated by commas. Multiple references under different names are separated by semicolons. For references having more than one author, the quoted name is that of the first author. For instance, [Labeyrie, 1, 2: Fienup, 4] refers to the first and second of the items under the name of Labeyrie, and the fourth item under Fienup's name – see the 'References' collected at the end of the book. They are listed alphabetically by the first authors' names. The item number, in square brackets, precedes the name – e.g. the above reference to Fienup is listed as:

[4]   Fienup, J.R., Crimmins, T.R., and Holsztynski, W. 1982 "Reconstruction of the support of an object from the support of its autocorrelation" J. Opt. Soc. Am. **72,** 610.

Books, special issues of journals and theses, which are listed separately in the References at the end of this volume, are referred to as above except that B, S, or T prefixes the number – e.g. [Andrews, B1], [Proc. IEEE, S2] or [Gullberg, T1].

The literature encompassing the subject of image science is so extensive that it would be futile to attempt a detailed survey here. This also applies to the more restricted subjects of image restoration and reconstruction. Nevertheless, we think it useful to include in the References various sources which cross-reference much of the published material related to this book. The first source is books which overview aspects of image processing. These are either texts aimed at the (post)graduate level or are collections of related papers. The second source comprises the journals and annual review volumes in which papers on image processing, and in particular image restoration and reconstruction, are most often published. A third source is a list of special issues of some of the aforementioned journals containing up-to-date information and apposite review articles. In recent years, conference proceedings have been a major means of disseminating image processing research results. Many of the conferences, whose proceedings are listed amongst the fourth source, are regular events and much of the material does not get published elsewhere. A fifth source is the series of technical reports published by many institutions with image processing laboratories. Selected reports are usually available on request. It is worth noting that such reports frequently contain

valuable practical information which is only published elsewhere in sum-
mary form if at all. This often applies to documentation of computer
programs or image processing systems. A sixth source is the list of specific
references of direct relevance to this book. Finally, we list a few theses
containing several worthwhile results, the full details of which cannot be
found anywhere else. The seven lists of references are far from exhaustive
but they do indicate where, in our experience, the image processing
literature can be located.

We emphasize that in Chapters II–IX all the quoted references are
collected in the introductory comments to the chapters. References
pertaining to §§2–5 are quoted in the two following paragraphs.

For §2, many overviews of image processing, both optical [Bracewell,
B1; Dainty, B1; Goodman, B1; Stroke, 1; Vander Lugt, 1] and digital
[Andrews, B1, B2, 3; Bernstein, B1; Castleman, B1; Hunt, 2, 4; Pratt,
B1; Rosenfeld, B1, B2] are available. See also [Rhodes, B1; Walkup,
B1]. The *Manual of Remote Sensing* [Greeves, 1] gives a particularly
broad overview of image processing and related subjects. Rosenfeld has
for some years published an annual literature review, such as [Rosenfeld,
2]. Sound introductions to rectification are given in [Bernstein, 1; Van
Wie, 1]. In addition to the above general treatments, enhancement is
overviewed in [Andrews, 2; Billingsley, 1]. Restoration is dealt with in
[Andrews, B3, 2; Huang, B1, 2; Sondhi, 1; Trussel, 1]. Good general
references for image reconstruction are [Herman, B1, B2, B3; IEEE, S2;
van Schooneveld, B1].

The notation and definitions introduced in §3 are discussed in [An-
drews, B3; Dainty, B1; Goodman, B1]. These references together with
[Huang, 2; Sondhi, 1] are relevant to §§4 and 5. Useful references for the
outline presented in §3 of image formation by an imaging instrument are
[Born, B1; Bates, 10].

The worked example (Example I) at the end of this chapter illustrates
the effects on a particular image of various kinds of degradation that
occur in important practical applications.

While details of notation and terminology are introduced whenever
appropriate and are summarized in the Glossary, it is convenient to
establish some basic conventions here. Equations are numbered consecutively within each section – e.g. (6.8) identifies the eighth equation in §6.
Any symbol, q say, which is invoked as an aid to mathematical analysis or
to represent a physical quantity is to be thought of as a *number, real* or
complex according to its definition. Consider *parameters* or *variables*, z_1,
z_2, through z_n, which are themselves numbers in the same sense that q is
a number. The notation $q = q(z_1, z_2, \ldots, z_n)$ implies that q is a *function*
of the variables – i.e. when one or more of the variables changes then q
itself varies, in general. The term 'amplitude' implies 'complex amplitude'
unless the word 'amplitude' is used in a technically conventional manner –

e.g. 'amplitude modulation'. Consequently, 'amplitude of q' is merely longhand for 'q', but such verbosity is sometimes apposite. Now consider a typical complex number W whose *real* and *imaginary parts* are U and V respectively (remember that both U and V are themselves real numbers). We adopt the following notation and terminology:

$$i = \text{the pure imaginary} = \sqrt{-1}; \tag{1.1}$$

$$W = U + i\,V = \text{complex amplitude of } W; \tag{1.2}$$

$$W^* = U - i\,V = \text{complex conjugate of } W; \tag{1.3}$$

$$|W| = (U^2 + V^2)^{\frac{1}{2}} = \text{magnitude of } W; \tag{1.4}$$

$$|W|^2 = W^*W = U^2 + V^2 = \text{intensity of } W; \tag{1.5}$$

$$\text{phase } \{W\} = \text{arc tan}(V/U) = \text{phase of } W. \tag{1.6}$$

Note that

$$W = |W| \exp(i\,\text{phase}\{W\}). \tag{1.7}$$

Abstract notation is abjured throughout most of this book, but it is invoked at times to ensure that some of the more complicated analyses can be concise. Use is made of the notation:

$$\in \text{ denotes membership (in a set or class);} \tag{1.8}$$

$$\cup \text{ denotes union (of sets);} \tag{1.9}$$

$$\cap \text{ denotes intersection (of sets);} \tag{1.10}$$

$$\varnothing \text{ denotes the null set;} \tag{1.11}$$

$$\{t_l; l = l_a, l_a + 1, \ldots, l_b\} \text{ denote the set of quantities } t_l, \text{ each of}$$
which is indexed by the integer l that runs from l_a to l_b. $\tag{1.12}$

The abbreviations LHS and RHS stand for '(the) left and right, respectively, hand sides (of)'.

2 Image science

Since our eyes gather more information than any of our other sense organs, and since the visual cortex seems to dominate the brain, images play a central role in our existence. We perceive space in depth, so that the images formed by the 'mind's eye' are (apparently) three-dimensional. For obvious technical reasons, but also because of deficiencies in our present understanding of the nature of perception, there has been little useful manipulation of three-dimensional images in scientific and technological applications. What catches the popular fancy about holography is its capability for reproducing the juxtaposition of objects in space. However, we do not know of any technical use for this fascinating property (there is no sign that holographic movies are going to be a practical proposition in even the moderately near future). Even in holo-graphic interferometry, which certainly is useful, information is displayed two-dimensionally although it is the spatial displacement of objects that is being recorded.

Only two-dimensional images are discussed in this book.

It is useful to define the term *image science* to encompass the recogni-tion of patterns in, the coding of, and the processing of, two-dimensional images. *Pattern recognition* and *image coding* are not treated here.

Information, in particular that present in images, can suffer many kinds of distortion. The word 'distortion' is used descriptively throughout this book. It is also given precise meanings on occasion, as in §3 for instance. This need not lead to any confusion, because the word has numerous (well accepted) connotations in the engineering literature. We feel that the scientifically and technically trained reader should have no difficulty distinguishing our general and specific invocations of the useful portman-teau term 'distortion'.

Image processing is concerned with the manipulation of data which are inherently two-dimensional. It often involves removing as much distortion and extraneous contamination as possible from data that are to be displayed pictorially, in order that their valuable information be efficiently coded (for storage or transmission) or the patterns implicit therein can be more easily recognized. The processing can be subdivided into rectifica-tion, enhancement, restoration, and reconstruction. *Image rectification* is concerned with spatial transformations that can remove geometric distor-tions or permit images to be properly registered with respect to each other, for instance in photographic cartography when wide-angle lenses and moving platforms (e.g. aircraft or artificial satellites) are used. *Image enhancement* is concerned with improved presentation (e.g. by noise reduction, compensation for non-linearities of the recording medium, contrast adjustment, edge sharpening) of images whose essential informa-tion is clouded but is nevertheless visually apparent before processing. Enhancement procedures have been perfected to an extraordinary de-gree, especially for satellite imagery (of the Earth and the planets).

This book is primarily concerned with *image restoration* and *image reconstruction*. The former term is sometimes equated to image enhance-ment, which is a pity because it is useful to make a clear distinction between enhancement and restoration. Pictorial information is often made unrecognizable by unwanted distortion, the exact nature of which is unknown *a priori*. Image restoration involves estimating the parameters of the distortion and using them to refurbish the original information. Image reconstruction is concerned with recovering detail in severely blurred images, the causes of whose imperfections are known *a priori*.

Many practical applications of image restoration and reconstruction are quoted throughout this book. All the methods of recovering information are based on the simple 'convolutional model' introduced in §4. The physical basis of the model is discussed in §3.

It is to be understood that the techniques analysed in this book are implemented in digital computers. In principle, many of the techniques could be carried out on coherent and/or incoherent optical benches interfaced to suitable electronic imaging devices. There are enormous difficulties, however, in the way of realizing useful computational accuracies. This is unfortunate because 'optical computing' offers the ultimate in speed (i.e. signals travelling at the velocity of light) and permits 'parallel processing' – e.g. a simple lens 'computes' the whole of a two-dimensional Fourier transform in nanoseconds. Nevertheless, it has not yet proved convenient to carry out other than preliminary computational procedures – e.g. averaging large sequences of two-dimensional data – on optical benches. It is worth distinguishing between 'optical bench computers' and 'electro-optic computers', the latter being composed of digital elements activated by light signals (these may well predominate in the 'guts' of future computers).

To be of practical use, restoration and reconstruction techniques need to be incorporated into a general purpose image processing system. These techniques then become part of a library of compatible image processing tools. The completion of an image processing task involves the selection and utilization of the appropriate tools. A restoration or reconstruction task rarely involves only a single algorithm or technique. It usually consists of several stages and is dependent on integrating a number of enhancement and rectification techniques into an overall systems plan.

The design and contents of an image processing system are thus of direct relevance to our main subject, which is the restoration and reconstruction of images. For this reason we discuss image processing system design in Chapter VII, and program or algorithm categories in Chapter VIII. Our intention is to emphasize that complicated image processing tasks can be carried out successfully with comparatively simple tools, provided they are used in appropriate combinations. Further comments on the purposes and the contents of other chapters are made in §§3 and 5.

3 Image distortion and how it arises

Consider quantifiable information, represented by the symbol f, existing in a plane which is called *image space*. An arbitrary point in this space is identified by the position vector $\mathbf{x}$. The functional dependence of f on $\mathbf{x}$ is written as

$$f = f(\mathbf{x}). \tag{3.1}$$

The functional dependences of all quantities existing in image space are written similarly.

Now suppose the information f is blurred by some *time-invariant distortion* represented by the general function $h = h(\mathbf{x}, \boldsymbol{\xi})$, where the functional dependence implies that f at $\boldsymbol{\xi}$ is spread out over image space according to the formula $h(\mathbf{x}, \boldsymbol{\xi})$. Only linear distortion is envisaged, so that the blurred information, denoted by $b = b(\mathbf{x})$, can be expressed quite generally as

$$b(\mathbf{x}) = \int\limits_{-\infty}^{\infty}\!\!\int f(\boldsymbol{\xi}) h(\mathbf{x}, \boldsymbol{\xi})\, d\sigma(\boldsymbol{\xi}), \qquad (3.2)$$

where the notation $d\sigma(\mathbf{x})$ denotes the element of area centred on the point in image space identified by the position vector $\mathbf{x}$. There are two integral signs because image space is two-dimensional. The infinite limits merely imply that all of the information is being considered.

In those situations for which the blurring is of such a general kind that no specialization or simplification of RHS (3.2) is permissible, it is only rarely that f can be successfully recovered from b. In fact, widely applicable restoration and reconstruction procedures have only been developed for distortions either of the kind called *point spread invariant*, for which the spreading is the same for all points $\mathbf{x}$, or of the sort that can be made effectively point spread invariant by one of two methods. The first of these methods involves using image rectification to transform a point spread variant image into a point spread invariant one. In the second method, a point spread variant image is segmented into a number of sub-images, each of which may be treated as point spread invariant. Both of these methods are discussed in §15.

Point spread invariance implies that the general function representing the distortion reduces to the specialized form

$$h(\mathbf{x}, \boldsymbol{\xi}) = h(\mathbf{x} - \boldsymbol{\xi}). \qquad (3.3)$$

When (3.3) is substituted into (3.2), the right-hand side of the latter assumes the form of what is called a *convolution integral*. The symbol $\bigodot$ is used throughout this book to denote the convolution operation. It provides a convenient shorthand for (3.2), when (3.3) applies:

$$b(\mathbf{x}) = f(\mathbf{x}) \bigodot h(\mathbf{x}). \qquad (3.4)$$

Even when the distortion is point spread invariant, there are no *a priori* constraints on the form of $h(\mathbf{x})$. However, certain forms often arise in practice. Four are listed in Table I.1 (see Example I at the end of this chapter). *Linear blur* is so called because it occurs when a photographed object moves in a straight line during the exposure (equivalently, the camera may swing involuntarily while the object remains stationary). The inset profile (see Table I.1) characterizes the motion of the object during the exposure (the sharp cut-off of the profile at each end indicates that the camera shutter opens and closes very rapidly). When the height of the profile is constant throughout the exposure, the distortion is said to be *uniform linear blur*. Another common cause of photographic blurring is a camera being *out-of-focus*, in which case $h(\mathbf{x})$ has a form very close to that of a circular disc (this can be deduced from simple geometrical optics because it describes the intersection with the focal plane of the cone of rays, emitted from a far-off point in the photographic field, that would converge to a point in that plane which would be the focal plane if the camera was in focus). When an object is viewed through a turbulent medium with a high-resolution imaging instrument, the blurring of a short exposure (i.e. rapid enough that the medium remains effectively still during it) is often well represented by an $h(\mathbf{x})$ having the form of a distribution of *random impulses*. On the other hand, the form of $h(\mathbf{x})$ tends to be *Gaussian* for a long exposure of an object viewed through a turbulent medium. Although there are many ways in which these four forms for $h(\mathbf{x})$ can arise, the above are perhaps the most usual.

We now outline image formation by an instrument viewing through a distorting medium. Our purpose is merely to reach (3.10) as expeditiously as possible. Detailed treatments can be found in the references quoted in §1. An arbitrary point in a plane on which radiation is falling is identified by the position vector $\mathbf{z}$. If the field existing at each point $\mathbf{z}$ is merely a (amplitude and phase) modulated version of the field which would exist at $\mathbf{z}$ in the absence of distortion, the distortion is then said to be *isoplanatic*. Although simple, this concept is of considerable practical importance, so that it is worth looking at it from another point of view. Consider the ray which emanates from an arbitrary point ζ within the source of the radiation and arrives at the point $\mathbf{z}$. Denote the attenuation and delay, introduced into this ray by the distortion, by the magnitude and phase, respectively, of the complex number $\mathfrak{D} = \mathfrak{D}(\zeta, \mathbf{z})$. The isoplanatic condition applies when $\mathfrak{D}$ is independent of ζ, i.e.

$$\mathfrak{D} = \mathfrak{D}(\mathbf{z}) \qquad \text{implies isoplanatism.} \qquad (3.5)$$

It is worth emphasizing that $\mathfrak{D}$ varies markedly with $\mathbf{z}$ in many practical situations involving isoplanatic distortion. The larger the physical dimensions of the source of radiation, the less likely is (3.5) to hold for any particular distorting medium. Also, for (3.5) to remain valid, the sizes of the 'cells' or 'blobs' of matter that introduce the distortion must exceed some minimum value set by the geometry of the source and the medium. This leads to the notion of an *isoplanatic patch*, which is the largest 'effective size' of the aforementioned source. It is useful to express the dimensions of the isoplanatic patch as angles. If, at all points $\mathbf{z}$, the source of the radiation subtends angles less than the dimensions of the isoplanatic patch then the distortion is isoplanatic.

At any instant t, denote the radiation field at $\mathbf{z}$ by $G(\mathbf{z}, t)$. Write its two-dimensional spatial Fourier transform (refer to §6) as $g(\mathbf{x}, t)$. Suppose that **z** lies in the pupil plane (i.e. aperture) of an imaging instrument (e.g. telescope, ultrasonic transducer, radio antenna). If the focal surface of the instrument is identified with the image space introduced in the first paragraph of this section then $g(\mathbf{x}, t)$ is the 'instantaneous image' formed by the instrument.

It is appropriate here to introduce the concept of *analytic signal*. Its defining characteristic is that it possesses no negative temporal frequencies. It is necessarily complex, with its imaginary part being the *Hilbert transform* of its real part. The latter part is conveniently taken to represent what can actually be measured in the real-world. The simplest instance of an analytic signal is the familiar $\exp(i[\omega t + \phi])$, where ω is the constant *angular frequency* and ϕ is a constant phase factor. The real signal corresponding to this is $\cos(\omega t + \phi)$. Analytic signals only make brief appearances in this book, and so we do not think it worthwhile to give them more than passing mention (the interested reader can find satisfactory treatments of them in references quoted in §1). However, whenever we introduce a signal-like quantity depending explicitly upon t, it is to be understood as being complex and devoid of negative temporal frequencies.

The nature of the 'recorded image' formed by the instrument depends upon whether the source of the radiation possesses a high or a low degree of *spatial coherence*. As far as the recorded image is concerned, the *degree of spatial coherence* is characterized by the dependence upon $|\mathbf{x} - \mathbf{x}'|$ of

$$\overline{g(\mathbf{x}, t)g^*(\mathbf{x}', t)} = (1/T) \int_0^T g(\mathbf{x}, t)g^*(\mathbf{x}', t)\, dt, \qquad (3.6)$$

where T represents a time which is *long* in whatever sense is apposite in the envisaged scientific/technical application. Full *coherence* arises when LHS(3.6) has significant value for any $|\mathbf{x} - \mathbf{x}'|$ for which both $|g(\mathbf{x}, t)|$ and $|g(\mathbf{x}', t)|$ are appreciable. For spatial *incoherence*, LHS(3.6) vanishes for $|\mathbf{x} - \mathbf{x}'|$ larger than the least linear dimension of the smallest detail which can be resolved by the imaging instrument.

We point out to the reader that a bar over an expression explicitly dependent upon time always implies a time-average in this book.

There are so few useful applications, of radiation emanating from sources having a moderate degree of spatial coherence, that further consideration is here given only to full spatial coherence and complete spatial incoherence. Neither of these extremes appertain exactly in practice, of course, but they are approached nearly enough – e.g. by reflecting and refracting structures illuminated by beams emanating from radio and microwave transmitters, ultrasonic transducers, and lasers on the one hand, and by the great majority of naturally occurring sources of radiation on the other. So, it makes sense to discuss only these two limiting coherence conditions.

When considering spatial coherence it is usually convenient to examine individual spectral components of images and radiations. This implies that we treat such quantities as monochromatic – e.g. we analyse the behaviour of an instantaneous image $g(\mathbf{x}, t)$ which is of the form $(g(\mathbf{x}) \exp(i\omega t))$. The *ideal recorded image*, here written as $r(\mathbf{x})$, is thus defined in terms of $g(\mathbf{x}, t)$ by

$$r(\mathbf{x}) = \begin{cases} g(\mathbf{x}, t) \exp(-i\omega t) & \text{for full spatial coherence} \\ \overline{|g(\mathbf{x}, t)|^2} & \text{for spatial incoherence} \end{cases}. \qquad (3.7)$$

It is worth emphasizing that the time-average indicated in (3.7) must be over many periods of the centre frequency of the field impinging on the focal surface of the imaging instrument. The time interval implied for this average is usually a miniscule fraction of the duration of the actual recording process (e.g. exposing a film, scanning one element of a photoelectric array device, obtaining a detectable signal from a microwave receiver). Note that the duration of a million periods of a visible light signal is only of the order of a nanosecond, and for most of the microwave band an interval of one microsecond encompasses more than one thousand periods. From an image processing point of view, the essential differences between spatial coherence and spatial incoherence are summarized by

$$r(\mathbf{x}) \quad \text{is} \quad \begin{cases} \text{complex for spatial coherence} \\ \text{non-negative real for spatial incoherence} \end{cases}. \qquad (3.8)$$

Mainly because of the practical difficulties associated with 'optical computing' (refer to §2), image processing appropriate for spatially coherent fields is largely neglected in this book. Except when explicitly stated otherwise, it is assumed from now on that

$$r(\mathbf{x}) \quad \text{is non-negative real.} \qquad (3.9)$$

Neglecting the inevitable recording noise which contaminates all real-world images, and assuming the blurring can be taken to be of the ideal isoplanatic kind, $r(\mathbf{x})$ is seen to be effectively the same as the quantity $b(\mathbf{x})$, as defined by (3.4). This is a consequence of the convolution theorem for Fourier transforms (see §7 and also §8, in which the imaging of spatially incoherent sources is discussed further). In keeping with (3.9) it is assumed throughout this book, except when explicitly stated otherwise, that

$$f = f(\mathbf{x}) \quad \text{and} \quad h = h(\mathbf{x}) \quad \text{are both non-negative real.} \qquad (3.10)$$

It must be emphasized that $r(\mathbf{x})$ is *diffraction-limited* because the aperture (or pupil) diameter D of any imaging instrument is necessarily finite. If λ is the centre wavelength of the radiation, the instrument cannot resolve detail in the actual source distribution that subtends angles less than λ/D. Super-resolution (see §11) is not impossible but it is only practicable if the resolved detail in the original image is appreciably larger than a pixel.

The kinds of blurring so far discussed in this section can be compensated for by the methods described in Chapters III and VI. The techniques introduced in Chapters VII–IX are not only relevant in this context, they are also useful for effecting rectification and enhancement (as defined in §2).

Image degradation does not only arise because of the non-ideal nature of a propagation medium or the inherent imperfections or inadequate adjustment of an imaging instrument. It is sometimes due to certain crucial data being unmeasureable or unavailable, as is the case for the problems considered in Chapter IV. In other instances it can be caused by the basic measurement procedure, which though effectively ideal introduces distortion which must be corrected before the processed images can be of practical use, as in the applications discussed in Chapter V.

4 Basic model for the distorted image

Several of the quantities introduced in §3 are now combined to form a theoretical model that serves as the basic premise from which the techniques described in this book are developed. The undistorted information, identified by the symbol $f = f(\mathbf{x})$ in §3, is referred to from now on as the *true image.* The blurred information $b = b(\mathbf{x})$, in the form defined by (3.4), is here called the *ideal blurred image.* The time-invariant, point-spread-invariant distortion $h = h(\mathbf{x})$, in the form defined by (3.3), is called the *point spread function* (psf).

In any practical situation the distortion is never exactly time-invariant. However, the only generally effective way of handling time-varying distortions is to record individual blurred images so rapidly that the psf can be assumed time-invariant during each recording (or 'exposure', if the images are captured photographically). Neither the distortion nor the recording process are ever perfectly linear or point spread invariant. Furthermore, *recording noise* is always present in any real-world image. It is, however, appropriate to lump together all imperfections, other than the ideal blurring, into a composite degradation which is here called the contamination $c = c(\mathbf{x})$.

The class of images which is most easily displayed pictorially is that which can be represented by two-dimensional arrays of *positive* (i.e. real and non-negative) numbers. Two-dimensional arrays of complex numbers can of course be classed as images on occasion. Images representing the amplitudes of coherent wave motions (e.g. ultrasound, electromagnetism, tsunami) are usually complex, and we consider such in some sections of this book. It is, nevertheless, probably fair to state that most of the images processed in important scientific applications are positive. One of the main reasons for this is that virtually all naturally occurring sources of radiation are effectively spatially incoherent. When such a source is viewed with an imaging instrument through a distorting medium, the blurred image $b(\mathbf{x})$ is equivalent to $r(\mathbf{x})$ as defined in (3.7) and (3.8), for the case of spatial incoherence.

While we give almost all of our attention to positive images, we do countenance real-and-partly-negative and complex images occasionally. In every case we write the observable information, which we prefer to call the *recordable image* $\mathfrak{r} = \mathfrak{r}(\mathbf{x})$, in the form

$$\mathfrak{r}(\mathbf{x}) = b(\mathbf{x}) + c(\mathbf{x}) = f(\mathbf{x}) \,\unicode{x24B8}\, h(\mathbf{x}) + c(\mathbf{x}). \qquad (4.1)$$

It is useful to distinguish between information that is recordable in principle and the image which is *actually recorded*. The latter is denoted by $\mathfrak{a} = \mathfrak{a}(\mathbf{x})$. It is also necessary to distinguish the actually recorded image from the form $a = a(\mathbf{x})$ it assumes after it has been preprocessed prior to being subjected to whatever restoration procedure is deemed appropriate. The quantities $\mathfrak{a}$ and a are given precise expression in §14.

The general problem considered in this book can be posed thus: given a particular recorded image, denoted by $\mathfrak{a}$, or a sequence of recorded images, denoted by $\{\mathfrak{a}_m\}$, to which f is common, and given h, or relevant characteristics of the point spread functions distorting the members of $\{\mathfrak{a}_m\}$, recover f.

The definition (4.1) is perfectly general, but it is not always useful. If the contamination swamps the ideal blurred image then only an inferior estimate of f can be recovered. However, in many applications of scientific and technical interest, (4.1) is found to usefully describe images that are actually recordable – which is why this book has been written!

Image restoration and reconstruction are Arts as well as Sciences. The challenge is to recover as much as possible of the true image, as free as possible of those artefacts that are not only visually annoying but are also misleading on occasion (this is especially so in medical applications). An *artefact* is defined to be a feature, apparent in a restored or a reconstructed image, that is not present in the true image. The contamination can introduce artefacts. This is obvious. What may not be immediately so obvious is that an inappropriate reconstruction procedure can result in a higher than necessary level of artefacts (a spectacular example of this is discussed in §33).

5   Some general comments

As explained in §2, the psf is known *a priori* for image reconstruction, but it has to be determined before the true image can be recovered by image restoration.

We think it appropriate to comment on our uses of the word 'known' –
it appears in the first sentence of this section, for instance. It is another of
those portmanteau terms which is just as useful when employed vaguely
as when it is precisely defined. It sometimes refers to what is 'given', in
which case its meaning is definite. Conversely, it occasionally intimates
that implicit information is available which, while it cannot be formulated
unambiguously, must nevertheless be taken into account in some explicit
fashion. As with the word 'distortion' (refer to the fourth paragraph of
§2), we feel that the wide range of meanings we ascribe to 'known' should
not cause any confusion, because the scientifically and technically trained
reader is preconditioned to such multiple simultaneous invocations of the
word.

In practical situations, *a priori* information, albeit imperfect and incomplete, is always available concerning the psf. Probably the commonest
method of estimating h given its general shape (e.g. whether it is due to
out-of-focus blur, or uniform linear blur) but not its extent, is to examine
the forms of f recovered by assuming a number of different point spread
functions, successive ones being little different from each other. That
form for h which results in the most 'satisfactory' form for f is assumed to
be the 'correct' one. As with all non-trivial human activities, image
processing relies much on subjective judgments. These are not examined
in this book but the reader is warned (one would hope this is unnecessary
for the scientific reader!) that excessive reliance on subjectivity can lead
to the 'creation' of unwarranted detail in the restored image. This is often
merely amusing; but it can be dangerous (in medical applications, for
instance).

The form of the psf can often be estimated using one of several widely
applicable techniques, which can be almost entirely objective provided
certain *a priori* information is available. Determining the psf is itself a
restoration problem and is thus best discussed after Chapter III. It would
not have been inappropriate to do so immediately after that chapter, but
we decided instead it was more suitable first to examine phase problems,
image reconstruction from projections and speckle imaging; especially
since the latter is a type of image processing in which restoration and
psf-determination are inextricably intertwined. When planning the final
three chapters we found it appropriate to introduce system and program
considerations before ultimately turning in Chapter IX to the details of
techniques for estimating the psf. Although the form of the psf is assumed
known *a priori* throughout Chapter III, we emphasize that image restoration can often be successfully accomplished without prior detailed knowledge of the psf.

Throughout most of this book the psf is taken to be *positive* (i.e. real
and non-negative) and point spread invariant (or isoplanatic). In some
sections of Chapter V, however, the partial isoplanaticity (i.e. non-perfect

point spread invariance) and partial non-positivity (i.e. it can be negative in parts of image space) of the psf have to be explicitly accounted for. A complex psf is envisaged in parts of Chapter VI.

It is appropriate here to introduce the concept of a picture element or *pixel*, because it underlies the development of Chapter II from the Fourier transform in §6 to sampling in §11 and the discrete Fourier transform in §12. A digital image is a rectangular grid of pixels. Each pixel is itself a rectangle (or more normally a square). The location of a pixel is its centre. A digital image represents a continuously variable image within a rectangular recording frame by assigning a discrete value to each pixel. Each pixel value represents the value (such as radiance or complex amplitude) of the continuously variable image within the pixel and is treated as a sample at the pixel. Typical recording instruments (which are discussed in §43) attempt to record the average radiance or intensity within each pixel. Because of the practical need to reduce recording noise, the aim is to record the average intensity within each pixel rather than the actual value at the pixel centre.

If the pixel size (or, equivalently, the pixel spacing) effectively equals the diffraction limit, the imaging instrument and its associated recording apparatus together perform as efficiently as possible from an information-gathering viewpoint. In some applications, of course, it can be vital for the pixel size to be appreciably smaller than the diffraction limit, even though information capacity is thereby seemingly squandered. The recorded information may be incomplete in some way (as for the problems studied in Chapter IV, for example). There is the further point that it may be possible to effect noise reduction, without subsequent loss of resolution, by averaging over neighbouring pixels. It is worth noting, on the other hand, that it is the pixel size, rather than the diffraction limit, which determines the overall information content when the pixel size exceeds the diffraction limit. This often occurs, for instance, whenever we wish to record the whole of a scene without regard for the actual resolution of whatever instrument (a vidicon camera, say) is available.

Example I – Blurred images

Figure Ia is a sub-scene (composed of 256×256 pixels) of an image of New Zealand formed from data gathered by the LANDSAT satellite. The particular radiation, from which the data were generated, was infra-red in the 800 nm to 1100 nm wavelength band. The major feature in the sub-scene is Lake Tarawera, which is centrally located in the northern part of the North Island. The image shown in Fig. Ia has been rectified to the New Zealand Map Grid projection, with the spacing of adjacent pixels equivalent to 50 m on the ground. The image has also been enhanced by histogram equalization (see §45).

Figures Ib through Ig are versions of Fig. Ia blurred by three different species of point-spread-invariant psf: linear blur, out-of-focus and Gaussian; as depicted in

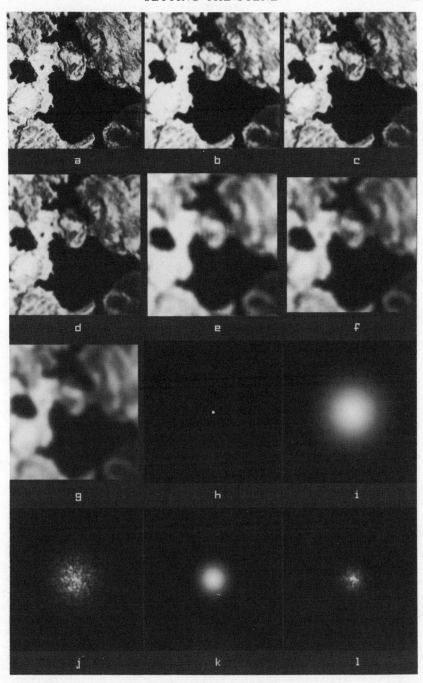

Fig. I.

Table I.1   Four particularly common forms for $h(\mathbf{x})$.

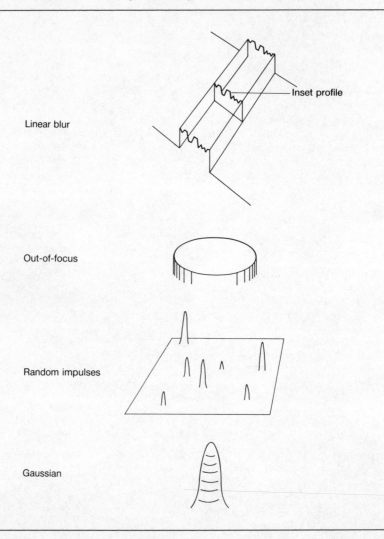

Table I.1 and discussed in §3, in the paragraph following that containing (3.3) and (3.4).

The psfs for Figs. Ib through Id are, respectively, horizontal uniform linear blur (5 pixels wide), out-of-focus (5 pixels diameter) and Gaussian (5 pixels effective width, which we take as the width when the amplitude of $h(\mathbf{x})$ is half its maximum value). The degradation apparent in each of these images is much the same in both degree and character. However, if we wish to recover Fig. Ia from any of these blurred images, it is vital to base the restoration procedure (e.g. whichever of those described in Chapter III seems most appropriate) on the correct form for the psf. Figures IIIn and IIIo in Example III emphasize how important this is.

The psfs for Figs. Ie through Ig are the same as for Figs. Ib through Id, respectively, but the effective widths are three times greater (i.e. 15 instead of 5). The increase in the degradation is very noticeable.

Figures Ih through Il illustrate blurring by the 'random impulses' psf depicted in Table I.1. This psf closely approximates that which commonly degrades any short exposure of a bright object viewed with a large optical astronomical telescope (see §34). Figure Ih shows what the image of a point object (e.g. a single star of small enough angular diameter that it is not resolved by the telescope) would look like in the absence of any distortion (i.e. the medium through which the object is viewed is perfectly homogeneous and the imaging instrument is free of aberrations). Figure Ii illustrates the form of a long-exposure image (obtained by viewing the object for much longer than the characteristic fluctuation time of the medium). Figure Ii is, in fact, typical of the 'seeing discs' observed by optical astronomers (see §34). If the exposure is short (i.e. its duration is significantly less than the aforesaid fluctuation time) the image has the characteristic mottled appearance shown in Fig. Ij.

The random impulses were spread over a roughly circular area of image space of approximate diameter 100 pixels for the psf which caused the blurring shown in Fig. Ij. Figure Ii can be regarded as the average of many statistically independent short exposures all of which are statistically akin to Fig. Ij.

Figures Ik and Il show how the long and short exposures differ when the medium through which the object is viewed fluctuates less wildly. The random impulses in the psf which caused the blurring in Fig. Il were spread over an area of approximate diameter 50 pixels.

Figures Ij and Il have the character of 'speckle images', which are of central concern in Chapter VI. Each of the speckle images included in Example VI has the typical appearance of an image captured with an electronic camera, whereas Figs. Ij and Il look like images recorded on photographic plates. The reader with astronomical leanings may wish to note the visual differences.

II

FOURIER THEORY

The image restoration and reconstruction techniques described in this book are based almost entirely on the properties of the Fourier transform. This chapter introduces the transform itself, investigates some of its idiosyncracies and derives a number of theorems related to its use. These results are invoked throughout the rest of the book, which can be thought of as a testament to the practical power of Fourier theory.

Most of the books listed in the References cover aspects of what is discussed in the sections (§§6–13) which comprise this chapter. Particularly relevant are [Bracewell, B1; Born, B1; Campbell, B1; Dainty, B1; Goodman, B1; Isaacson, B1; Scaife, B1].

In §6 we introduce the Fourier transform [Bracewell, B1], the 'sinc' function and the delta function as well as the concepts of Fourier space, spatial frequency spectrum and energy conservation. The definition of convolution and correlation in §7 leads to the convolution, correlation, and autocorrelation theorems, and finally to a discussion of image sizes and image frames. The consequences of an image being real and non-negative are investigated in §8. In addition, the relationships between the focal plane and image space, and the pupil plane and Fourier space are described. The concepts of the equivalent planar source distribution, the illumination taper and the optical transfer function are also introduced.

Projections are defined in §9, which presents the projection theorem [Lewitt, 1, 2, 3], before introducing the Kronecker delta and the Hankel transform [Lewitt, 1] via the agency of trigonometric Fourier series. We discuss the latter in §10 in some detail, which leads us to a treatment of the sampling theorem and of periodic images.

Sampling [Bracewell, B1], interpolation [Isaacson, B1], and extrapolation [Scaife, B1] are concepts central to many of the methods developed in later chapters of this book. They are discussed in §11 with emphasis on Lagrangian interpolation [Davis, B1] and Gerchbergian extrapolation [Gerchberg, 1, 2]. There is an enormous literature on interpolation in one dimension, as quickly becomes apparent on perusing any pertinent references [Isaacson, B1; Milner, T1; Prenter, B1; Scaife, B1]. While two-dimensional interpolation, which is what is of chief interest here, has of course been systematically studied [Franke, 1], not all that much is known about how to effect it smoothly and to perform it accurately and efficiently. It is for this reason that we pay particular attention to Lagrange polynomials of a complex variable, whose virtues are not yet widely appreciated [Bates, 18].

In this book many techniques are analysed and described in terms of quantities which are (piecewise at least) continuous and differentiable. The implementation of these techniques in the computer is necessarily digital and discrete, however. Our bridge between continuous and discrete analysis – i.e. the Discrete Fourier Transform – is developed in §12, where we introduce two useful concepts: the point spread frame and the periodic overlapped ideal blurred image. Finally, we outline the practical implementation of the Discrete Fourier Transform by means of the Fast Fourier Transform algorithm [Bergland, 1; Brigham, B1; Cochran, 1; IEEE, B2] (this is examined further in §48).

The concepts of the complex Fourier plane and complex zeros, which have been recurring themes in our work at Canterbury [Bates, 1, 2, 3, 6, 17; McKinnon, 1], are discussed in §13.

Example II illustrates ways in which the Fast Fourier Transform algorithm (now commonly referred to as the F.F.T.) can be used with advantage. It is worth emphasizing that the F.F.T. is almost always invoked nowadays when Fourier transforms have to be evaluated numerically.

6   The Fourier transform

When the functions $f = f(x)$ and $F = F(u)$, of the single variables x and u respectively, are the members of a Fourier transform pair, they are here defined in terms of each other by

$$F(u) = \int_{-\infty}^{\infty} f(x) \exp(i\, 2\pi ux)\, dx \quad \text{and} \quad f(x) = \int_{-\infty}^{\infty} F(u) \exp(-i\, 2\pi xu)\, du.$$

$$(6.1)$$

The variables x and u are termed *conjugate* in this context.

The consistency of the two simultaneous integral equations in (6.1) is formally established by substituting one into the other. It is worth noting that for multiple Fourier integrals, and indeed for multiple combinations of Fourier series and integrals, there is rarely any need to think before changing the order in which integrals and summations are evaluated, provided the functions f and F are of the kind that can usefully represent physical processes. The mathematicians have done the spadework, leaving us to forge ahead willy-nilly. There are, of course, occasions on which this abandoned behaviour leads to trouble, but it is usually easily recognized *post facto*, and then is time enough to retrace one's steps and seek the cause of the difficulty. In this book, integrals and summations are interchanged without comment. It follows from (6.1) that

$$F(u) = \iint_{-\infty}^{\infty} F(\alpha) \exp(i\, 2\pi[u - \alpha]x)\, dx\, d\alpha.$$

$$(6.2)$$

It is now convenient to introduce the function

$$\delta(\beta) = \int_{-\infty}^{\infty} \exp(i\,2\pi\beta y)\,dy = \lim_{T\to\infty} \int_{-T/2}^{T/2} \exp(i\,2\pi\beta y)\,dy$$

$$= \lim_{T\to\infty} T\,\text{sinc}(T\beta), \tag{6.3}$$

where the 'sinc' function is defined by

$$\text{sinc}(t) = (\sin(\pi t))/\pi t. \tag{6.4}$$

It is seen that

$$\delta(-\beta) = \delta(\beta), \qquad \delta(0) = \infty, \qquad \int_{-\infty}^{\infty} \delta(\beta)\,d\beta = 1, \tag{6.5}$$

and, because $T \to \infty$ in (6.3),

$$\int_{0^-}^{0^+} \delta(\beta)\,d\beta = 1, \tag{6.6}$$

where the useful notation

$$\xi^{\pm} = \lim_{\varepsilon\to 0}(\xi \pm \varepsilon), \text{ where } \varepsilon \text{ is real and positive}, \tag{6.7}$$

has been invoked. When $\beta \neq 0$, the value of $\delta(\beta)$ is clearly indeterminate; but that does not matter because first, $\delta(\beta)$ is an oscillatory function whose adjacent half-cycles are infinitesimally narrow and of the same shape but opposite sign, and second, $\delta(\beta)$ only figures in analysis under the integral sign which means that the contributions from adjacent half-cycles cancel except near $\beta = 0$. So, $\delta(\beta)$ is, of course, equivalent to Dirac's famous *delta function* – it is sometimes called the *impulse function*. The delta function is useful because of its property of 'collapsing' integrals. For instance, consider the integral of the product of $\delta(\beta - \alpha)$ with a function $g(\beta)$ which is continuous in β. It follows from (6.5), (6.6) and the above discussion that the integral can be reduced in the following steps (provided α is real):

$$\int_{-\infty}^{\infty} g(\beta)\,\delta(\beta - \alpha)\,d\beta = g(\alpha) \int_{-\infty}^{\infty} \delta(\beta - \alpha)\,d\beta$$

$$= g(\alpha) \int_{-\infty}^{\infty} \delta(\beta)\,d\beta = g(\alpha). \tag{6.8}$$

It further follows, given that Γ denotes a segment (not necessarily simple-connected) of the real β-axis, that

$$\int_\Gamma g(\beta)\,\delta(\beta-\alpha)\,d\beta = g(\alpha) \qquad \text{if } \alpha \text{ lies within } \Gamma,$$

$$= 0 \qquad \text{if } \alpha \text{ lies outside } \Gamma. \qquad (6.9)$$

Substituting (6.3) into (6.2), and making use of (6.4) through (6.8), gives

$$F(u) = \int_{-\infty}^{\infty} F(\alpha)\,\delta(u-\alpha)\,d\alpha = F(u), \qquad (6.10)$$

which is comforting. This way of checking the consistency of integral representations occurring in Fourier theory and practice can always be invoked, and it is only rarely that any special care need be taken. It is worth remarking that (6.8) and (6.9) are usually valid even when $g(\beta)$ is only piecewise continuous. The most constructive attitude for an applied scientist to take is to assume the general validity of (6.8) and (6.9) and only worry about them when a difficulty actually occurs.

Complementary to the image space introduced in §3, we now introduce a *Fourier space*, an arbitrary point in which is identified by the position vector $\mathbf{u}$. It is convenient to write the Cartesian coordinates of $\mathbf{x}$ and $\mathbf{u}$ as (x, y) and (u, v) respectively. It is also convenient to introduce the respective cylindrical polar coordinates $(r; \theta)$ and $(\rho; \phi)$. Semicolons inside parenthesis always precede an angular coordinate. The functional dependences on $\mathbf{x}$ in image space, and $\mathbf{u}$ in Fourier space, are conveniently written respectively in the following different ways:

$$\left.\begin{array}{l} f = f(\mathbf{x}) = f(x, y) = f(r; \theta) \\ F = F(\mathbf{u}) = F(u, v) = F(\rho; \phi) \end{array}\right\}. \qquad (6.11)$$

When an imaging instrument is involved, so that $F(\mathbf{u})$ is related to the field in the pupil plane (the exact nature of this relation depends upon whether the sources of the field are spatially coherent or incoherent – as discussed in §§3 and 8), then u and v are 'direction cosines' divided by the mean wavelength of the radiation field. It is consequently useful to think of u and v as *spatial frequencies*, by analogy with signals and their spectra for which the conjugate independent variables are 'time' and 'frequency'. It is also useful to introduce the concept of the 'effective spatial bandwidth' which is the range of $|\mathbf{u}|$ over which $|F(\mathbf{u})|$ is appreciable.

When f and F are members of a two-dimensional Fourier transform pair, they are here defined in terms of each other by

$$F(\mathbf{u}) = \int\int_{-\infty}^{\infty} f(\mathbf{x}) \exp(\mathrm{i}\,2\pi\mathbf{u}\cdot\mathbf{x})\,d\sigma(\mathbf{x}),$$

$$= \int\int_{-\infty}^{\infty} f(x, y) \exp(\mathrm{i}\,2\pi[ux + vy])\,dx\,dy,$$

$$= \int_{0}^{\infty}\int_{0}^{2\pi} f(r;\theta) \exp(\mathrm{i}\,2\pi\rho r \cos(\phi - \theta))r\,d\theta\,dr, \qquad (6.12)$$

and, denoting by $d\Sigma(\mathbf{u})$ the element of area in Fourier space centred on the point $\mathbf{u}$,

$$f(\mathbf{x}) = \int\int_{-\infty}^{\infty} F(\mathbf{u}) \exp(-\mathrm{i}\,2\pi\mathbf{x}\cdot\mathbf{u})\,d\Sigma(\mathbf{u}),$$

$$= \int\int_{-\infty}^{\infty} F(u, v) \exp(-\mathrm{i}\,2\pi[xu + vy])\,du\,dv,$$

$$= \int_{0}^{\infty}\int_{0}^{2\pi} F(\rho;\phi) \exp(-\mathrm{i}\,2\pi r\rho \cos(\theta - \phi))\rho\,d\phi\,d\rho. \qquad (6.13)$$

Substituting (6.13) into (6.12) and invoking (6.3), (6.5), (6.8), and (6.11) gives

$$F(u, v) = \int\int\int\int_{-\infty}^{\infty} F(\alpha, \beta) \exp(\mathrm{i}\,2\pi[(u - \alpha)x + (v - \beta)y])\,dx\,dy\,d\alpha\,d\beta,$$

$$= \int\int_{-\infty}^{\infty} F(\alpha, \beta)\,\delta(u - \alpha)\,\delta(v - \beta)\,d\alpha\,d\beta = F(u, v), \qquad (6.14)$$

which is just as comforting as (6.10).

It is sometimes convenient to use the shorthand notations

$$F = \mathsf{F}\{f\} \quad \text{and} \quad f = \mathsf{F}\{F\} \qquad (6.15)$$

for either (6.12) or the first of equations (6.1) and, respectively, either (6.13) or the second of equations (6.1). No confusion is caused by the failure to distinguish between 'forward' and 'reverse' transforms. Consistency relations, such as (6.10) and (6.14), then become

$$F = \mathsf{F}\{\mathsf{F}\{F\}\} \quad \text{and} \quad f = \mathsf{F}\{\mathsf{F}\{f\}\}. \qquad (6.16)$$

When the function, g say, that is being transformed depends upon two variables, ξ and η say, but the transformation is one-dimensional, it is necessary to indicate which variable figures in the transformation. This is done by subscripting g with the relevant variable, ξ say, in parentheses and subscripting F with the number 1 followed by the conjugate variable figuring in the transform, e.g.

$$\mathsf{F}_{(1\alpha)}\{g_{(\xi)}\} = \int_{-\infty}^{\infty} g(\xi, \eta) \exp(\mathrm{i}\, 2\pi\alpha\xi)\, d\xi. \tag{6.17}$$

When one-dimensional and two-dimensional Fourier transforms figure in a series of operations, it can be convenient to identify the dimensions of any particular transform, which is done by subscripting F with the number 1 or the number 2 – either number being in parentheses.

Whether F is one-dimensional or two-dimensional it is appropriate to call it the *spatial frequency spectrum*, or just the *spectrum*, of f.

It is instructive to introduce the *two-dimensional delta function* $\delta(\mathbf{x})$, characterized by

$$\delta(-\mathbf{x}) = \delta(\mathbf{x}), \qquad \delta(0) = \infty \quad \text{and} \quad \iint_{-\infty}^{\infty} \delta(\mathbf{x})\, d\sigma(\mathbf{x}) = 1. \tag{6.18}$$

It seems intuitively obvious that $\delta(\mathbf{x})$ should be required to be circularly symmetric. However, inspection of (6.3) and (6.4) then shows that

$$\delta(\mathbf{x}) \neq \delta(x)\, \delta(y), \tag{6.19}$$

but there is no need to be upset by this because the two sides of (6.19) are 'operationally' equivalent:

$$\iint_{-\infty}^{\infty} g(\boldsymbol{\xi})\, \delta(\mathbf{x}-\boldsymbol{\xi})\, d\sigma(\boldsymbol{\xi}) = g(\mathbf{x}) = g(x, y), \tag{6.20}$$

and

$$\iint_{-\infty}^{\infty} g(\xi, \eta)\, \delta(x-\xi)\, \delta(y-\eta)\, d\xi\, d\eta = g(x, y). \tag{6.21}$$

Finally, note from (6.11) and (6.13) that

$$\iint_{-\infty}^{\infty} |f(\mathbf{x})|^2\, d\sigma(\mathbf{x}) = \iint_{-\infty}^{\infty} |f(x, y)|^2\, dx\, dy,$$

$$= \iiiint_{-\infty}^{\infty}\!\!\!\!\int\!\int F(u', v')F^*(u, v)$$

$$\times \exp(\mathrm{i}\, 2\pi[(u'-u)x + (v'-v)y])\, dx\, dy\, du'\, dv'\, du\, dv,$$

$$= \iiiint_{-\infty}^{\infty} F(u', v')F^*(u, v)\, \delta(u-u')\, \delta(v-v')\, du'\, dv'\, du\, dv,$$

$$= \iint_{-\infty}^{\infty} |F(u, v)|^2\, du\, dv = \iint_{-\infty}^{\infty} |F(\mathbf{u})|^2\, d\Sigma(\mathbf{u}), \tag{6.22}$$

which is here called the *energy conservation theorem* for Fourier transforms, which is a more graphic name than the several more usual ones, because it seems to make good physical sense to think of $|f(x, y)|^2$ and $|F(u, v)|^2$ as the energy densities in image space and Fourier space respectively. The one-dimensional form of this theorem is

$$\int_{-\infty}^{\infty} |f(x)|^2 \, dx = \int_{-\infty}^{\infty} |F(u)|^2 \, du. \tag{6.23}$$

7 Convolution and correlation

The symbols $\odot$ and $*$ are used to denote *convolution* and *correlation* respectively. In one dimension, for functions $g = g(x)$ and $h = h(x)$, these operations are defined by

$$g \odot h = \int_{-\infty}^{\infty} g(\xi) h(x - \xi) \, d\xi = h \odot g, \tag{7.1}$$

and

$$g * h = \int_{-\infty}^{\infty} g(\xi) h(x + \xi) \, d\xi. \tag{7.2}$$

Note that $g * h$ considered as a function of x is equal to $h * g$ considered as a function of $-x$. Writing the Fourier transforms of g and h as $G = G(u)$ and $H = H(u)$, respectively, invoking the definitions (6.1) and manipulating delta functions in the manner introduced in §6, indicates that the Fourier transform of (7.1) can be expressed as

$$\int_{-\infty}^{\infty} (g \odot h) \exp(i \, 2\pi u x) \, dx,$$

$$= \iiiint_{-\infty}^{\infty} G(\alpha) H(\beta) \exp(i \, 2\pi[ux - \alpha\xi - \beta x + \beta\xi]) \, dx \, d\xi \, d\alpha \, d\beta,$$

$$= \iint_{-\infty}^{\infty} G(\alpha) H(\beta) \, \delta(u - \beta) \, \delta(\beta - \alpha) \, d\alpha \, d\beta,$$

$$= \int_{-\infty}^{\infty} G(\beta) H(\beta) \, \delta(u - \beta) \, d\beta = G(u) H(u), \tag{7.3}$$

which is the *convolution theorem*. Treating (7.2) similarly gives

$$\int_{-\infty}^{\infty} (g \ast h) \exp(i\, 2\pi u x)\, dx = G(-u)H(u), \qquad (7.4)$$

which is the *correlation theorem*.

In two dimensions, the convolution and correlation operations are defined in terms of functions $g = g(\mathbf{x})$ and $h = h(\mathbf{x})$ by

$$g \odot h = \iint_{-\infty}^{\infty} g(\boldsymbol{\xi})h(\mathbf{x}-\boldsymbol{\xi})\, d\sigma(\boldsymbol{\xi}) = \iint_{-\infty}^{\infty} g(\xi, \eta)h(x-\xi, y-\eta)\, d\xi\, d\eta, \quad (7.5)$$

and

$$g \ast h = \iint_{-\infty}^{\infty} g(\boldsymbol{\xi})h(\mathbf{x}+\boldsymbol{\xi})\, d\sigma(\boldsymbol{\xi}) = \iint_{-\infty}^{\infty} g(\xi, \eta)h(x+\xi, y+\eta)\, d\xi\, d\eta. \quad (7.6)$$

The *convolution* and *correlation theorems* become, respectively,

$$\iint_{-\infty}^{\infty} (g \odot h) \exp(i\, 2\pi \mathbf{u} \cdot \mathbf{x})\, d\sigma(\mathbf{x}) = G(\mathbf{u})H(\mathbf{u}) = \mathsf{F}\{g \odot h\}, \qquad (7.7)$$

and

$$\iint_{-\infty}^{\infty} (g \ast h) \exp(i\, 2\pi \mathbf{u} \cdot \mathbf{x})\, d\sigma(\mathbf{x}) = G(-\mathbf{u})H(\mathbf{u}) = \mathsf{F}\{g \ast h\}. \qquad (7.8)$$

For a function $g = g(\mathbf{x})$ that can assume complex values, it is convenient to call the correlation of g^* with g the

$$\text{autocorrelation of } g = g^* \ast g. \qquad (7.9)$$

Inspection of (7.8) then establishes the *autocorrelation theorem*:

$$\iint_{-\infty}^{\infty} (g^* \ast g) \exp(i\, 2\pi \mathbf{u} \cdot \mathbf{x})\, d\sigma(\mathbf{x}) = |G(\mathbf{u})|^2 = \mathsf{F}\{g^* \ast g\}, \qquad (7.10)$$

which is, of course, also valid in one dimension.

A useful exercise in manipulating Fourier integrals is to confirm that convolution is commutative and associative, i.e.

$$p \odot q \odot r = (p \odot q) \odot r = p \odot (q \odot r) = p \odot r \odot q = r \odot q \odot p$$
$$= q \odot r \odot p = q \odot p \odot r = r \odot p \odot q. \qquad (7.11)$$

Almost any image of practical interest is of *finite size*, meaning that its intensity is everywhere zero outside some finite region of image space.

For any image $\bar{f} = f(\mathbf{x})$, the rectangle (with its sides parallel to the x- and y-axes) which just encloses the aforesaid finite region is here denoted by Ω_f and is called the *image frame*, or just the *frame* when no confusion is caused thereby. The lengths of the sides of Ω_f in the x- and y-directions are called, respectively, the *x-extent* and the *y-extent*, which are collectively called the *extents* of f (and of its frame). In one dimension the terms *finite length* and *image segment* (or just segment) are used instead of finite size and image frame. The length of the segment, which is always in fact its x-extent in this book, is called its *extent*.

Denoting by $L^{(f)}$ either the x-extent or the y-extent of Ω_f, it follows immediately from inspection of (7.5) and (7.6) that

$$L^{(g \odot h)} = L^{(g * h)} = L^{(g)} + L^{(h)}, \tag{7.12}$$

provided both g and h are of finite size. It is convenient to refer to the result embodied in (7.12) as the *extent of convolution theorem* or the *extent of correlation theorem*. When $g = h^*$, the latter theorem is conveniently rechristened the *extent of autocorrelation theorem*.

Suppose that an image f satisfies some relation that bears on its size, but does not fix it. Then the *most compact* image consistent with this relation is here defined as the one for which $(L_1^{(f)} L_2^{(f)})$ is a minimum, where $L_1^{(f)}$ and $L_2^{(f)}$ are the image's x-extent and y-extent respectively.

If one is given the value of some $L^{(\mathfrak{z})}$, where $\mathfrak{z}$ is the autocorrelation of an image which is denoted for convenience by h, inspection of (7.6), with g set equal to h^*, immediately reveals that

$$L^{(\bar{h})} = L^{(\mathfrak{z})}/2, \tag{7.13}$$

where $\bar{h}$ is the most compact image whose autocorrelation is $\mathfrak{z}$. The point is that images with larger extents can have the same autocorrelation because the oscillations of the integrand of (7.6), with $g = h^*$, can be such that there is perfect cancellation over the whole integral for all $\mathbf{x}$ outside $\Omega_{\mathfrak{z}}$. The result embodied in (7.13) is here called the *minimum-extent of autocorrelation theorem*.

It follows from definitions introduced in §6 that

$$\iint\limits_{-\infty}^{\infty} f(\mathbf{x}) \, d\sigma(\mathbf{x}) = \iiint\limits_{-\infty}^{\infty} F(\mathbf{u}) \exp(-i \, 2\pi \mathbf{u} \cdot \mathbf{x}) \, d\Sigma(\mathbf{u}) \, d\sigma(\mathbf{x})$$

$$= \iint\limits_{-\infty}^{\infty} F(\mathbf{u}) \, \delta(\mathbf{u}) \, d\Sigma(\mathbf{u}) = F(0). \tag{7.14}$$

When $f(\mathbf{x})$ is of finite size, it is a mere matter of definition that

$$\iint\limits_{-\infty}^{\infty} f(\mathbf{x}) \, d\sigma(\mathbf{x}) = \iint\limits_{\Omega_f} f(\mathbf{x}) \, d\sigma(\mathbf{x}). \tag{7.15}$$

Consequently,

$$F(0) = \int\limits_{-\infty}^{\infty}\!\!\int F(\mathbf{u})\, \delta_f^*(\mathbf{u})\, d\Sigma(\mathbf{u}), \tag{7.16}$$

where the *limited delta function* $\delta_f(\mathbf{u})$ is defined by

$$\delta_f(\mathbf{u}) = \int\limits_{\Omega_f}\!\!\int \exp(\mathrm{i}\, 2\pi\mathbf{u} \cdot \mathbf{x})\, d\sigma(\mathbf{x}). \tag{7.17}$$

8 Positive, compact images

For the reasons given in §4, the images considered in this book are, except when it is explicitly stated otherwise, constrained by (3.10):

$$f^* = f \quad \text{and} \quad f \geq 0. \tag{8.1}$$

Substituting (8.1) into (6.12), with $\mathbf{u}$ replaced by $-\mathbf{u}$, shows by inspection that

$$F(-\mathbf{u}) = \int\limits_{-\infty}^{\infty}\!\!\int f^*(\mathbf{x})\exp(-\mathrm{i}\, 2\pi\mathbf{u} \cdot \mathbf{x})\, d\sigma(\mathbf{x}) = F^*(\mathbf{u}). \tag{8.2}$$

Combining (7.6) with (8.1) indicates that the autocorrelation of f is not only real, so that

$$f^* \ast f = f \ast f, \tag{8.3}$$

but it is also an even function of $\mathbf{x}$.

Coherent, monochromatic fields existing instantaneously in the pupil and focal planes of an imaging instrument are each others' Fourier transforms. It is, nevertheless, appropriate here to make a clear distinction between the pupil plane (in which an arbitrary point is identified by $\mathbf{z}$) of a (generalized) imaging instrument and Fourier space (in which an arbitrary point is identified by $\mathbf{u}$), since so many images of practical importance are of objects whose emissions are uncorrelated, in the sense that there is negligible coherence between radiations emanating from arbitrary closely spaced points. The image of such an object is the Fourier transform of the autocorrelation of its pupil field. It is conversely not nearly so necessary to distinguish between the focal plane and image space because the image of a spatially incoherent source distribution is merely the time-average of the intensity of the field existing in the image plane. The following exposition should clarify all this. The formulas (8.10) and (8.13) typify, respectively, the equivalence of the focal plane and image space and the non-equivalence of the pupil plane and Fourier space.

It is appropriate then to refer back to §3 and consider in greater detail the imaging of spatially incoherent source distributions. It is convenient to introduce a *source plane*. This can be situated anywhere in between the actual sources and the pupil plane. However, it is only useful if it is parallel to the pupil plane and is positioned as close as possible to the sources without actually intersecting them. The sources must lie on the opposite side of the source plane to the pupil plane. The radiation field $G(\mathbf{z}, t)$ appearing in the instrument's pupil can be expressed with equal exactness in terms of either the actual sources or equivalent sources situated on the source plane.

It is important to understand that, even if the actual sources are spatially incoherent, the equivalent sources may not be. Consider an arbitrary point, identified by the position vector ξ, in the source plane. The complex amplitude $s(\xi, t)$ of the equivalent source density at this point is given by the summation of radiations impinging upon it from throughout the whole of the actual source distribution. It follows that, for any two points ξ and ξ' in the source plane, there must be correlation between $s(\xi, t)$ and $s(\xi', t)$ because each of these amplitudes are integrals over all of the actual sources. This is a long way from being the whole story, however, as far as the field in the pupil plane is concerned. At each point $\mathbf{z}$ in the pupil, $G(\mathbf{z}, t)$ can be regarded as the sum of contributions from every point in the actual source distribution (at least, every point lying within the field of view of the imaging instrument). All the rays from any single point in the actual source distribution impinging upon the pupil intersect a region, $\Re$ say, of the source plane whose size depends on the ratio of the pupil diameter to the distance from the pupil to the source plane. If this ratio is very small, which it is in the great majority of imaging applications, the linear dimensions of $\Re$ are likely to be less than the resolution limit (or diffraction limit) of the instrument. However, if the actual sources are spatially incoherent, the moiety of the equivalent sources which contributes to the pupil plane field cannot exhibit any finite degree of coherence over distances larger than the linear dimensions of $\Re$. Consequently, when the aforesaid ratio is small enough, the part of the equivalent source distribution which contributes to the pupil plane field is effectively spatially incoherent. From now on we imply this part of the distribution when we employ the notation $s(\xi, t)$, which we call the *equivalent planar source distribution*.

The above discussion takes no account of any distortion $\mathfrak{D}(\zeta, \mathbf{z})$ introduced by the propagation medium in between the source and pupil planes – refer to the paragraph containing (3.5). Furthermore, in order to calculate the detailed form of the image produced by the instrument, the latter's aberrations and its *pupil function* must be allowed for. It is appropriate to lump in the aberrations with $\mathfrak{D}(\zeta, \mathbf{z})$. The pupil function, sometimes called the *apodization* function in the optical literature, is the same as what is referred to in radio-engineering as the *illumination taper* – it is discussed further in §15. It is written here as $\Re = \Re(\mathbf{z})$. Its primary defining characteristic is

$$\mathfrak{K}(\mathbf{z}) = 0 \qquad \text{for points } \mathbf{z} \text{ outside the pupil.} \tag{8.4}$$

So that the quantity $g(\mathbf{x}, t)$ introduced in §3 can represent the instantaneous field in the focal plane of an apodized instrument viewing through a distorting medium, it is convenient to extend the definition of $G(\mathbf{z}, t)$ to include the propagation distortion, the pupil function and the aberrations. On defining

$$S(\mathbf{z}, t) = \mathsf{F}\{s(\boldsymbol{\xi}, t)\} \tag{8.5}$$

and requiring the distortion to be isoplanatic, as expressed by (3.5), it is seen that $G(\mathbf{z}, t)$ is now given by

$$G(\mathbf{z}, t) = \mathfrak{D}(\mathbf{z})\mathfrak{K}(\mathbf{z})S(\mathbf{z}, t), \tag{8.6}$$

on the understanding that the distance from the source plane is so large, compared to both the shortest significant wavelength of the radiation and the largest linear dimension of the pupil, that the pupil can be considered in the far-field (Fraunhofer region) of the part of the source plane in the instrument's field of view (this condition holds in perhaps most imaging applications).

Invoking the definition $g(\mathbf{x}, t) = \mathsf{F}\{G(\mathbf{z}, t)\}$ introduced in §3, making double use of the convolution theorem (7.7), and noting that convolution is associative – as indicated by (7.11) – we find from (8.5) and (8.6) that

$$g(\mathbf{x}, t) = \mathfrak{d}(\mathbf{x}) \mathbin{\text{\textcircled{0}}} \mathfrak{f}(\mathbf{x}) \mathbin{\text{\textcircled{0}}} s(\mathbf{x}, t), \tag{8.7}$$

where $\mathfrak{d} = \mathsf{F}\{\mathfrak{D}\}$ and $\mathfrak{f} = \mathsf{F}\{\mathfrak{K}\}$. Because of (8.4), the quantity

$$\tilde{s}(\mathbf{x}, t) = \mathfrak{f}(\mathbf{x}) \mathbin{\text{\textcircled{0}}} s(\mathbf{x}, t) \tag{8.8}$$

is the instantaneous image of the equivalent planar source distribution $s(\boldsymbol{\xi}, t)$ – i.e. it is the field which would appear in the focal plane of the instrument in the absence of aberrations and propagation distortion.

If $s(\boldsymbol{\xi}, t)$ is effectively spatially incoherent, it follows from the reasoning presented in the paragraph containing (3.6) that

$$\overline{s(\boldsymbol{\xi}, t)s^*(\boldsymbol{\xi}', t)} = \iota(\boldsymbol{\xi})\, \delta(\boldsymbol{\xi} - \boldsymbol{\xi}'), \tag{8.9}$$

where $\delta(\mathbf{x})$ is the two-dimensional delta function introduced in (6.18) and $\iota(\boldsymbol{\xi})$ is the average intensity of $s(\boldsymbol{\xi}, t)$. Since RHS (8.8) represents the diffraction-limiting effect of the imaging instrument – because $\mathfrak{f}(\mathbf{x})$ is the Fourier transform of the quantity $\mathfrak{K}(\mathbf{z})$ which is zero outside the pupil, as indicated by (8.4) – we see from (8.9) that the time-average of the product of $\tilde{s}(\mathbf{x}, t)$ and $\tilde{s}^*(\mathbf{x}', t)$ reduces to the true image (as defined in §3) of the equivalent planar source distribution at the point $\mathbf{x} = \mathbf{x}'$. So

$$\overline{\tilde{s}(\mathbf{x}, t)\tilde{s}^*(\mathbf{x}', t)} = f(\mathbf{x})\, \delta(\mathbf{x} - \mathbf{x}') \tag{8.10}$$

As is remarked in the final sentence of the third paragraph of §4, the two quantities $r(\mathbf{x})$ and $b(\mathbf{x})$ introduced in §3 are equivalent. So, substituting (8.7) and (8.8) into the lower relation in (3.7) leads to the following manipulations, in which (7.5) and (8.10) are invoked:

$$b(\mathbf{x}) = \overline{|\mathfrak{d}(\mathbf{x}) \odot \tilde{s}(\mathbf{x}, t)|^2}$$

$$= \int\!\!\int\!\!\int\!\!\int_{-\infty}^{\infty} \overline{\mathfrak{d}^*(\mathbf{x}-\boldsymbol{\eta})\tilde{s}^*(\boldsymbol{\eta}, t)\mathfrak{d}(\mathbf{x}-\boldsymbol{\xi})\tilde{s}(\boldsymbol{\xi}, t)}\; d\sigma(\boldsymbol{\eta})\, d\sigma(\boldsymbol{\xi})$$

$$= \int\!\!\int\!\!\int\!\!\int_{-\infty}^{\infty} \mathfrak{d}^*(\mathbf{x}-\boldsymbol{\eta})\mathfrak{d}(\mathbf{x}-\boldsymbol{\xi})\; \overline{\tilde{s}^*(\boldsymbol{\eta}, t)\tilde{s}(\boldsymbol{\xi}, t)}\; d\sigma(\boldsymbol{\eta})\, d\sigma(\boldsymbol{\xi})$$

$$= \int\!\!\int\!\!\int\!\!\int_{-\infty}^{\infty} \mathfrak{d}^*(\mathbf{x}-\boldsymbol{\eta})\mathfrak{d}(\mathbf{x}-\boldsymbol{\xi})\; \delta(\boldsymbol{\eta}-\boldsymbol{\xi})f(\boldsymbol{\xi})\, d\sigma(\boldsymbol{\eta})\, d\sigma(\boldsymbol{\xi})$$

$$= \int\!\!\int_{-\infty}^{\infty} |\mathfrak{d}(\mathbf{x}-\boldsymbol{\xi})|^2 f(\boldsymbol{\xi})\, d\sigma(\boldsymbol{\xi}) = |\mathfrak{d}(\mathbf{x})|^2 \odot f(\mathbf{x}), \qquad (8.11)$$

which is not only an important result, it is also a good example of the power of Fourier analysis. Comparing (3.4) and (8.11) shows that

$$h(\mathbf{x}) = |\mathfrak{d}(\mathbf{x})|^2. \qquad (8.12)$$

The Fourier transform $H = H(\mathbf{u})$ of h characterizes the time-averaged effect of the distorting medium on the field in the pupil plane of the imaging instrument. The name often given to $H(\mathbf{u})$ – i.e. the *optical transfer function* (OTF) – is therefore as graphic as the name psf for $h(\mathbf{x})$. Reference to the autocorrelation theorem (7.10) and to (8.12) shows that

$$H(\mathbf{u}) = \mathscr{D}^* \bigstar \mathscr{D} = \int\!\!\int \mathscr{D}^*(\mathbf{z})\mathscr{D}(\mathbf{z}+\mathbf{u})\, d\sigma(\mathbf{z}). \qquad (8.13)$$

There is a final, very important point worth making about real and non-negative images of finite size. The minimum-extent of auto-correlation theorem (see §7) ensures that the extents of the most compact image (or images if there can be more than one – this question is discussed at length in Chapter IV) $\mathfrak{f}$ compatible with a given $(\mathfrak{f}^* \bigstar \mathfrak{f})$ are necessarily one-half of the respective extents of $(\mathfrak{f}^* \bigstar \mathfrak{f})$. However, when (8.1) holds, the only image(s) f compatible with a given $(f \bigstar f)$ is(are) necessarily the most compact one(s), as is clear from inspection of the integral, over the frame Ω_f, defining the autocorrelation of the real and non-negative f:

$$f \bigstar f = \int\!\!\int_{\Omega_f} f(\boldsymbol{\xi})f(\mathbf{x}+\boldsymbol{\xi})\, d\sigma(\boldsymbol{\xi}). \qquad (8.14)$$

Since f is never negative, the integrand of (8.14) cannot oscillate, so that it is impossible for the integral to be zero for values of $\mathbf{x}$ for which $(f(\boldsymbol{\xi})f(\mathbf{x}+\boldsymbol{\xi})) \neq 0$ over a finite region of the $\boldsymbol{\xi}$-plane. Consequently,

$$L^{(f)} \equiv L^{(f*f)}/2, \tag{8.15}$$

where the notation $L^{(f)}$ is defined in §7. The result embodied in (8.15) is conveniently called the *extent–constraint theorem for positive images* (where *positivity* here implies being real and non-negative).

9 Projections

A number of somewhat isolated results are developed in this section. While they are invoked mainly in Chapter V, they are also needed in various parts of the rest of this book.

Consider Cartesian coordinates (ξ, η) rotated in image space by an angle ϕ with respect to the Cartesian coordinates (x, y). Also consider Cartesian coordinates (α, β) rotated in Fourier space by the same angle ϕ with respect to the Cartesian coordinates (u, v). Elementary analytic geometry shows that

$$ux + vy = \alpha\xi + \beta\eta \quad \text{and} \quad \iint\limits_{\Omega_f} \mathfrak{f}\, dx\, dy = \iint\limits_{\Omega_f} \mathfrak{f}\, d\xi\, d\eta, \tag{9.1}$$

where $\mathfrak{f}$ is any image existing within the frame Ω_f, and

$$u = \alpha\, \cos(\phi) - \beta\, \sin(\phi) \quad \text{and} \quad v = \alpha\, \sin(\phi) + \beta\, \cos(\phi). \tag{9.2}$$

Combining (9.1) and (9.2) with (6.11) and (6.12), and then setting $\beta = 0$, gives

$$F(\alpha\, \cos(\phi), \alpha\, \sin(\phi)) = \int\limits_{-\infty}^{\infty}\!\!\int f(x, y)\, \exp(\mathrm{i}\, 2\pi\alpha\xi)\, d\eta\, d\xi, \tag{9.3}$$

since $d\sigma(\mathbf{x})$ can be written as $d\xi\, d\eta$, just as well as $dx\, dy$.

The name *projection* (at angle ϕ) is given to the quantity

$$p = p(\xi; \phi) = \int\limits_{-\infty}^{\infty} f(x, y)\, d\eta. \tag{9.4}$$

The name is exceptionally apposite when f is required to satisfy (3.10) because, for any particular value of ξ, the right side represents the integrated image intensity along a ray parallel to the η-axis distant ξ from the coordinate origin – i.e. $p(\xi; \phi)$ is the projected intensity in the η-direction. Substituting (9.4) into (9.3), and replacing α by the polar coordinate ρ, gives

$$F(\rho \cos(\phi), \rho \sin(\phi)) = F(\rho; \phi) = \int_{-\infty}^{\infty} p(\xi; \phi) \exp(i\, 2\pi\rho\xi) \, d\xi, \quad (9.5)$$

which is the *projection theorem* in two dimensions. Making use of the notation introduced in the paragraph containing (6.15) through (6.17), this theorem can be alternatively expressed as

$$F(\rho; \phi) = \mathbf{F}_{(1\rho)}\{p_{(\xi)}(\xi; \phi)\} \quad \text{and} \quad p = \mathbf{F}_{(1\xi)}\{F_{(\rho)}\}. \quad (9.6)$$

It is convenient to express F and p as trigonometric Fourier series (these are discussed in some detail in §10) in the angle ϕ:

$$F(\rho; \phi) = \sum_{l=-\infty}^{\infty} F_l(\rho) \exp(i\, l\phi) \quad \text{and} \quad p(\xi; \phi) = \sum_{l=-\infty}^{\infty} p_l(\xi) \exp(i\, l\phi), \quad (9.7)$$

where the notations $F_l = F_l(\rho)$ and $p_l = p_l(\xi)$ are introduced for later use. On substituting (9.7) into either (9.5) or (9.6), it is seen, for each integer l, that

$$F_l = \mathbf{F}\{p_l\}, \quad (9.8)$$

because, for any integers m and n,

$$(1/2\pi) \int_{0}^{2\pi} \exp(i\,[m-n]\theta)\, d\theta = \delta_{m,n} = 1 \qquad \text{if } m = n$$

$$= 0 \qquad \text{if } m \neq n, \quad (9.9)$$

where $\delta_{m,n}$ is called the *Kronecker delta*. Note that the subscripts attached to F and p in (9.8) do not refer to transformation variables, because these subscripts are not enclosed in parentheses. It is further convenient to express f in terms of polar coordinates in image space and to write it as yet another trigonometric Fourier series:

$$f(r; \theta) = \sum_{l=-\infty}^{\infty} f_l(r) \exp(i\, l\theta). \quad (9.10)$$

Before continuing with the main argument the following formula must be recalled:

$$\exp(i\, a \cos(\alpha)) = \sum_{l=-\infty}^{\infty} i^l J_l(a) \exp(i\, l\alpha), \quad (9.11)$$

where $J_l(\cdot)$ denotes the *Bessel function* of the first kind of order l. On recalling (6.11), on substituting (9.10) and the first relation in (9.7) into (6.12), and on invoking (9.9) and (9.11), it is seen that

$$F_l(\rho) = 2\pi i^l \int_{0}^{\infty} f_l(r) J_l(2\pi\rho r) r \, dr. \quad (9.12)$$

Making use of (6.13) in a similar fashion shows that

$$f_l(r) = 2\pi(-i)^l \int_0^\infty F_l(\rho)J_l(2\pi\rho r)\rho \, d\rho. \tag{9.13}$$

The two integral formulas (9.12) and (9.13) are called *Hankel transforms* of order *l*. On substituting (9.12) into (9.13), it must follow, by analogy with (6.2) and (6.10), that

$$4\pi^2 \int_0^\infty J_l(2\pi\rho r)J_l(2\pi\alpha r)r \, dr = (\delta(\rho-\alpha))/\rho, \tag{9.14}$$

which is indeed found to be so – equation (8) of §5.11 of [Watson, B1] allows the above integral to be evaluated immediately. The contribution from the lower limit vanishes. When the asymptotic formulas for Bessel functions are invoked, the contribution from the upper limit becomes

$$\operatorname*{Lim}_{T\to\infty} (T/\rho) \operatorname{sinc}(T[\rho-\alpha]).$$

Reference to (6.3) then confirms the correctness of (9.14).

The definition (9.4) ensures that

$$p(-\xi; \phi+\pi) = p(\xi; \phi), \tag{9.15}$$

which, when substituted into the second formula in (9.7), gives

$$p_l(-\xi) = (-1)^l p_l(\xi). \tag{9.16}$$

It then follows from (9.8) and the definitions (6.1) that

$$F_l(-u) = (-1)^l F_l(u). \tag{9.17}$$

Consequently, the even and odd Fourier coefficients in the summations introduced in (9.7) are themselves even and odd functions, respectively, of their arguments.

It is useful to define the *integrated image* $\mathbf{I}_f$ by

$$\mathbf{I}_f = \int\!\!\int_{-\infty}^\infty f(x, y) \, dx \, dy. \tag{9.18}$$

It follows from (9.1) and (9.4) that

$$\mathbf{I}_f = \int_{-\infty}^\infty p(\xi; \phi) \, d\xi, \tag{9.19}$$

which is here referred to as the *primary consistency condition* for projections.

10 Fourier series and sampling theorem

When the true image $f(\mathbf{x})$ is of finite size (see §7), it is often convenient to place the centre of its frame Ω_f (whose x-extent and y-extent are here denoted by L_1 and L_2 respectively) at the origin of image space:

$$f(\mathbf{x}) = f(x, y) = 0 \qquad \text{for } |x| > L_1/2,\ |y| > L_2/2. \qquad (10.1)$$

Within its frame, $f(\mathbf{x})$ can be represented by the *trigonometric Fourier series*:

$$f(\mathbf{x}) = (1/L_1 L_2) \sum_{l=-M_1}^{M_1} \sum_{m=-M_2}^{M_2} F_{l,m} \exp(-\mathrm{i}\, 2\pi[lx/L_1 + my/L_2])$$

$$\text{for } \mathbf{x} \in \Omega_f, \qquad (10.2)$$

where the $F_{l,m}$ are constants, called *Fourier coefficients*, which satisfy

$$F_{-l,-m} = F_{l,m}^* \qquad (10.3)$$

because of (3.10). The integers M_1 and M_2 are both infinite in general. However, real-world images are never free of contamination. It is pointless to attempt to represent detail that is irretrievably lost in whatever 'noise' is present. Since the $|F_{l,m}|$ must decrease fairly rapidly with increasing $|l|$ and increasing $|m|$, for $|l|$ and $|m|$ larger than particular integers (say, M_1 and M_2 respectively), only a finite number of them are meaningful. Anyway, only a finite number of anything is computable! But it is worth remembering that M_1 and M_2 usually increase as the contamination decreases – the higher the quality of the image, the larger is the number of Fourier coefficients that it is worth taking into account.

On substituting (10.1) and (10.2) into (6.12), and recalling the definition (6.4), it is seen that

$$F(\mathbf{u}) = F(u, v) = \sum_{l=-M_1}^{M_1} \sum_{m=-M_2}^{M_2} F_{l,m} \operatorname{sinc}(L_1 u - l) \operatorname{sinc}(L_2 v - m). \qquad (10.4)$$

Inspection of (6.4) indicates that, for any two different integers j and k,

$$\operatorname{sinc}(t - j) = 1 \qquad \text{for } t = j,$$
$$= 0 \qquad \text{for } t = k, \qquad (10.5)$$

which shows that

$$F(l/L_1, m/L_2) = F_{l,m} \qquad (10.6)$$

follows from (10.4). The formula (10.6) is the two-dimensional *sampling theorem* – its one-dimensional form is conveniently expressed by setting $F_{l,m} \equiv 0$ for either $l \neq 0$ or $m \neq 0$. The significance of this theorem is that an image of finite size can be exactly represented in Fourier space by samples (i.e. values assumed by $F(\mathbf{u})$ at discrete points) spaced by the reciprocals of the lengths of the sides of any rectangular domain which encloses the image. It is often (but by no means always – refer to §11) convenient for L_1 and L_2 to assume their smallest allowable values. In any practical situation, a finite number of samples suffices. It is worth remarking that the samples referred to above are of the kind called point samples in §11. The sample spacings, implicit in (10.6), of $1/L_1$ parallel to the u-axis and $1/L_2$ parallel to the v-axis are called the *Nyquist frequencies* in the u-direction and the v-direction respectively.

It is here convenient to introduce the *periodic image* $\text{p}(\mathbf{x})$, which repeats itself throughout image space within identical contiguous rectangular frames, whose extents are N times those of Ω_f, where N is any particular positive integer. The l, mth frame, here denoted by $\Omega_{\text{p},l,m}$, occupies the domain $0 < |x - lNL_1| < NL_1/2$, $0 < |y - mNL_2| < NL_2/2$ where the integers l and m range from $-\infty$ to ∞. It is further convenient to introduce the frames $\Omega_{f,l,m}$ which occupy the domains $0 < |x - lNL_1| < L_1/2$, $0 < |y - mNL_2| < L_2/2$. Note that each $\Omega_{f,l,m}$ occupies the central rectangle, of extents L_1 and L_2, of each $\Omega_{\text{p},l,m}$. Now $\text{p}(\mathbf{x})$ is defined by

$$\text{p}(\mathbf{x}) = f(x - lNL_1, y - mNL_2) \qquad \text{for } \mathbf{x} \in \Omega_{f,l,m},$$
$$= 0 \qquad \text{for } \mathbf{x} \in \bar{\Omega}_{f,l,m}, \qquad (10.7)$$

where $\bar{\Omega}_{f,l,m}$ is the part of $\Omega_{\text{p},l,m}$ *not* occupied by $\Omega_{f,l,m}$. It follows that

$$\text{p}(\mathbf{x}) = (1/N^2L_1L_2) \sum_{l=-M_1'}^{M_1'} \sum_{m=-M_2'}^{M_2'} \mathfrak{P}_{l,m} \exp(-i\, 2\pi[lx/L_1 + my/L_2]/N) \qquad (10.8)$$

for $\mathbf{x} \in \Omega_{\text{p},0,0}$, where the integers M_1' and M_2' are large enough that RHS (10.8) describes $\text{p}(\mathbf{x})$ as accurately as $f(\mathbf{x})$ is described by RHS (10.2), and

$$\mathfrak{P}_{l,m} = \int_{-L_1/2}^{L_1/2} \int_{-L_2/2}^{L_2/2} f(x, y) \exp(i\, 2\pi[lx/L_1 + my/L_2]/N)\, dx\, dy. \qquad (10.9)$$

It follows further that

$$\mathfrak{P}(\mathbf{u}) = \mathsf{F}\{\text{p}(\mathbf{x})\},$$

$$= (1/N^2L_1L_2) \sum_{l=-M_1'}^{M_1'} \sum_{m=-M_2'}^{M_2'} \mathfrak{P}_{l,m}$$

$$\times \int\!\!\!\int_{-\infty}^{\infty} \exp(i\, 2\pi[(u - l/L_1)x + (v - m/L_2)y]/N)\, dx\, dy,$$

$$= (1/N^2L_1L_2) \sum_{l=-M_1'}^{M_1'} \sum_{m=-M_2'}^{M_2'} \mathfrak{P}_{l,m}\, \delta(u - l/NL_1)\, \delta(v - m/NL_2), \qquad (10.10)$$

where (6.3) has been invoked. Note that $\mathfrak{P}(\mathbf{u})$ exists effectively only at the points $(l/L_1, m/L_2)$ in Fourier space.

11 Sampling, interpolation, and extrapolation

Before a measured quantity (or a mathematical function) can be subjected to any digital computational operation, it must be sampled. To be specific, consider sampling the spectrum in Fourier space. The 'resolution' of the sampling procedure is characterized by what is here called the *instrument function* of the apparatus with which the sampling is effected. Since the sensitivity of the apparatus can depend upon the position in Fourier space of the *sample point* $\mathbf{u}_k$, it is appropriate to write the instrument function as $\mathrm{inst}_k(\mathbf{u})$. The *sample* of $F(\mathbf{u})$ at $\mathbf{u}_k$ is here defined by

$$F'_k = F(\mathbf{u}) \,\text{\textcircled{$\ominus$}}\, \mathrm{inst}_k(\mathbf{u}). \tag{11.1}$$

Note that RHS (11.1) depends only on the point $\mathbf{u}_k$ because of the way $\mathrm{inst}_k(\mathbf{u})$ has been defined. The quantity

$$F_k = F(\mathbf{u}_k) \tag{11.2}$$

is here called a *point sample* because it corresponds to an instrument having perfect resolution – i.e. it is sensitive only to the amplitude of $F(\mathbf{u})$ at $\mathbf{u}_k$, implying that $\mathrm{inst}_k(\mathbf{u}) = (\delta(u - u_k)\,\delta(v - v_k))$ where u_k and v_k are the Cartesian components of $\mathbf{u}_k$. Note that the $F_{l,m}$, introduced in (10.2), are point samples.

In many practical applications it is both satisfactory and convenient to employ *separable* instrument functions, whose u and v dependences factorize thus:

$$\mathrm{inst}_k(\mathbf{u}) = \mathrm{inst}_k^+(u)\,\mathrm{inst}_k^-(v), \tag{11.3}$$

where the $\mathrm{inst}_k^\pm(\cdot)$ are called *one-dimensional instrument functions*, useful examples of which have the (idealized) forms $((4/\pi\sigma^2)^{\frac{1}{2}}\exp(-4u^2/\sigma^2))$, $(\sigma^{-1}\,\mathrm{rect}(u/\sigma))$ and $((2/\sigma)\,\mathrm{tri}(u/\sigma))$, where σ is the 'effective width' of each function. All three of these reduce to $\delta(u)$ as $\sigma \to 0$. Note that

$$\mathrm{rect}(u) = \mathrm{tri}(u) = 0 \qquad \text{for } |u| > \tfrac{1}{2}, \tag{11.4}$$

and

$$\left.\begin{array}{l} \mathrm{rect}(u) = 1 \\ \mathrm{tri}(u) = 1 - 2u \end{array}\right\} \quad |u| < \tfrac{1}{2}. \tag{11.5}$$

Given samples of some function, say $F(\mathbf{u})$, at a set $\{\mathbf{u}_k;\ k = 1, 2, \ldots, K\}$ of sample points, one often wants to estimate the amplitudes of the function at other points. If there exists, or can be found, a set $\{\mathrm{samp}_k(\mathbf{u});\ k = 1, 2, \ldots, K\}$ of suitable 'interpolation functions', or *sampling functions* as they are called here, then $F(\mathbf{u})$ can be estimated throughout Fourier space:

$$F(\mathbf{u}) \simeq \sum_{k=1}^{K} F_k\,\mathrm{samp}_k(\mathbf{u}). \tag{11.6}$$

When sufficient *a priori* information concerning $F(\mathbf{u})$ is available, (11.6) can on occasion be an exact formula, rather than merely an estimate. Suppose, for instance, that the data points lie on a rectangular grid. It is convenient to take the u- and v-axes parallel to the lines of the grid, whose spacings are here denoted by $1/L_1'$ and $1/L_2'$ in the u- and v-directions respectively. Suppose further that the image $f(\mathbf{x})$ is known to satisfy (10.1) with

$$L_1 \le L_1' \quad \text{and} \quad L_2 \le L_2'. \tag{11.7}$$

It then follows from the first two paragraphs of §10 that

$$F(\mathbf{u}) = \sum_{l=-M_1'}^{M_1'} \sum_{m=-M_2'}^{M_2'} F_{l,m}' \operatorname{sinc}(L_1'u - l) \operatorname{sinc}(L_2'v - m) \tag{11.8}$$

is exact provided the $F_{l,m}'$ are point samples, and every member of $\{F_k; k = 1, 2, \ldots, K\}$ corresponds to one and only one of the $F_{l,m}'$ where $l/L_1' = u_k$ and $m/L_2' = v_k$, and each non-zero $F_{l,m}'$ corresponds to one and only one F_k, and

$$F(l/L_1', m/L_2') = 0 \quad \text{for } |l| > M_1' \text{ and/or } |m| > M_2'. \tag{11.9}$$

In practice, there is virtually never enough *a priori* information to permit an exact formula for $F(\mathbf{u})$ to be constructed. However, it is sometimes known how closely the samples should be spaced to ensure that $|F(\mathbf{u})|$ is simply convex or concave in between most of the sample points. When the sample spacing satisfies this condition, interpolation can usually be effected to a useful accuracy.

For many applications it is necessary to handle enormous numbers of samples, processing them sequentially in groups. The formula (11.6) is then only useful if there is a straightforward way of substituting the given samples into it. The formula is particularly convenient and computationally efficient if, for each k,

$$\operatorname{samp}_k(\mathbf{u}_k) = 1 \quad \text{and} \quad \operatorname{samp}_j(\mathbf{u}_k) = 0 \quad \text{for } j \ne k, \tag{11.10}$$

because RHS (11.6) then reduces to F_j when $\mathbf{u} = \mathbf{u}_j$.

It is only useful to incorporate all K of the given samples into a single interpolation formula, such as RHS (11.6), if $|\operatorname{samp}_k(\mathbf{u})|$ decreases monotonically, or exhibits damped oscillations about a falling mean, as $|\mathbf{u} - \mathbf{u}_k|$ increases. The sinc functions, defined by (6.4) and most commonly occurring in formulas such as (10.4), behave in this way. However, it is only justifiable to invoke *sinc interpolation* when one is sure that the function being interpolated is *band-limited* – i.e. its Fourier transform is identically zero outside some particular frame. One also needs to possess fairly good estimates of the extents of this frame (refer to §7) if the interpolation is to be successful.

For most sampling functions, best results are obtained by processing the samples sequentially in small groups. The optimum number, here denoted by J for a one-dimensional formula and M $(\simeq J^2)$ for a two-dimensional formula, of samples to include in a group depends upon the sampling functions being used and on the type of quantity being interpolated. It is seldom, however, that J can be permitted to exceed 7 and a value of 4 or 5 is often best.

Probably the simplest way of ensuring that the sampling functions satisfy (11.10) is to separate their u and v dependencies:

$$\mathrm{samp}_k(\mathbf{u}) = \mathrm{samp}_j^+(u)\,\mathrm{samp}_k^-(v), \tag{11.11}$$

with

$$\mathrm{samp}_k^\pm(\alpha) = 1 \quad \text{and} \quad \mathrm{samp}_j^\pm(\alpha_k) = 0 \qquad \text{for } j \neq k, \tag{11.12}$$

where $\alpha_k = u_k$ and $\alpha_k = v_k$ correspond, respectively, to the $+$ and $-$ superscripts. Unfortunately, (11.12) is only compatible with (11.10) if the sample points lie on a rectangular grid. Note that the pair of sinc functions inside the summation in (11.8) are characterized precisely by (11.11) if u_k and v_k are identified with l/L_1' and m/L_2' respectively. The use of sinc functions is only unexceptional when (10.1) and (11.7) are known to be true. If no *a priori* estimate of the size of $f(\mathbf{x})$ is available, or if $F(\mathbf{u})$ is merely some function of any two parameters (called u and v purely for convenience) whose form one wishes to estimate from given samples, then it can be more appropriate to employ Lagrange polynomials. Consider a subset $\{\mathbf{u}_{j,k}; j = 1, 2, \ldots, J;$ one value of $k\}$ of the set of sample points, where the Cartesian components of $\mathbf{u}_{j,k}$ are u_j and v_k – i.e. all of the $\mathbf{u}_{j,k}$ lie on a straight line parallel to the u-axis and distant v_k from it. An estimate of the form of $F(\mathbf{u})$ along this line is given by

$$F(u, v_k) \simeq \sum_{j=1}^{J} F_{j,k}\,\mathrm{lagrange}_j(u), \tag{11.13}$$

where $F_{j,k}$ is the given sample corresponding to the sample point $\mathbf{u}_{j,k}$, and where the jth *Lagrange polynomial* is here defined by

$$\mathrm{lagrange}_j(u) = \prod_{\substack{l=1 \\ l \neq j}}^{J} (u - u_l)/(u_j - u_l)$$

$$= 1 \qquad \text{for } u = u_j$$

$$= 0 \qquad \text{for } u = u_l \text{ when } l \neq j. \tag{11.14}$$

The right-hand side of (11.13) often provides a usefully accurate estimate of $F(\mathbf{u})$ along the segment of the line $v = v_k$ lying between the least and greatest of the u_j – i.e. it is a useful *interpolation* formula – but it is usually dangerous to use it outside this segment – i.e. it is *not* a useful *extrapolation* formula.

Lagrange interpolation is also straightforward when the sample points lie along a set of arbitrarily inclined straight lines that criss-cross each other. One selects subsets of the sample points on J of the lines closest to the desired interpolation point (α, β), and employs Lagrange interpolation along each line to estimate $F(\mathbf{u})$ at the J points having the same v-coordinate as the interpolation point. Because these J points lie themselves along a straight line, $F(\alpha, \beta)$ can be estimated by a further application of conventional Lagrange interpolation.

The enormous versatility of Lagrange interpolation, as well as its computational attractiveness, resides in its ability to handle sample points that are irregularly spaced along straight or even curved lines – for the latter, the variable u on RHS (11.13) becomes arc length along a curve. These advantages evaporate, however, when sample points are arbitrarily distributed over a plane. The trouble is that there do not seem to be any tricks, for describing smooth curves through irregularly positioned points, that are significantly cheaper computationally than interpolation based on *global polynomials*, a typical one of which contains of the order of J^2 terms (where J has the same meaning as in the previous three paragraphs), e.g.

$$F(\mathbf{u}) = \sum_{m=0}^{J} \sum_{n=0}^{m} A_{m,n} u^{m-n} v^n, \tag{11.15}$$

where the $A_{m,n}$ are constants. On RHS (11.15) it is to be understood that the origin of the u, v-coordinates is shifted to the vicinity of the interpolation point. Before $F(\alpha, \beta)$ can be estimated, the $A_{m,n}$ must be evaluated by substituting into (11.15) on LHS and RHS, respectively, the given samples and the coordinates of the sample points (denoting Cartesian coordinates of each $\mathbf{u}_k$ by u_k and v_k):

$$F(\mathbf{u}_k) = \sum_{m=0}^{J} \sum_{n=0}^{m} A_{m,n} u_k^{m-n} v_k^n \tag{11.16}$$

for $M = (J+2)(J+1)$ values of the integer k. The $A_{m,n}$ are then found by elimination or matrix inversion, procedures that are very expensive computationally when many interpolations must be performed.

Before describing how the M given samples closest to the desired interpolation point can be incorporated into two-dimensional formulas which are both accurate and computationally convenient, we think it is worth remarking that subtractive deconvolution (§17) is a kind of interpolation technique. The point is that subtractive deconvolution is particularly useful when data are gathered in Fourier space, over a frame large enough to guarantee the desired resolution, but with appreciable gaps in the coverage of this frame. The algorithm outlined in §17 can be regarded as species of nonlinear processing for filling these gaps. The maximum entropy method discussed in §18 can be similarly viewed.

Much of the computational convenience of conventional Lagrange interpolation can be retained, when the sample points are irregularly positioned, by mapping the real plane one-to-one onto a complex plane. The real position vector $\mathbf{u}$ is replaced in (11.6) by a complex variable, w say. Having chosen the $M \simeq J^2$ sample points closest to the desired interpolation point, one divides them into J groups, with the points in each group lying as closely as the given sample points allow to a straight line parallel to, say, the v-axis. One point, near the middle of this 'line', is chosen (temporarily) as the origin of the w-plane. In the vicinity of the kth of these 'lines', an estimate of $F(\mathbf{u})$ is given by (11.13) with the $F_{j,k}$ now being the given samples at the sample points on the kth 'line'. For each k, we require that

$$\text{Imaginary part of } F(w_{j,k}) = 0 \tag{11.17}$$

for all J values of j, where $w_{j,k} = u_{j,k} + \mathrm{i}\, v_{j,k}$. One next uses (11.13) to estimate the imaginary part of $F(u + \mathrm{i}\,\beta)$ for values of u close to the 'average' u-coordinate of the kth 'line', choosing that value α_k for which

$$\text{Imaginary part of } F(\alpha_k + \mathrm{i}\,\beta) = 0. \tag{11.18}$$

By doing this for all J 'lines', estimates are obtained for J values of $F(u + \mathrm{i}\,\beta)$ whose imaginary parts are zero. Their real parts can be considered samples of $F(u, v)$ at new sample points (α_k, β). Since the latter have been forced to lie on a straight line, an estimate of the amplitude of $F(\mathbf{u})$ at the desired interpolation point (α, β) is immediately given by conventional Lagrange interpolation.

Up to now we have been concentrating on how to interpolate as accurately and as efficiently as possible, because in many situations of scientific and technical importance it is necessary to handle huge amounts of data. When one is interpolating as part of a structural or fabricational design, for example, or as an intermediate image processing step, it is usually necessary to achieve considerable accuracy, to ensure the quality of the finished product or to prevent errors contaminating the results of subsequent steps. There are other instances, however, when rapid, relatively crude techniques are satisfactory. It is sometimes possible (as in the application discussed at length in §33) to order the steps in a large image processing procedure in such a way that simple interpolation methods are adequate. Two particular methods, which are described in detail within a specific practical context in §33, are there called *nearest neighbour interpolation* and *linear interpolation*. Problems associated with implementing algorithms for interpolating onto rectangular grids are discussed in §§45 through 47.

While interpolation can often be successfully implemented without too much difficulty, *extrapolation* tends to be an unstable operation unless tightly constrained. Very rarely is it possible to introduce constraints straightforwardly. One must usually resort to some iterative procedure. A good and useful example is the Gerchberg algorithm, the essentials of which are described below. In imaging contexts, extrapolation is often referred to as *super-resolution* (it is mentioned further in §15).

Suppose $F(\mathbf{u})$ is given within a restricted region R_- *of Fourier space and it is desired to estimate* $F(\mathbf{u})$ *throughout a larger region* R *which encloses* R_-. It is convenient to define

$$F(\mathbf{u}) = F_g(\mathbf{u}) \qquad \text{for } \mathbf{u} \in \mathsf{R}_-,$$
$$= F_e(\mathbf{u}) \qquad \text{for } \mathbf{u} \in \mathsf{R}_+, \tag{11.19}$$

where R_- and R_+, which together constitute R, do not intersect:

$$\mathsf{R} = \mathsf{R}_- \cup \mathsf{R}_+ \quad \text{and} \quad \mathsf{R}_- \cap \mathsf{R}_+ = \varnothing, \tag{11.20}$$

where $\varnothing$ denotes the empty set. The subscripts g and e introduced in (11.19) imply, respectively, what is 'given' and what is to be 'estimated'.

Gerchberg's algorithm is predicated upon $F_g(\mathbf{u})$ (which is here taken to equal $F_g^*(-\mathbf{u})$ because the image is real – see §8) spanning sufficient of Fourier space to permit the image frame (see §7) Ω_f to be estimated to a useful accuracy. A *preliminary estimate* $f_e^{(p)}(\mathbf{x})$ of the true image is defined by

$$f_e^{(p)}(\mathbf{x}) = \mathsf{F}\{F_g(\mathbf{u})\}. \tag{11.21}$$

All points in image space where $f_e^{(p)}(\mathbf{x})$ exceeds a pre-set threshold (chosen to suit the expected uncertainty due to noise, the incompleteness of the data, etc.) are identified. The smallest rectangle, with its sides parallel to the x- and y-axes, which just encloses these points is defined to be Ω_f.

The *first estimate* $f_e(\mathbf{x})$ of the image is the non-negative part of $f_e^{(p)}(\mathbf{x})$ within Ω_f. The first estimate of $F_e(\mathbf{u})$ is now defined by

$$F_e(\mathbf{u}) = \mathsf{F}\{f_e(\mathbf{x})\} \qquad \text{for } \mathbf{u} \in \mathsf{R}_+, \tag{11.22}$$

which is then added to the given data – i.e. $F_g(\mathbf{u})$ – to provide a second preliminary estimate – also denoted by $f_e^{(p)}(\mathbf{x})$ to avoid multiplication of notation – of the image by Fourier transformation:

$$f_e^{(p)}(\mathbf{x}) = \mathsf{F}\{F_g(\mathbf{u} \in \mathsf{R}_-) + F_e(\mathbf{u} \in \mathsf{R}_+)\}. \tag{11.23}$$

The second estimate – again denoted by $f_e(\mathbf{x})$ – of the true image itself is the part of $f_e^{(p)}(\mathbf{x})$ which is both within Ω_f and above the aforementioned threshold. An iterative loop has thus been set up because the new $f_e(\mathbf{x})$ can be substituted into (11.22) to provide a new $F_e(\mathbf{u})$ which is itself substituted into (11.23), etc. This procedure is continued until the quantity

$$\left(\iint\limits_{\Omega_f} |f_{e,m}(\mathbf{x}) - f_{e,m-1}(\mathbf{x})|^2 \, d\sigma(\mathbf{x}) \right) \Big/ \iint\limits_{\Omega_f} |f_{e,m}(\mathbf{x})|^2 \, d\sigma(\mathbf{x})$$

is less than some pre-set level directly related to the threshold, where $f_{e,m}(\mathbf{x})$ is the mth estimate of the true image. The convergence of this algorithm can often be accelerated by refinements of the kind discussed in a slightly different context in §23; but the above outline covers the essential features of the algorithm.

The basic constraint here is forcing each $f_{e,m}(\mathbf{x})$ to be positive and to lie inside Ω_f. Without this constraint the algorithm is useless. What is critical, therefore, is whether $F_g(\mathbf{u})$ really does span sufficient of Fourier space to permit Ω_f to be estimated usefully. This is hardly surprising when one realizes that extrapolation can only be expected to be successful when the data are adequate enough.

There are situations of practical interest for which the image frames are known *a priori*. Gerchberg's algorithm is then markedly more powerful and can sometimes provide remarkably large extrapolations from comparatively meagre data. This is discussed further in §15.

12 Discrete and fast Fourier transforms

The *discrete Fourier transform* (D.F.T.) is a trigonometric Fourier series representation which relates point samples of an image of finite size to point samples of its Fourier transform. The sampling points must lie on a rectangular grid, spaced by Δ_1 say in the x-direction and Δ_2 say in the y-direction. If $f(x, y)$ is the image then its samples are necessarily given by

$$f_{l,m} = \int_{-L_1/2}^{L_1/2} \int_{-L_2/2}^{L_2/2} f(x, y)\, \delta(x - l\,\Delta_1)\, \delta(y - m\,\Delta_2)\, dx\, dy. \qquad (12.1)$$

It is, of course, impossible to record point samples from actual data. What is implied by (12.1) from a practical point of view is that the 'effective spatial bandwidth' – this concept is defined in §6 in the paragraph following (6.11) – is considerably less than $1/\Delta$, where Δ is the smaller of Δ_1 and Δ_2. Since $f(x, y)$ is constrained to be of finite size, it can be characterized by (10.1) and (10.2), so that the image samples are given in terms of the Fourier samples by

$$f_{l,m} = (1/L_1 L_2) \sum_{l'=-M_1}^{M_1} \sum_{m'=-M_2}^{M_2} F_{l',m'}$$
$$\times \exp(-i\, 2\pi[ll'\, \Delta_1/L_1 + mm'\, \Delta_2/L_2]). \qquad (12.2)$$

It would clearly be advantageous to find a formula, similar to RHS (12.2) but giving each $F_{l,m}$ in terms of a summation over the $f_{l,m}$. Such a formula can, in fact, be developed provided L_1 and L_2 are chosen to be integral multiples of Δ_1 and Δ_2 respectively; e.g. set

$$L_1/\Delta_1 = 2M_1 + 1 \quad \text{and} \quad L_2/\Delta_2 = 2M_2 + 1. \tag{12.3}$$

The point of this is that

$$(1/(2M+1)) \sum_{l=-M}^{M} \exp(i\, 2lm\pi/(2M+1))$$

$$= (\sin(m\pi))/(2M+1) \sin(m\pi/[2M+1]) = \delta_{0,m-(2M+1)l} \tag{12.4}$$

for all integers l, where the Kronecker delta is defined in (9.9). Consequently, the left side of (12.4) is a 'discrete delta function', which plays the same role for trigonometric Fourier series as $\delta(\cdot)$ does for Fourier integrals. So, multiplying $f_{l,m}$ by $\exp(i\, 2\pi[ll'/(2M_1+1) + mm'/(2M_2+1)])$ and summing from $-M_1$ to M_1 over l and from $-M_2$ to M_2 over m gives

$$F_{l,m} = (1/\Delta_1 \Delta_2) \sum_{l'=-M_1}^{M_1} \sum_{m'=-M_2}^{M_2} f_{l',m}$$

$$\times \exp(i\, 2\pi[ll'/(2M_1+1) + mm'/(2M_2+1)]), \tag{12.5}$$

as follows directly from (12.2) and (12.4).

It is now instructive to apply, in combination, the techniques introduced in this section and in §10 to $b(\mathbf{x})$, which is called the ideal blurred image in §4. Consider (3.4). Any practical psf is (effectively) of finite size, in the sense that its amplitude is negligible outside a finite region of image space. The *point spread frame*, denoted by Ω_h (invoking the notation introduced in §7), is here defined to be the rectangle (with its sides parallel to the x- and y-axes) which is just large enough to enclose the aforesaid region. It follows that Ω_b can be constructed by placing the centre of Ω_h on the perimeter of Ω_f and moving it all round this perimeter. The outer edges of Ω_h then generate the perimeter of Ω_b. Now think of image space covered by an infinity of contiguous, non-overlapping rectangles each of the same size as Ω_f. Superimposed upon each of these rectangles is a frame the same size as Ω_b with its centre coinciding with that of the rectangle. The ideal blurred image is repeated within each of these frames. Note that the outer parts of adjacent versions of $b(\mathbf{x})$ overlap because adjacent frames overlap. The resulting periodic image, denoted by $\mathfrak{p}_b(\mathbf{x})$ and called the *periodic overlapped ideal blurred image*, can be written as

$$\mathfrak{p}_b(\mathbf{x}) = \sum_{l,m=-\infty}^{\infty} b(x - lL_1, y - mL_2), \tag{12.6}$$

the spatial periods of which are $(1/N)$ of those of $\mathfrak{p}(\mathbf{x})$, as defined in §10. On denoting by $\Omega_{l,m}$ the l, mth of the aforementioned contiguous non-overlapping rectangles, and invoking the definitions (6.12) and (6.15), it is seen that the following formula can be derived from (12.6):

$$\mathfrak{P}_b(\mathbf{u}) = \mathsf{F}\{\mathfrak{p}_b(\mathbf{x})\} = \sum_{l,m=-\infty}^{\infty} \iint_{\Omega_{l,m}} b(x - lL_1, y - mL_2)$$

$$\times \exp(\mathrm{i}\, 2\pi[ux + vy])\, dx\, dy$$

$$= \sum_{l,m=-\infty}^{\infty} \exp(\mathrm{i}\, 2\pi[lL_1 u + mL_2 v]) \iint_{\Omega_f} b(x, y)$$

$$\times \exp(\mathrm{i}\, 2\pi[ux + vy])\, dx\, dy, \tag{12.7}$$

where, in each term in the second summation, x and y equal, respectively, $(x - lL_1)$ and $(y - mL_2)$ in the first summation, so that each $\Omega_{l,m}$ in the first summation becomes Ω_f in the second summation. Now refer to (12.4) and consider what happens when it is multiplied through by $(2M + 1)$ and the limit $M \to \infty$ is taken. It is seen that the Kronecker delta becomes a Dirac delta function – e.g.

$$\sum_{m=-\infty}^{\infty} \exp(\mathrm{i}\, 2m\pi\alpha) = \sum_{m=-\infty}^{\infty} \delta(\alpha - m), \tag{12.8}$$

which, when substituted into (12.7), together with the definition

$$B(\mathbf{u}) = \mathsf{F}\{b(\mathbf{x})\}, \tag{12.9}$$

gives

$$\mathfrak{P}_b(\mathbf{u}) = B(\mathbf{u}) \sum_{l,m=-\infty}^{\infty} \delta(L_1 u - l)\, \delta(L_2 v - m). \tag{12.10}$$

Since (3.4), (12.9) and the convolution theorem (7.7) combine to give

$$B(\mathbf{u}) = F(\mathbf{u})H(\mathbf{u}), \tag{12.11}$$

where $F(\mathbf{u})$ and $H(\mathbf{u})$ are defined in §§6 and 8 respectively, it follows that (12.10) reduces to

$$\mathfrak{P}_b(\mathbf{u}) = \sum_{l,m=-\infty}^{\infty} B_{p,l,m}\, \delta(L_1 u - l)\, \delta(L_2 v - m), \tag{12.12}$$

where it is seen from (6.4), (10.4), and (12.11) that

$$B_{p,l,m} = F_{l,m} H(l/L_1, m/L_2) \tag{12.13}$$

This formula is discussed further in §14.

The formulas (12.2) and (12.5), which correspond to the Fourier integral formulas (6.13) and (6.12) respectively, are the relations that define the D.F.T. Note that the point samples of the image appearing in these formulas are the point samples of the true image itself provided it is effectively contained within the rectangle $|x| < L_1/2$, $|y| < L_2/2$ and the $F_{l,m}$ are effectively zero for $|l| > M_1$ and $|m| > M_2$. So, the D.F.T. can be just as 'exact' as the conventional Fourier transform.

The D.F.T. is the theoretical starting point for the development of the Fast Fourier Transform (F.F.T.) algorithm, which has revolutionized many spectral computational procedures. The way in which the F.F.T. arises from the D.F.T. is explained later in this section. Its computational efficiency is such that, when one is required to compute $f \odot g$, it is advisable to invoke the convolution theorem (7.7) and carry out the computational procedure: $\mathsf{F}\{\mathsf{F}\{f\}\mathsf{F}\{g\}\}$. For the computation of $f \ast g$ to be efficient, it should be performed similarly, after invoking the correlation theorem (7.8). It is also worth noting that most F.F.T. program packages require the number of samples to be a power of 2, in which case the summations in the D.F.T. formulas can no longer be symmetrical, as in (12.2) and (12.5), – this usually causes some initial difficulty for the user, but the algorithm's idiosyncracies tend to become transparent after a brief acquaintance.

In many practical situations it is useful to be able to vary the resolution to which spectra are computed. Suppose it is known (or thought) to be sufficient (in a particular application) to sample a given function $f(x)$ at 2^N points spaced by Δ. Invoking only these samples when employing the D.F.T. constraints $F(u)$ to be computed only at 2^N points spaced by $1/L$, where $L = (2^N - 1)\Delta$. How can one arrange matters so as to decrease the spacing of the points on the u-axis at which $F(u)$ is computed? It is obvious that f must somehow be made to occupy a longer segment of the x-axis. But suppose f is effectively zero outside the segment of length L. The only answer is to take extra samples of zero amplitude outside the segment of length L. In fact, a new segment of length $\hat{L}$ must be taken, where $\hat{L}/L$ must be a power of 2 if a standard F.F.T. program package is to be used. The extra samples must still be spaced by Δ. It follows that by employing this neat trick, which has come to be generally known as 'packing f with zeros', the computed samples of $F(u)$ can be spaced by $1/2L$, or $1/4L$, etc., along the u-axis. The same trick can be used in two, or more, dimensions. A given $f(\mathbf{x})$ must be packed with zeros (i.e. samples of zero amplitude) outside the rectangular region within which it has sensible value.

In order to implement (12.2) and (12.5) by means of the F.F.T., we replace (12.3) by

$$L_1/\Delta_1 = N_1 \quad \text{and} \quad L_2/\Delta_2 = N_2, \tag{12.14}$$

and take the unit of length in the x- and y-directions to be Δ_1 and Δ_2 respectively. So, (12.2) and (12.5) become

$$f_{l,m} = (1/N_1 N_2) \sum_{l'=0}^{N_1-1} \sum_{m'=0}^{N_2-1} F_{l',m'} \exp(-i \, 2\pi[ll'/N_1 + mm'/N_2]), \tag{12.15}$$

and

$$F_{l,m} = \sum_{l'=0}^{N_1-1} \sum_{m'=0}^{N_2-1} f_{l',m'} \exp(i \, 2\pi[ll'/N_1 + mm'/N_2]), \tag{12.16}$$

respectively. For the purposes of the F.F.T. both $f_{l,m}$ and $F_{l,m}$ are considered to be periodic, with N_1 and N_2 being the periods in l and m respectively. Thus

$$f_{l,m} = f_{l+jN_1, m+kN_2} \quad \text{and} \quad F_{l,m} = F_{l+jN_1, m+kN_2}, \tag{12.17}$$

where j and k are arbitrary integers. Any selection of summation ranges in (12.15) and (12.16), for the purpose, say, of fitting in with some particular computing convention, is therefore acceptable. This implies that f as represented in Ω_f by the $f_{l,m}$ can be taken to be *circular* in Ω_f. By this we mean that the top and bottom edges of Ω_f can be considered the same, as can the left and right edges of Ω_f. An identical argument applies to F within its corresponding frame Ω_F in Fourier space.

It should be noted that if f is real, (8.2) implies that $F_{0,0}$, $F_{N_1/2,m}$ and $F_{l,N_2/2}$ are all real. Furthermore, the signs of the exponents in (12.15) and (12.16) are interchanged in some texts. This does not matter so long as data are organized consistently with whatever program package is being used.

The crucial advantage of the F.F.T. over direct implementation of the D.F.T. is its speed. For an $N \times N$ pixel complex image, the direct D.F.T. requires of the order of $4N^4$ complex operations (consisting of multiplications and additions), whereas the corresponding number for the F.F.T. is $(8N^2 \log_2(N))$. For instance, when $N = 256$, the improvement factor is 4096. There are two reasons for this speed increase. First, the $N \times N$ two-dimensional F.F.T. can be treated as $2N$ one-dimensional F.F.T.s each of length N. Second, if $N = 2^n$, each one-dimensional F.F.T. of length N can be broken down into $N/2$ sets of one-dimensional F.F.T.s of length 2. To make clear why this second consideration leads to increased computational efficiency, it is convenient to express (12.16) in one-dimensional form:

$$F_l = \sum_{l'=0}^{N-1} f_{l'} \exp(-i\, 2\pi l l'/N). \tag{12.18}$$

Evaluating RHS (12.18) as it stands would involve $2N$ complex operations. So, $2N^2$ such operations would be needed to calculate all N of the F_l. However, we can split the summation into its even and odd parts, i.e.

$$F_l = A_l^+ + A_l^- \exp(-i\, 2\pi l/N), \tag{12.19}$$

where

$$A_l^{\pm} = \sum_{j=0}^{N/2-1} f_{2j+k(1)} \exp(-i\, 4\pi j l/N), \tag{12.20}$$

with $k(1)$ being 0 for A_l^+ and 1 for A_l^-. Note that each $A_l^{\pm}$ is of period $N/2$. There are now 2 summations, but each involves only N complex operations. To calculate all of the $A_l^{\pm}$, therefore, we must perform

$((2N) \times (N/2)) = N^2$ operations. In addition, $2N$ operations are needed to calculate all the F_l defined by (12.19). This gives a total of $(N^2 + 2N)$ operations. So the number of necessary operations has been approximately halved. This trick can be repeated:

$$A_l^+ = A_l^{++} + A_l^{+-} \exp(-i\, 4\pi l/N), \tag{12.21}$$

and

$$A_l^- = A_l^{-+} + A_l^{--} \exp(-i\, 4\pi l/N), \tag{12.22}$$

where

$$A_l^{\pm\pm} = \sum_{j=0}^{N/4-1} f_{4j+k(2)} \exp(-i\, 8\pi j l/N), \tag{12.23}$$

with $k(2)$ being 0 for A_l^{++}, 2 for A_l^{+-}, 1 for A_l^{-+}, and 3 for A_l^{--}. So only $(N^2/2)$ complex operations are now needed to calculate the required coefficients $A_l^{\pm\pm}$. This gives a total of $(N^2/2 + 4N)$ operations when those implied by (12.21), (12.22), and (12.19) are included. This procedure can be performed $((\log_2 N) - 1)$ times resulting finally in $N/2$ summations each containing 2 terms. The total number of complex operations is then

$$2N^2/(2^{(\log_2 N)-1}) + 2N((\log_2 N) - 1) = 2N(1 + \log_2 N), \tag{12.24}$$

which is, in fact, the number of operations needed to evaluate all of the F_l by the procedure outlined above. When implemented as software (as is discussed in §48), this procedure constitutes the F.F.T. algorithm. The reduction in the number of operations, from the $(2N^2)$ required to calculate RHS (12.18) as it stands, is of course considerable when N is appreciable. The reduction factor is then approximately $(N/\log_2 N)$. Hence the wide interest in, and the rapidly increasing incorporation into computational practice of, the F.F.T. algorithm.

We emphasize that the reduction factor $(N/\log_2 N)$ applies to a one-dimensional D.F.T. of length N (used when a function of a single variable, represented by N point samples, is to be Fourier-transformed). For an $N \times N$ D.F.T. (needed, for instance, when an $N \times N$ pixel image is to be Fourier-transformed), the reduction factor is seen, from the paragraph containing (12.18), to be $(N^2/2 \log_2 N)$, which of course represents an even more striking improvement.

13 Complex Fourier plane

It is sometimes useful to generalize a one-dimensional Fourier domain to a complex plane. Consider $g = g(x)$ and $G = G(u)$ where

$$G = \mathsf{F}\{g\} = \mathsf{F}_{(1u)}\{g_{(x)}\} \quad \text{with } g = 0 \text{ for } x \geq L/2, \tag{13.1}$$

where the notation introduced in (6.17) has been invoked. The real constant L satisfies

$$0 < L < \infty, \tag{13.2}$$

so that g is a one-dimensional image of finite length whose image segment is centred at the origin (note that the terminology introduced in §7 has been invoked). It is convenient to express g, which is restricted to being real, as a trigonometrical Fourier series:

$$g(x) = (1/L) \sum_{m=-M}^{M} G_m \exp(-i\, 2m\pi x/L) = g^*(x), \tag{13.3}$$

where M can always be taken as finite for the reasons presented in the paragraph containing (10.3).

It is now convenient to generalize the real variable u to the complex variable

$$w = u + i\, v, \tag{13.4}$$

where v, the imaginary part of w, is not to be confused with the Cartesian coordinate v in Fourier space introduced in the paragraph containing (6.11). The functional dependence of G is also generalized:

$$G = G(w) = \int_{-L/2}^{L/2} g(x) \exp(i\, 2\pi wx)\, dx$$

$$= \sum_{m=-M}^{M} G_m \operatorname{sinc}(Lw - m), \qquad \text{with } G_{-m} = G_m^*, \tag{13.5}$$

where (13.3) and the conjugate symmetry characterized by (8.1) and (8.2) have been invoked, and the same algebra that produces (10.4) from (10.2) has been used. Inspection of RHS (13.5) indicates that $G(w)$ is analytic for all finite values of w. In fact, $G(w)$ is said to be an entire (integral) function of exponential type. On recalling the definition (6.4) and the formula

$$\sin(\pi t) = \pi t \prod_{m=1}^{\infty} (1 - t^2/m^2), \tag{13.6}$$

it is seen that (13.5) can be re-expressed as

$$G(w) = K \prod_{m=M+1}^{\infty} (1 - L^2 w^2/m^2) \prod_{m=1}^{M} (w - w_m)(w + w_m^*), \tag{13.7}$$

where K is a real positive constant. The set $\{w_m;\ m = 1, 2, \ldots, M\}$ of constants is called the set of right-half-plane *complex zeros* of $G(w)$. It is convenient to define

$$w_m = u_m + i\, v_m, \qquad \text{with } u_m > 0. \tag{13.8}$$

Inspection of (13.7) shows that every zero of $G(w)$ in the right-half-plane is reflected in the imaginary axis, so that all of the zeros are characterized by the right-half-plane zeros, of which there is a denumerable infinity that are necessarily real (these are M/L for $m > M$) and M that may be complex (these are the w_m for $0 < m \le M$).

Yet another way of writing (13.5) is

$$G(w) = (Q(w)/\pi) \sin(\pi L w), \tag{13.9}$$

where

$$Q(w) = \sum_{m=-M}^{M} (-1)^m G_m/(Lw - m). \tag{13.10}$$

By direct (though tedious!) manipulation, (13.10) can be re-expressed as

$$Q(w) = P(w, M) \Big/ \prod_{m=-M}^{M} (Lw - m), \tag{13.11}$$

where the coefficients of the polynomial $P(w, M)$, which is of order $2M$, are themselves expressed in terms of the G_m. Standard 'root-finding' computer programs can then calculate the complex zeros. When M is too large for this to be convenient, the w_m can be computed by submitting the right side of (13.10) to a 'gradient search'.

For later convenience, a concise notation for sets of zeros is introduced. For a function, say g, varying in only one dimension in image space and having G for its one-dimensional Fourier transform, the symbol $\mathbf{Z}_g$ represents the set of right-half-plane zeros characterizing G in the complex plane. The sets of *necessarily real zeros* and *zeros that may be complex* are denoted by $\mathbf{Z}_g^r$ and $\mathbf{Z}_g^c$ respectively:

$$\mathbf{Z}_g = \mathbf{Z}_g^r \cup \mathbf{Z}_g^c. \tag{13.12}$$

Finally note that, given $\mathbf{Z}_g$, it is straightforward to compute $G(u)$, because (13.4), (13.6), and (13.7) combine to give

$$G(u) = (K/\pi L u) \sin(\pi L u) \prod_{m=1}^{M} ((u - w_m)(u + w_m^*)/(1 - L^2 u^2/m^2)). \tag{13.13}$$

In many applications it is unimportant to know the value of K; but note that the latter is given immediately from (13.7) if $G(0)$ is known (which it is often is).

It is useful to consider the spectrum $\mathfrak{G} = \mathfrak{G}(u)$ defined by

$$\mathfrak{G} = ((u - \zeta^*)/(u - \zeta))G \qquad \text{with } \zeta = \xi' + i\,\eta, \tag{13.14}$$

where ξ' and η are both real. Note that

$$|\mathfrak{G}| = |G| \tag{13.15}$$

Fourier transforming (13.14), invoking the convolution theorem (7.3) and defining $g(x) = g = F\{\mathfrak{G}\}$, gives

$$g = g + (\zeta - \zeta^*)g \odot t, \qquad \text{with } t = t(x) = F\{1/(u - \zeta)\}. \qquad (13.16)$$

To be specific, take $\eta > 0$. Invoking Cauchy's theorem to evaluate $t(x)$ by closing the contour of the Fourier integral at infinity in the upper and lower halves of the complex plane according as $x < 0$ and $x > 0$, it is seen that

$$\begin{aligned} t(x) &= i\, 2\pi \exp(-i\, 2\pi\zeta x) && \text{for } x < 0, \\ &= 0 && \text{for } x > 0, \end{aligned} \qquad (13.17)$$

which when substituted into (13.16) gives

$$g(x) = g(x) + i\, 2\pi(\zeta - \zeta^*)G(\zeta)\exp(i\, 2\pi\zeta x)\int_x^\infty g(\xi)\exp(i\, 2\pi\zeta\xi)\, d\xi. \qquad (13.18)$$

Reference to (13.1) then confirms that

$$\begin{aligned} g(x) &= i\, 2\pi(\zeta - \zeta^*)G(\zeta)\exp(i\, 2\pi\zeta x) && \text{for } x < -L/2, \\ &= 0 && \text{for } x > L/2. \end{aligned} \qquad (13.19)$$

Also, $G(\zeta) = 0$ by definition if ζ is any one of the w_m or if $\zeta = m/L$ for any integer $|m| > M$. So, it is seen from (13.19) and the definition of Z_g that

$$g(x) \quad \begin{matrix} \text{is of finite extent } L \text{ when } \zeta \in Z_g \\ \text{is of infinite extent when } \zeta \notin Z_g \end{matrix} \bigg\}, \qquad (13.20)$$

which result is unchanged if $\eta < 0$; if $\eta = 0$ then $g \equiv g$ because $(\zeta - \zeta^*) = 0$.

The one-dimensional form of the minimum-extent of autocorrelation theorem – refer to (7.13) – is nicely illustrated by (13.20). The images $g(x)$ that are of extent L are those to which the theorem applies. If $g(x)$ itself is merely real, which is all that is demanded by (13.3), then its extent can be either L or ∞. However, if $g(x)$ is required to be positive, the one-dimensional form of the extent–constraint theorem – refer to (8.15) – then ensures that its extent must be L. Within its image segment, which is of extent $2L$ necessarily, the autocorrelation of $g(x)$ can be written as

$$g(x) \divideontimes g(x) = (\tfrac{1}{2}L)\sum_{m=-M'}^{M'} \Gamma_m \exp(-i\, m\pi x/L), \qquad (13.21)$$

where the Γ_m are constants and the positive integer M' is large enough that RHS (13.21) is as accurate a representation of the autocorrelation as RHS (13.3) is of $g(x)$. It follows from (13.15) and the autocorrelation theorem (7.10) that

$$|G(u)|^2 = F\{g(x) \divideontimes g(x)\} = F\{g(x) \divideontimes g(x)\}, \qquad (13.22)$$

so that the set of zeros characterizing $|G(u)|^2$ – which it is worth noting is the limit as $v \to 0$ of $(G^*(w^*)G(w))$ – must be written as $\mathbf{Z}_{g*g}$ to be consistent with the notation introduced earlier in this section. Consequently, by replacing $g(x)$, $G(w)$, L, and G_m by $(g(x) \divideontimes g(x))$, $(G^*(w^*)G(w))$, $2L$, and Γ_m, respectively, in the second and third paragraphs of this section, it is seen that $\mathbf{Z}_{g*g}$ can be computed when $|G(u)|^2$ is given as readily as $\mathbf{Z}_g$ can be computed when $G(u)$ is given. Note further from (13.9) through (13.11) that

$$G^*(w^*)G(w) = \left| (\sin(\pi L w)) \middle/ \pi \prod_{m=-M}^{M} (Lw - m) \right|^2 R(w), \quad (13.23)$$

where $R(w)$ is a polynomial of order $4M$ whose zeros necessarily form quartets symmetrically disposed about the origin of the complex w-plane:

$$R(w) = \prod_{m=1}^{M} (w - w_m)(w - w_m^*)(w + w_m)(w + w_m^*). \quad (13.24)$$

Denoting by $\mathbf{Z}^*$ the set whose members are the complex conjugates of the members of $\mathbf{Z}$, it is seen that

$$\mathbf{Z}_{g*g} = \mathbf{Z}_g^* \cup \mathbf{Z}_g. \quad (13.25)$$

Example II – Using the F.F.T.

We illustrate several facets of Fourier transformation in this example. All the pictures displayed here comprise 256×256 pixels. For each picture which represents a Fourier transform of some other picture included in the example, the pixels were calculated with the aid of the F.F.T. The centres of all pictures have been arranged to coincide with the origin of either image space or Fourier space, as appropriate.

Figure IIa shows a uniform circular disc of diameter 15 pixels. It could be thought of, for instance, as a pictorial representation of the out-of-focus psf shown in Table I.1 (p. 16). Figures IIb–IId illustrate ways of emphasizing different characteristics of Fig. IIa's spectrum – i.e. its Fourier transform: refer to paragraph following that containing (6.15) through (6.17). Figure IIb is the histogram-equalized (see §45) logarithm of the positive part of the spectrum. The maximum amplitude of the logarithm was scaled to 255. The processed logarithm is displayed as black wherever the scaled logarithm is less than 1. This type of display has the advantage that it emphasizes the locations in Fourier space of the phase reversals of the spectrum. Figure IIc was processed similarly, but for the modulus, rather than the positive real part, of the spectrum. The nulls of the spectrum, and its sidelobe structure, are more apparent than in Fig. IIb but it does not reveal whether or not a particular null is associated with a phase reversal. Figure IId displays the spectrum's phase (black and white represent 0 and π respectively).

The inverse relationship between extents in image space and the spacings of nulls in Fourier space is emphasized by Figs. IIe and IIf, which are similar to Figs. IIa and IIb, respectively, except that the disc in Fig. IIe is only 5 pixels in diameter.

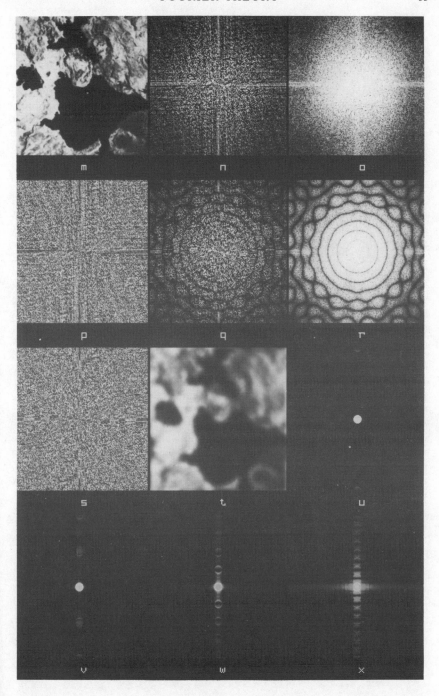

Fig. II.

The three pairs of images Figs. IIg,h, IIi,j, and IIk,l correspond to Figs. IIa,b, except that the circular disc is replaced, respectively, by a horizontal bar (of width 5 pixels), a Gaussian and a cross (comprising two bars of the kind shown in Fig. IIg, inclined at ±45°). The length of each bar and the effective width of the Gaussian are both 15 pixels. The variations between the four images Figs. IIb,h,j,l are appreciable. Even though the disc and the Gaussian (Figs. IIa and IIi respectively) look much the same, the spectrum of the former possesses oscillatory side lobes (of phase 0 and π successively) whereas that of the latter is wholly positive, being negligible outside a comparatively small circle centred on the origin of Fourier space. Figures IIb and IIj display these properties of the two respective spectra.

Figure IIm is the same as Fig. Ia. Figures IIn,o,p bear the same relation to Fig. IIm as Figs. IIb,c,d do to Fig. IIa.

We next demonstrate how the spectrum of an image alters when the image is blurred (by an out-of-focus psf – see Table I.1 – of diameter 15 pixels; Fig. IIa is a pictorial representation of such a psf, as has already been noted). Since convolution in image space corresponds to multiplication in Fourier space (see §7), the spectrum of the blurred version of Fig. IIm is obtained by multiplying its spectrum by the spectrum of Fig. IIa. Figures IIq,r,s bear the same relation to the blurred image as do Figs. IIn,o,p to Fig. IIm. The effect of the blurring on the displays of the spectral characteristics is as marked as on the image itself (Fig. IIt is the blurred image). It is worth remarking here that Fig. If (see Example I) is a truncated version of Fig. IIt.

It is instructive to see the pictorial effects of erroneous Fourier processing. A particularly common and important error (often quite unwilful, of course, and frequently unavoidable) is to undersample Fourier space. This implies that the spacing of the samples is not close enough to satisfy the sampling theorem (10.6). Figure IIu looks very similar to Fig. IIa but it is, in fact, an aliased version of the latter. It was produced by, first, undersampling the spectrum of Fig. IIa by a factor of 2 in both the u- and v-directions, second, resampling by linear interpolation – refer to paragraph containing (33.3) and (33.4) – and, finally, computing the inverse F.F.T. Figures IIv,w,x were generated in the same way as Fig. IIu except that the undersampling of the spectrum of Fig. IIa was by factors of 4, 8, and 16 respectively. The distortion of the circular disc becomes markedly severer as the spectrum is increasingly undersampled.

III

DECONVOLUTION

Having established our mathematical notation and basic theory in Chapters I and II, we are now ready to discuss applications. This chapter describes the approaches to two-dimensional deconvolution which we think are useful in practice.

The term deconvolution encompasses what is perhaps the most important and widely used collection of image processing techniques. The need for deconvolution arises in all areas of science concerned with measurement. It consequently permeates physics and engineering, and it is often invoked in chemical, biological, and medical contexts. There are many comprehensive treatments of deconvolution methods, some very specialized and others wide-ranging [Andrews, B1, B3; Bates (J.H.T.), T1; Bates, 10; Bernstein, B1; Carson, 1; Hawkes, B1; Huang, 2, B1; McDonnell, T1; Rabiner, B1; Rosenbaum, 1; Rosenfeld, B2; Sage, B1; Saxton, B1; Sondhi, 1].

In §14 we discuss the nature of the deconvolution problem, and point out that it is desirable to transform the given blurred image into the form of a consistent convolution [Bates, 2, 17; McKinnon, 1]. It is also noted that there is a tendency for the problem to be complicated by the truncation of the blurred image by the recording frame [McDonnell, 7].

The deconvolution problem may be solved in several ways, which are discussed in §§16–19. The technique that is most appropriate depends upon a number of factors including the form and extent of the psf, the nature of the original image, and the severity of the truncation by the recording frame.

Whichever technique is used it is almost always necessary to preprocess the given blurred image into a form suitable for deconvolution. This crucial step in the overall deconvolution process is discussed in §15. We find it convenient to divide preprocessing into five categories – enhancement, sectioning, windowing, edge extension, and super-resolution. These are considered in turn. Enhancement in this context is concerned with reducing the contamination [McDonnell, 5, 14] and we defer discussion of it until Chapters VIII and IX. Sectioning [Huang, 2; McDonnell, 7, 8, T1; Trussel, 3, 4] involves dividing an image with a spatially varying psf into sections each of which may be treated as having an approximately spatially invariant psf. Windowing [Harris, 1] is a useful technique for reducing the effect of the truncation by the recording frame. However, it can have shortcomings when compared to the edge-extension method which we have usefully invoked in several contexts [Bates, 9; Lewitt, 2; McDonnell, 3, 5, 7, 8, 14, 15, T1]. There are two kinds of edge extension called simple and overlapped. The second is usually superior because it takes advantage of what is referred to in §14 as the consistency of periodic convolutions. We think this is worth emphasizing because it is yet another example [Bates, 19] of the efficiency of a numerical technique improving markedly when it accords as closely as possible with the underlying mathematical physics of the problem being studied [Bates, 20]. Super-resolution [Gerchberg, 1, 2; Harris, 2; Rushforth, 1; Slepian, 3] is considered under the heading of preprocessing because in its most general sense it allows the contamination to be reduced. Also relevant here is our attempt to both super-resolve and restore an optically resolved image of the star Betelgeuse [McDonnell, 9] because it illustrates how constraints on the form and symmetry of an image and its psf may be utilized to good advantage.

In §16 we consider multiplicative deconvolution, using the Wiener filter [Andrews, B3; Helstrom, 1; Huang, 2; Hunt, 2; Slepian, 1] and the equivalent homomorphic filters [Oppenheim, 1; Stockham, 2]. Multiplicative deconvolution is the most widely used method of restoring an image which is a consistent convolution. A blurred image which is not consistent must be converted into one that is, as discussed in §15.

In §17 we describe the generalized approach to subtractive deconvolution recently developed at Canterbury [Bates (J.H.T.), 1, 2, 3]. This stems from the CLEAN algorithm [Högbom, 1; van Schooneveld, B1] which was devised for refining maps of the radio heavens formed from data gathered with synthesis telescopes. It is worth noting the similarity of our 'subtractive reblurring' to the van Cittert iterative method of deconvolution [Hill, 1]. The final two paragraphs of §17 summarize some recent findings [Bates (J.H.T.), 4; Cornwell, 1].

Subtractive deconvolution is especially useful when the imperfections apparent in the recorded image are due not so much to loss of resolution as to obscuration of small detail – e.g. when the psf has a main lobe as narrow as the detail we wish to resolve but possesses a wide skirt of appreciable amplitude or exhibits high side-lobes. Subtractive techniques can be readily adapted to deal with point spread variant degradations, although it must be realized that they then tend to become computationally expensive.

In §18 we introduce and compare, with each other and with the methods treated in §§16 and 17, a variety of approaches to deconvolution. These comprise non-recursive [Arguello, 1; Frieden, 5; Honda, 1, 2; McDonnell, 1, 2, T1; Riemer, 1; Saleh, 1] and recursive [Hall, 1; Huang, 2; IEEE, B1; Rabiner, B1; Shanks, 1; Zimmerman, 1] techniques in image space, direct matrix methods [Andrews, B3], and maximum entropy and maximum likelihood techniques [Frieden, 3, 6, 7; Nityananda,

1]. This section is included mainly to give the reader an overview of those important approaches which we do not ourselves greatly favour. The length of the section would become inordinate if we attempted any detailed descriptions. Consequently, we merely outline the various techniques and comment on their advantages and disadvantages.

Consistent deconvolution, which has grown out of our studies of complex zeros in Fourier space [Bates, 1, 17; McKinnon, 1], is primarily a one-dimensional technique which can be invoked as widely (if not always as conveniently) as multiplicative deconvolution. In §19 we remark on its two-dimensional applications [Napier, 1] which are intimately connected with projection theory (see §9 and Chapter V).

One of the most important practical deconvolution techniques is that called *blind deconvolution*. We do not discuss it in this chapter because it is more appropriately dealt with in Chapter IX (in §52). It is worth noting, as is remarked again in the introductory comments to Chapter VI, that all speckle imaging methods can be regarded as species of blind deconvolution.

Besides the prosaic, standard, and of course very useful, applications of deconvolution, various exotic employments are sometimes found for it. One of the more remarkable is the recovery (effected by invoking blind deconvolutional techniques) of the (hopefully) pristine voices of famous singers from ancient gramophone records [Oppenheim, 1; Stockham, 2]. Some of our own more out of the way forays have been into forensics [McDonnell, 8, 14, T1], reconstruction of 'star spots' [McDonnell, 9], analysis of photographs of supposedly unidentified flying objects [McDonnell, 11] and the restoration of Voyager I images of Jupiter's moon Io [McDonnell, 5].

Example III is devoted entirely to illustrations of Wiener filtering, which we hold to be by and large the most widely applicable deconvolutional technique in practice. The reader should note that an illustration of Cleaning (see §17) is included in Example VI.

14 The deconvolution problem

It is here convenient to repeat (3.4):

$$b(\mathbf{x}) = f(\mathbf{x}) \ominus h(\mathbf{x}). \tag{14.1}$$

The convolution theorem (7.7) confirms that the Fourier transform of (14.1) is

$$B(\mathbf{u}) = F(\mathbf{u})H(\mathbf{u}), \tag{14.2}$$

where the above three quantities are defined respectively in (12.9), (6.12) and the paragraph preceding (8.13).

We pose the *idealized finite deconvolution problem* as: given $b(\mathbf{x})$ and

$h(\mathbf{x})$, reconstruct $f(\mathbf{x})$; on the understanding that all three quantities are of finite size (see §7).

Inspection of (14.2) reveals that the above problem is solved trivially:

$$f(\mathbf{x}) = \mathsf{F}\{(B(\mathbf{u}))/H(\mathbf{u})\}. \qquad (14.3)$$

The division operation inside the braces on RHS (14.3) is here called *simple inverse filtering*. This invocation of the 'filtering' concept is by analogy with the terminologies of classical circuit theory and modern signal processing. The classical filter is a device which alters the temporal frequency content of a signal. The spectrum $B(\mathbf{u})$ is a function of spatial frequency (see §6). The OTF $H(\mathbf{u})$ alters the spatial frequency content of $B(\mathbf{u})$ by the aforesaid division operation.

Because processed images are usually stored as quantized pixels in computer memory, image processing tends to rely on *digital filters* rather than the classical *analogue filters*. We here define a digital filter as a discrete array of possibly complex numbers whose effect is to alter the spatial frequency content of an image during some processing operation. The filtering operation can be in either image space or Fourier space. Consequently, both $h(\mathbf{x})$ in (14.1) and $H(\mathbf{u})$ in (14.2) can be regarded as filters (and they are manifested digitally in most applications). The standard classification of digital filters was developed in signal processing contexts (i.e. for one-dimensional images, which are what signals can be regarded as if one wishes to treat signal processing and image processing as two branches of a single technical science). We now convert this terminology into appropriate two-dimensional language. Note that the signal processor's term 'sample' translates into 'pixel' in image processing contexts. It is worth emphasizing that both pixels and samples must be quantized in amplitude before they can be processed digitally. The image that is to be filtered is here called the *given image*, which is said to consist of *given pixels*. The filtered image is said to consist of *output pixels*. For a *non-recursive digital filter*, each output pixel is a weighted sum of given pixels. For a *recursive digital filter*, each output pixel is a weighted sum of given pixels and previously computed output pixels. All practically realizable digital filters, of course, comprise *finite filter arrays*, whose extents in image space and/or Fourier space are conveniently expressed in terms of the notation introduced in §7, in the four paragraphs following (7.11) (in one dimension we often refer to a finite filter as being *short*). We say that digital filters are *direct* or *spectral* if they are applied, respectively, in image space or Fourier space. A *causal* filter or psf is one-sided in the sense that its response always follows its input (this tends to be somewhat artificial in two dimensions but it is, of course, crucial for the one-dimensional filtering operations which characterize signal processing). Causal filters are almost always direct. A *multiplicative digital filter* is a spectral filter for which each output pixel consists of the product of a given pixel and a single element of the filter array.

We give much consideration to multiplicative filters in this chapter, especially in §§15 and 16. Recursive and non-recursive filters are discussed at some length in §18.

If (14.3) characterized all essential aspects of practical deconvolution problems, the contents of this chapter could be comfortably squeezed into §7. Deconvolution presents plenty of practical difficulties, however! They arise because data are always corrupted in some sort of way in the real-world.

Before posing the practical deconvolution problem, it is instructive to examine certain consistency properties of convolutions.

The one-dimensional form of (14.2) is

$$B(w) = F(w)H(w), \tag{14.4}$$

where the real variable u had been generalized to w, as in (13.4). When $f(x)$ and $h(x)$ are of finite length – so that $b(x)$ is also of finite length (refer to the extent of convolution theorem in §7) – their spectra are characterized by sets of zeros in the complex w-plane (see §13). So, invoking the notation introduced in the paragraph containing (13.12) and (13.13),

$$\mathbf{Z}_b = \mathbf{Z}_f \cup \mathbf{Z}_h. \tag{14.5}$$

It follows that the one-dimensional deconvolution problem is *inconsistent* unless all the zeros of $H(w)$ are also zeros of $B(w)$. Consequently, the quantities $b(x)$ and $h(x)$ cannot be specified independently, but must be known *a priori* to satisfy (14.1). The same consideration applies to two-dimensional convolutions.

Now return to the periodic overlapped ideal blurred image $p_b(\mathbf{x})$, defined by (12.6), and its spectrum $\mathfrak{P}_b(\mathbf{u})$, as expressed in (12.12). The latter can be rewritten with aid of (12.13) as

$$\mathfrak{P}_b(\mathbf{u}) = \sum_{l,m=-\infty}^{\infty} F_{l,m}H_{l,m}\delta(L_1 u - l)\,\delta(L_2 v - m), \tag{14.6}$$

where the $F_{l,m}$ are the Fourier coefficients of the true image $f(\mathbf{x})$, as defined by (10.2), and are also the point samples of $F(\mathbf{u})$ that figure in the sampling theorem – these point samples lie at the sample points $(l/L_1, m/L_2)$ in Fourier space, as can be seen from (10.4). The $H_{l,m}$ introduced in (14.6) are the point samples of the OTF (see §8) $H(\mathbf{u})$ at the same sample points:

$$H_{l,m} = H(l/L_1, m/L_2), \tag{14.7}$$

where l and m are arbitrary integers.

It is now convenient to pose the *idealized periodic deconvolution problem* as: given $p_b(\mathbf{x})$ and $h(\mathbf{x})$, reconstruct $f(\mathbf{x})$; on the understanding that $f(\mathbf{x})$ and $h(\mathbf{x})$ are of finite size but $p_b(\mathbf{x})$ is periodic, as defined by (12.6).

From the given $b(\mathbf{x})$ one computes $B(\mathbf{u})$ and immediately finds the

$$B_{p,l,m} = B(l/L_1, m/L_2),\qquad(14.8)$$

as follows from (12.10) and (12.12). One similarly evaluates the point samples $H_{l,m}$ of the OTF. It is seen from (14.6) that each $F_{l,m}$ is given by the division operation $(B_{p,l,m}/H_{l,m})$, which can always be effected except for those $H_{l,m}$ that are actually zero. This simple approach is adequate, for $b(\mathbf{x})$ and $h(\mathbf{x})$ chosen quite independently, because $\mathfrak{P}_b(\mathbf{u})$ effectively exists only at the aforementioned sample points, as (14.6) confirms. This approach is inadmissable for the idealized finite problem because $B(\mathbf{u})$ is then a continuous function of $\mathbf{u}$.

Remarkably, therefore, the sole consistency condition on periodic convolutions, such as RHS (12.6), is that $H_{l,m}$ can only be allowed to be zero for those values of l and m for which $B_{p,l,m} = 0$. This is here called the *consistency of periodic convolutions*. It is worth emphasizing that no $H_{l,m}$ can ever be exactly zero when $h(\mathbf{x})$, or equivalently $H(\mathbf{u})$, is measured, so that periodic convolutions are always consistent in practice (they are, of course, extremely 'noisy' when many of the $H_{l,m}$ are 'small' for values of l and m for which the amplitudes of the $B_{p,l,m}$ are significant). This is discussed further in §§15 and 18.

Practical imperfections are realistically modelled by the recordable image $r(\mathbf{x})$, introduced in §4, except that RHS (4.1) does not explicitly acknowledge that any *actually recorded image* $\mathfrak{a}(\mathbf{x})$ must exist within a finite frame.

The *practical deconvolution problem* is here posed as: given $\mathfrak{a}(\mathbf{x})$ and $h(\mathbf{x})$, reconstruct $f(\mathbf{x})$; on the understanding that $\mathfrak{a}(\mathbf{x})$ is a truncated version of the $r(\mathbf{x})$ defined by (4.1).

One of the golden rules of image reconstruction is to avoid processing data which exhibit any species of 'glitches' – i.e. anything having the character of discontinuties, of which truncations are among the more objectionable – because spurious detail (often called *artefacts*, especially in medical contexts) is then almost always generated. It is thus generally desirable to *preprocess* $\mathfrak{a}(\mathbf{x})$ in order to compensate as far as is possible for whatever glitches, or other ameliorable imperfections, afflict it. Useful preprocessing techniques are described in §15.

Any kind of preprocessing is, of course, likely to add to the distortion of $f(x)$ already present in $r(\mathbf{x})$. If the glitches are not removed, however, the resulting artefacts are likely to swamp any extra contamination introduced by preprocessing. The massaged form of $\mathfrak{a}(\mathbf{x})$ is here denoted by $a(\mathbf{x})$ and is called the *preprocessed recorded image*. Even though the three quantities on RHS (4.1) must be altered by the preprocessing, there is seldom any means of estimating by how much, so that it is usually pointless to worry about whatever differences may exist between $a(\mathbf{x})$ and $r(x)$. These two images are henceforth treated as identical, at least within the frame Ω_a (i.e. the region of image space) enclosing the preprocessed form of $a(\mathbf{x})$. It is assumed therefore that

$$a(\mathbf{x}) = f(\mathbf{x}) \ominus h(\mathbf{x}) + c(\mathbf{x}). \tag{14.9}$$

This assumption is never really offensive because the contamination $c(\mathbf{x})$ includes the effects of any added distortion due to the preprocessing.

It is now appropriate to introduce the *recoverable true image* $\hat{f}(\mathbf{x})$ which is the estimate of $f(\mathbf{x})$ obtainable from $a(\mathbf{x})$ by invoking a particular reconstruction method (e.g. any one of those described in §§16 through 19).

Any useful approach to solving the practical deconvolution problem starts with forming $a(\mathbf{x})$ from the given $a(\mathbf{x})$. A suitable deconvolution procedure (see §§16 through 19) is then chosen for operating with $h(\mathbf{x})$ on $a(\mathbf{x})$ to obtain $\hat{f}(\mathbf{x})$. Some of these procedures can be interpreted as forming a *modified point spread function* $\hat{h}(\mathbf{x})$ which is related to the preprocessed recorded image and the recoverable true image by

$$a(\mathbf{x}) = \hat{f}(\mathbf{x}) \ominus \hat{h}(\mathbf{x}). \tag{14.10}$$

It is convenient to denote the Fourier coefficients (as defined in §10) of $\hat{f}(\mathbf{x})$ as $\hat{F}_{l,m}$ and to identify the spectra of $a(\mathbf{x})$, $c(\mathbf{x})$, $\hat{f}(\mathbf{x})$ and $\hat{h}(\mathbf{x})$ by their corresponding upper-case letters, with or without circumflex accents as appropriate.

If there is any fear that the differences between $\hat{f}(\mathbf{x})$ and $f(\mathbf{x})$ may be significantly amplified by the lack of consistency between $a(\mathbf{x})$ and $h(\mathbf{x})$ when they are taken to be explicitly finite, the periodic image $p_a(\mathbf{x})$ can be formed, as defined by RHS (12.6) with a replacing b. The spectrum $\mathfrak{P}_a(\mathbf{u})$ of the convolution is then given by RHS (14.6) but with $\hat{F}_{l,m}$ and $\hat{H}_{l,m}$ replacing $F_{l,m}$ and $H_{l,m}$ respectively. Remember that periodic convolutions cannot suffer from the inconsistency which, as noted earlier in this section, can afflict convolutions of quantities of explicitly finite extents.

15 Preprocessing

In our experience the success of image restoration depends strongly upon the quality of the preprocessing by which $a(\mathbf{x})$ is produced from $a(\mathbf{x})$. We find it convenient to divide preprocessing into five categories – enhancement, sectioning, windowing, edge extension, and super-resolution – which are examined in turn below.

It is usually appropriate to *enhance* $a(\mathbf{x})$ in order to reduce the effects of the contamination as much as possible. While this may often be largely cosmetic, it can nevertheless be of more practical use. Remember that the quantity $c(\mathbf{x})$ introduced in (14.9) includes the consequences of such things as recording non-linearities, recording noise, bit errors, missing information (i.e. missing pixels or groups of pixels), saturation, and the dust, dirt, and scratches which tend to degrade photographic records. Enhancement can be regarded as a two-dimensional manifestation of the elementary signal processing notion that one must try to reject all noise whose spectral components lie outside the band of temporal frequencies spanned by the information carried by the particular communication channel which is under consideration. Much of the image contamination referred to above can be expected to be independent from pixel to pixel, whereas the interesting detail in an image is often correlated across several neighbouring pixels. This is the same as saying that the spatial frequency spectrum of the contamination is appreciably wider than that of the image, in which case one obviously gains by spatially filtering $a(\mathbf{x})$ so as to retain only those spectral components of $c(\mathbf{x})$ which are resolved to the same degree as detail in the true image.

Several enhancement algorithms are discussed in detail in Chapters VIII and IX. It is the contamination level *after* preprocessing, much more than the deconvolution technique which is actually used, that (in our experience, at least) limits the faithfulness of the restored image.

The elegant deconvolution techniques described in §§16–19 can only be invoked straightforwardly when the psf is spatially invariant. The loss of point spread invariance changes the nature of the deconvolution problem, greatly increasing the computational complexity and expense of even the techniques which are able to handle a spatially varying psf. In many practical situations, point spread variance arises, not so much due to the intrinsic nature of the psf, but because of geometric distortions occurring during the recording process (such distortions are often caused by the lenses in imaging instruments, for instance). We accordingly find it convenient to classify rectification under the general heading of enhancement. The rectification methods introduced in §47 can be invoked to compensate for geometric distortion, thereby producing a psf which is effectively spatially invariant. It is worth giving a specific example. Suppose a scene is being photographed from a rotating vehicle, to which the camera is rigidly fixed. The plane in which the camera film lies corresponds to what is called image space throughout this book. Knowing the geometrical relation of the scene to the vehicle, we can calculate the position of the *axial point* where the axis of rotation intersects image space. Even if the camera is properly focused, the recorded image is distorted by a spatially varying psf characterized at each point in the *rotationally blurred* image by an arc centred on the axial point. The angular extent of the arc is proportional to the product of the photographic exposure time and the rate of rotation of the vehicle. The appropriate rectification procedure is to transform each arc into a straight segment of constant length. The transformed psf is then spatially invariant, corresponding to a linear blur (see Table I.1 in Example I). After compensating for the blurring by whatever deconvolutional method seems most suitable, the original geometry is restored by the inverse to the rectification transformation.

Two approaches are available to handle a spatially varying psf which cannot be corrected by rectification. It may be possible to invoke one of the direct methods discussed in §18. They tend to be so computationally expensive, however, that they are only practicable for small images (128×128 say) or when the psf varies in only one dimension. The second, usually preferable approach, which we call *sectioning*, involves partitioning the recorded image into a number of contiguous, equal-sized sub-images. The blurring of each sub-image is taken to be due to the form of the actual psf at the centre of the sub-image. Any error in this assumption is included in the overall contamination of the sub-image, the size of which must be small enough to prevent the contamination becoming excessive. On the other hand, it is our experience that the size of a sub-image should be at least four to eight times the effective size of the psf. All of this assumes that the actual psf varies smoothly and slowly across the recorded image, which fortunately is often true in practice Sectioning thus allows restoration of an image characterized by a spatially varying psf to be treated as a sequence of what are defined in §14 as practical deconvolution problems. The complete restored image is generated by fitting together (we call this *mosaicing*, which is discussed in Chapter VIII) the restored versions of the individual sub-images.

Invoking the terminology introduced in §7, we should strictly denote by Ω_a the region of image space occupied by the given (but enhanced, as described above) recorded image. This accords with the definitions, introduced in §§3 and 14, connecting the recordable and actually recorded images, $r(\mathbf{x})$ and $a(\mathbf{x})$ respectively. However, $r(\mathbf{x})$ is truncated to form $a(\mathbf{x})$ in most practical applications, which is why the two are in general different. Remember that $a(\mathbf{x})$ is a contaminated version of $b(\mathbf{x})$. The latter is what we are really interested in because it represents the quantity which we would like to 'deconvolve'. Since $b(\mathbf{x})$ extends outside the above frame, it is helpful to think of $a(\mathbf{x})$ doing likewise. Note that $a(\mathbf{x})$ can still differ from $r(\mathbf{x})$. We therefore introduce another symbol, Γ in fact, to identify this frame. It only makes sense to attempt to recover those parts of $f(\mathbf{x})$ which affect the form of $b(\mathbf{x})$ within Γ. These are the parts of $f(\mathbf{x})$ originally within Γ and those parts which are spread, by the action of the psf, into Γ from outside of it. We denote by Ω the frame containing the sum of these parts. Reference to (14.1) confirms that Ω can be constructed by placing the centre of the point spread frame Ω_h on the perimeter of Γ and moving it all around the perimeter. The frame Ω is then the union of all points in Γ and all points encompassed by Ω_h on its passage around Γ (refer to the discussion relating to Fig. IIIn in Example III).

Since the parts of the true image lying outside Ω are irretrievably lost, we might as well restrict $f(\mathbf{x})$ to lie within Ω. It is to be understood from now on, therefore, that

$$f(x) = 0 \qquad \text{for } x \notin \Omega, \tag{15.1}$$

which, of course, implies that

$$\Omega_f \equiv \Omega. \tag{15.2}$$

Because $f(\mathbf{x})$ exists within Ω, it must have influence within a larger frame, due to the spreading action of $h(\mathbf{x})$. As this larger frame contains all the parts of $b(\mathbf{x})$ which are not irretrievably expunged by truncation of $r(\mathbf{x})$, we denote it by Ω_b. Since $a(\mathbf{x})$ is a contaminated version of $b(\mathbf{x})$, we now see that Ω_a is identical to Ω_b.

Even though we know $a(\mathbf{x})$ exists throughout Ω_b, it is only given within Γ. It is usually appropriate to carry out further preprocessing (besides enhancement) to compensate as far as possible for the effects of the truncation, and also for the inconsistency of convolution already noted in §14. Consequently, we require the preprocessed recorded image, introduced in (14.9), to satisfy

$$\begin{aligned} a(\mathbf{x}) &= \text{pre}\{a(\mathbf{x})\} &&\text{for } x \in \Omega_b \\ &= 0 &&\text{for } x \notin \Omega_b, \end{aligned} \tag{15.3}$$

where $\text{pre}\{\cdot\}$ represents the preprocessing operations described below.

The truncation of $a(\mathbf{x})$ is so significant from a practical point of view that it is essential to emphasize its consequences. To do so we must first specialize the shape of Γ. Only rectangular or circular frames are considered here because they are the ones which occur in the majority of practical applications. So, if L_1 and L_2 are the x and y extents, respectively, of a rectangular Γ, or if R is the radial extent of a circular Γ, then it follows from the definitions introduced in §§4 and 14 that either

$$a(\mathbf{x}) = r(\mathbf{x}) \, \text{rect}(x/L_1) \, \text{rect}(y/L_2) \tag{15.4}$$

or

$$a(\mathbf{x}) = r(\mathbf{x})\text{rect}(r/2R), \tag{15.5}$$

where the coordinate origin is chosen to lie at the centre of Γ. It is convenient to examine the two versions of $a(\mathbf{x})$ separately. On defining

$$\mathfrak{A} = \mathsf{F}\{a\} \quad \text{and} \quad \mathfrak{R} = \mathsf{F}\{r\} \tag{15.6}$$

it follows from (15.4) and the convolution theorem (7.7) that

$$\mathfrak{A}(\mathbf{u}) = L_1 L_2 \mathfrak{R}(\mathbf{u}) \circledcirc \text{sinc}(L_1 u) \, \text{sinc}(L_2 v), \tag{15.7}$$

where $\text{sinc}(u)$ – refer to the definition (6.4) – is the Fourier transform of $\text{rect}(x)$.

On turning now to (15.5), Fourier-transforming it and invoking the first of the two definitions introduced in (15.6), we see that

$$\mathfrak{A}(\mathbf{u}) = \pi R^2 \mathfrak{R}(\mathbf{u}) \ominus \text{jinc}(\rho R), \tag{15.8}$$

where ρ is the radial coordinate in Fourier space – refer to the paragraph containing (6.11) – and

$$\text{jinc}(t) = (J_1(2\pi t))/\pi t, \tag{15.9}$$

where $J_1(\cdot)$ denotes the Bessel function of the first kind of order 1.

Note that $\text{sinc}(t)$ is an oscillatory function, having a central peak (often called the *main lobe*) roughly of width unity, and an infinite succession of smaller peaks (sometimes called *side lobes*), each of effective width $\frac{1}{2}$, whose amplitudes fall off comparatively slowly (i.e. as $1/|t|$). These side lobes tend to give rise to unacceptable artefacts if, before it is appropriately preprocessed, $\mathfrak{a}(\mathbf{x})$ is subjected to a filtering operation (such as any of those described in §§16 through 19 below – refer also to the discussion of digital filters included in §14). Though these comments apply specifically to the $\mathfrak{a}(\mathbf{x})$ defined by (15.4), they are also germane for the $\mathfrak{a}(\mathbf{x})$ defined by (15.5). The jinc function introduced in (15.8) is similar to the sinc. In fact, its effect is roughly equivalent to that of the pair of sinc functions appearing in (15.7). Note that a typical filtering operation can be characterized by

$$\mathsf{F}\{\mathfrak{Z}\mathsf{F}\{\mathfrak{a}\}\} = \hat{\mathfrak{a}}, \tag{15.10}$$

where $\mathfrak{Z} = \mathfrak{Z}(\mathbf{u})$ is a multiplicative filter (see §14), designed to produce from $\mathfrak{a}$ an $\hat{\mathfrak{a}}$ having certain desired characteristics. The side lobes of the jinc, and pair of sinc, functions tend to distort the outer Fourier transform in (15.10), often generating ugly high-amplitude ripples across the region of image space where $\mathfrak{a}(\mathbf{x})$ is appreciable, thereby obscuring low-amplitude detail in the filtered image.

Since $\mathfrak{a}(\mathbf{x})$ is identically zero outside Γ, it is usually impracticable to realize (in a restored image) a resolution finer than that represented by the widths of the main lobes of the sincs in (15.7) or the jinc in (15.8). On the other hand it is often quite practicable to ameliorate the effects of the side lobes of the sinc and jinc functions by appropriate preprocessing.

When we know that the more interesting parts of $f(\mathbf{x})$ lie towards the centre of Γ, in cases for which its size is appreciably larger than that of Ω_h, a satisfactory type of preprocessing is *windowing*. It involves multiplying $\mathfrak{a}(\mathbf{x})$ by a *window function*, denoted here by $\mathfrak{w} = \mathfrak{w}(\mathbf{x})$, which falls smoothly to zero all around the perimeter of Γ, and is zero everywhere outside Γ. This has the effect of forcing Ω_b to equal Γ. Reference to (15.3) then shows that the preprocessed recorded image takes the form

$$a(\mathbf{x}) = \text{pre}\{\mathfrak{a}(\mathbf{x})\} = \mathfrak{w}(\mathbf{x})\mathfrak{a}(\mathbf{x}), \tag{15.11}$$

where $a(\mathbf{x})$ is understood to be an enhanced version of the actually recorded image.

Windowing results in an unavoidable loss of resolution, but this is often a small price to pay for the avoidance of the aforementioned artefacts. Standard texts list many window functions which exhibit satisfactory compromises between side-lobe reduction and resolution loss. We therefore feel it is sufficient to demonstrate some general properties of window functions, which we illustrate by analysing a particularly versatile window – the recth function defined by (15.18) – which has been largely neglected in the relevant literature.

Since the only forms for a considered here are characterized by (15.4) and (15.5), there seems little point in studying window functions that are neither circularly symmetric nor separable into a function of x multiplied to a function of y. It is sufficient, therefore, to investigate one-dimensional window functions, e.g. $\mathfrak{w}(x)$, where x can represent either x or y or r. It is convenient to take L to be the length in the x-direction of the (truncated) actual recorded image. On writing $\mathfrak{W}(u) = \mathfrak{W} = \mathbf{F}\{\mathfrak{w}\}$, and assuming $\mathfrak{w}$ to be analytic (i.e. 'continuously smooth' meaning that it is infinitely differentiable) throughout the interval $|x| < L/2$, it is seen that the Fourier integral defining $\mathfrak{W}$ can be integrated by parts to yield

$$\mathfrak{W}(u) = (\mathfrak{w}(L/2) \exp(\mathrm{i}\,\pi Lu) - \mathfrak{w}(-L/2) \exp(-\mathrm{i}\,\pi Lu)/\mathrm{i}\,2\pi u$$

$$+ (\mathrm{i}/2\pi u) \int_{-L/2}^{L/2} \mathfrak{w}^{(1)}(x) \exp(\mathrm{i}\,2\pi ux)\, dx, \qquad (15.12)$$

where the notation

$$\mathfrak{w}^{(n)}(x) = d^n \mathfrak{w}(x)/dx^n \qquad (15.13)$$

has been invoked. The integral in (15.12) can itself be integrated by parts, and the integral in the resulting expression can be integrated by parts, and so on, so that

$$\mathfrak{W}(u) = (1/\mathrm{i}\,2\pi u) \sum_{n=0}^{N} (\mathfrak{w}^{(n)}(L/2) \exp(\mathrm{i}\,\pi Lu)$$
$$- \mathfrak{w}^{(n)}(-L/2) \exp(-\mathrm{i}\,\pi Lu))/(-\mathrm{i}\,2\pi u)^n$$

$$- (-1/\mathrm{i}\,2\pi u)^{n+1} \int_{-L/2}^{L/2} \mathfrak{w}^{(n+1)}(x) \exp(\mathrm{i}\,2\pi ux)\, dx, \quad (15.14)$$

where the notation (15.13) has again been invoked. Provided the summation in (15.14) actually converges as $N \to \infty$, it provides a concise characterization of the asymptotic behaviour (as $|u| \to \infty$) of the side lobes of $\mathfrak{W}(u)$. To force the side lobes to fall off as rapidly as possible with increasing $|u|$, it is necessary to arrange that

$$\mathfrak{w}^{(n)}(\pm L/2) = 0 \quad \text{for} \quad n = 1, 2, \ldots, \tilde{N}, \tag{15.15}$$

where the positive integer $\tilde{N}$ is as large as can be managed. In practice, $\tilde{N}$ is seldom chosen to be greater than unity, because it is usually either too difficult to control the form of $\mathfrak{w}(x)$ throughout $|x| < L/2$ (e.g. when $\mathfrak{w}(x)$ represents the illumination taper of an antenna, or the apodization of an optical instrument) or the side-lobe reduction is sufficient with $\tilde{N} = 1$ (or the resolution loss is unnecessarily high if $\tilde{N} > 1$). For certain optimum compromises between resolution loss and side-lobe level none of the $\mathfrak{w}^{(n)}(\pm L/2)$ are forced to zero – they are merely made small enough to satisfy particular criteria.

Many standard window functions are of the form

$$\begin{aligned} \mathfrak{w}(x) &= \tilde{\mathfrak{w}}(L/2 + x) \quad &\text{for} \quad -L/2 < x < -L/2 + T \\ &= 1 \quad &\text{for} \quad |x| < L/2 - T \\ &= \tilde{\mathfrak{w}}(L/2 - x) \quad &\text{for} \quad L/2 - T < x < L/2, \end{aligned} \tag{15.16}$$

where $\tilde{\mathfrak{w}} = \tilde{\mathfrak{w}}(x)$ is chosen such that

$$\mathfrak{w}^{(n)}(\pm L/2) = \mathfrak{w}^{(n)}(\pm L/2 \mp T)) = 0 \quad \text{for} \quad n = 1 \text{ and } 2. \tag{15.17}$$

The side lobes fall off asymptotically as $|u|^{-3}$. Varying the parameter T allows a trade-off between resolution loss and side-lobe level.

It is sometimes convenient to construct a window function that is analytic throughout $|x| < L/2$ and possesses two independently variable parameters which permit advantageous compromises to be reached. Consider the particular window function $\text{recth}(x \mid \alpha, \beta)$ which is here defined to be

$$\text{recth}(x \mid \alpha, \beta) = (\tanh((x + \alpha/2)/\beta) - \tanh((x - \alpha/2)/\beta))/2. \tag{15.18}$$

Note that

$$\underset{\beta \to 0}{\text{Lim}} \text{recth}(x \mid \alpha, \beta) = \text{rect}(x/\alpha). \tag{15.19}$$

Note also that the integral

$$\int_{-\infty}^{\infty} \tanh(x) \exp(i\, 2\pi u x)\, dx$$

is an odd function of u, and its integrand has simple poles on the imaginary axis of the complex x-plane, at the points $i(2l+1)\pi/2$ where l is any integer. Since $\exp(i\, 2\pi u x)$ is analytic throughout the upper half of the complex x-plane when u is real and positive, the above integral can be evaluated by applying Cauchy's theorem to

$$\int_C \tanh(x) \exp(i\, 2\pi ux)\, dx \quad \text{for} \quad u > 0,$$

where C is the closed contour consisting of the real axis and the semicircle at infinity in the upper half of the complex x-plane. Consequently,

$$\int_{-\infty}^{\infty} \tanh(x) \exp(i\, 2\pi ux)\, dx = \mathfrak{P} + \mathfrak{S}, \tag{15.20}$$

where $\mathfrak{P} = \mathfrak{P}(u)$ is the contribution from the poles in the upper half plane and $\mathfrak{S} = \mathfrak{S}(u)$ is what is obtained from the integral along the infinite semicircle. Since $\cosh(x) = 1$ when $\sinh(x) = 0$, it follows that

$$\mathfrak{P} = 2\pi \sum_{l=0}^{\infty} \exp(-(2l+1)\pi^2 u)$$

$$= (2\pi \exp(-\pi^2 u))/(1 - \exp(-2\pi^2 u)). \tag{15.21}$$

On the infinite semicircle it is convenient to write $x = R \exp(i\,\theta)$ where R is an arbitrarily large, real, positive constant. Note that

$$\tanh(x) = \text{sgn}(\pi/2 - \Theta) \quad \text{when} \quad x = R \exp(i\,\Theta) \quad \text{and} \quad R \to \infty \tag{15.22}$$

where the signum function is defined by

$$\text{sgn}(\Theta) = \pm 1 \quad \text{when} \quad \Theta \gtrless 0. \tag{15.23}$$

It follows that

$$\mathfrak{S} = i\, R \int_0^{\pi} \text{sgn}(\pi/2 - \Theta) \exp(i\, 2\pi uR \exp(i\,\Theta)) \exp(i\,\Theta)\, d\Theta$$

$$= -R \int_{Q_1} \exp(i\, 2\pi uRt)\, dt + R \int_{Q_2} \exp(i\, 2\pi uRt)\, dt, \tag{15.24}$$

where Q_l denotes the lth quadrant of the unit circle in the complex t-plane. Since $\exp(i\, 2\pi uRt)$ is analytic inside and on this unit circle, each Q_l can be deformed to lie along the parts of the real and imaginary axes connecting the origin to the ends of the quadrant. So, (15.24) can be rearranged such that

$$\mathfrak{S} = R \int_{-1}^{1} \text{sgn}(t) \exp(i\, 2\pi uRt)\, dt - 2R \int_0^{i} \exp(i\, 2\pi uRt)\, dt$$

$$= i(\exp(-2\pi Ru) - \cos(2\pi Ru))/\pi u, \tag{15.25}$$

which, as expected, is an odd function of u. Note that, as $R \to \infty$, $\mathfrak{S}$ reduces to

$$\mathfrak{S} = (-i/\pi u) \cos(2\pi R u). \tag{15.26}$$

Reference to the convolution theorem (7.7) and to (15.11) indicates that the Fourier transform of the preprocessed image is the convolution of the Fourier transforms of the window function and the actual recorded image. The contribution of $\mathfrak{S}$ to any such convolution necessarily vanishes because the cosine function in (15.26) is of finite amplitude but infinite frequency – the integrated effects of successive half cycles must cancel. consequently, RHS (15.20) reduces to $\mathfrak{P}$. Then, on writing

$$\mathsf{F}\{\text{recth}\} = \text{sinch} = \text{sinch}(u \,|\, \alpha, \beta), \tag{15.27}$$

it is seen from (6.4) and (15.18) that

$$\text{sinch}(u \,|\, \alpha, \beta) = \alpha \mathsf{s}(\beta u) \, \text{sinc}(\alpha u), \tag{15.28}$$

where the *side-lobe reduction factor* $\mathsf{s}(\beta u)$ is given by

$$\mathsf{s}(\beta u) = (2\pi^2 \beta u \exp(-\pi^2 \beta u))/(1 - \exp(-2\pi^2 \beta u)). \tag{15.29}$$

Note that $\lim\limits_{\beta \to 0} \mathsf{s}(\beta u) = 1$, which together with (15.19) confirms that (15.18) and (15.28) are consistent.

The width of the main lobe of the sinch function defined by (15.28) varies with both α and β. However, the ratio of this width to that of the sinc function on RHS (15.28) stays close to unity for virtually all values of α and β that are useful in practice. Consequently, the resolution loss is expressed as the fraction α/L. Since the highest side-lobe of a sinc function is roughly 0.21 of its main-lobe height and occurs when its argument is approximately $3\pi/2$, the highest side-lobe of $\text{sinch}(u \,|\, \alpha, \beta)$ is roughly $0.21\mathsf{s}(3\pi\beta/2\alpha)$. Since the actual recorded image is truncated, the practical window function is not the recth function but rather $(\text{rect}(x/L) \, \text{recth}(x \,|\, \alpha, \beta))$. It follows from (15.12) then that, if the highest side-lobe of the practical window function is not to exceed the estimate obtained two sentences back,

$$\text{recth}(L/2 \,|\, \alpha, \beta) < 0.21\mathsf{s}(3\pi\beta/2\alpha), \tag{15.30}$$

which is the condition that fixes the compromise between resolution loss and side-lobe level – the point is that β/α sets the side-lobe level, but this level can actually be achieved only if α/L is small enough to permit the inequality (15.30) to be satisfied.

Windowing ameliorates the effect of truncating $\mathfrak{a}(\mathbf{x})$, but it destroys information, preventing the restoration of $f(\mathbf{x})$ within what we call the *border* of Γ. Suppose that $f(\mathbf{x})$ *can* be restored faithfully within a frame Γ_o, whose size is, of course, less than that of Γ. The border is the region, of effective width T_o say, between the perimeters of Γ_o and Γ. In order to be able to make further use of this concept, we say we are here talking of the *border between Γ_o and Γ*. Now, T_0 is the sum of the 'falloff' distance of the window – e.g. T in (15.16) – , plus half of the effective width $\hat{T}$ of the *inverse psf* $\hat{h}^{-1}(\mathbf{x})$. The latter is to be thought of as the operator which gives $\delta(\mathbf{x})$ when convolved with the modified psf $\hat{h}(\mathbf{x})$ introduced in (14.10). Note that $\hat{h}^{-1}(\mathbf{x})$ symbolically characterizes the deconvolution operation in the sense that $\hat{f}(\mathbf{x})$ is given by $(a(\mathbf{x}) \ominus \hat{h}^{-1}(\mathbf{x}))$. A very significant practical point is that $\hat{T}$ is not simply proportional to the reciprocal of the effective width of $h(\mathbf{x})$. In particular, for any given form of $h(\mathbf{x})$, $\hat{T}$ increases as the contamination level decreases and as the size of Ω_h increases. Furthermore, $\hat{T}$ is much less for a Gaussian blur than for blurring due to linear motion or out-of-focus imaging. The psf characterizing each of these three types of blurring is, respectively,

$$h(r) = (4/\pi\sigma^2) \exp(-4r^2/\sigma^2), \tag{15.31}$$

where $r = (x^2 + y^2)^{\frac{1}{2}}$,

$$h(x) = \sigma^{-1} \operatorname{rect}(x/\sigma), \tag{15.32}$$

and

$$h(r) = (4/\pi\sigma^2) \operatorname{rect}(r/\sigma), \tag{15.33}$$

where σ is the effective width of $h(\mathbf{x})$ in each case. It follows that, when the blurring is severe, the ratio of the sizes of Γ_o and Γ can on occasion be quite small.

The truncation of $a(\mathbf{x})$ can be countered, more sophisticatedly than by windowing, by extrapolating the contents of Γ throughout the frame Ω_b. The latter is introduced in the paragraph preceeding the one containing (15.13). There must be no thought of attempting to recover the form of $f(\mathbf{x})$ outside Γ. The only purpose of this kind of preprocessing is to replace the truncated image by one which:

(i) is free of sudden amplitude variations near to its perimeter,
(ii) is of the right size as regards the recoverable part of the true image (i.e. it exists within the frame Ω_b), and
(iii) retains all of the recorded information.

We give the name *edge extension* to this kind of preprocessing. It is characterized by

$$a(\mathbf{x}) = a(\mathbf{x}) \quad \text{for } \mathbf{x} \in \Gamma. \tag{15.34}$$

There are two types of edge extension. *Simple edge extension* involves continuing $a(\mathbf{x})$ from Γ out to the perimeter of Ω_b along straight lines perpendicular to the perimeter of Ω_b. This is, of course, straightforward when Γ and Ω_b are rectangular (which they are in most practical applications). Now refer to (15.3) and consider the border between Γ and Ω_b.

The operation $\text{pre}\{a(\mathbf{x})\}$ is defined by

$$\text{pre}\{a(\mathbf{x})\} = k_0 + k_1\xi + k_2\xi^2 + k_3\xi^3 \qquad (15.35)$$

along any of the abovementioned 'straight lines' (so the coordinate ξ represents either x or y). The values of the constants k_0 through k_3 are found by requiring the amplitude and first derivative of $\text{pre}\{a(\mathbf{x})\}$ to equal those of $a(\mathbf{x})$ at the perimeter of Γ and setting $\text{pre}\{a(\mathbf{x})\}$ to zero at the perimeter of Ω_b – a constraint on the first derivative at the perimeter of Ω_b is also usually employed.

While simple edge extension tends to be less satisfactory than windowing as regards the central region of the restored image, it always leads to the recovery of more of the information contained in the true image. It is also an effective means of compensating for the truncation of $a(\mathbf{x})$. It suffers, however, from the inconsistency of deconvolution noted in §14.

Overlapped edge extension avoids the inconsistency of convolution by treating the preprocessed image $a(\mathbf{x})$ as if it was periodic (refer to the relevant discussion in §14). One is to imagine image space being covered with contiguous frames each of the same size as Ω, which itself equals Ω_f from (15.2). We refer to these contiguous frames as *basic cells*. An *inner cell*, which is a frame congruent to Γ, is centred within each basic cell. The name *cell border* is here given to the border between a basic cell and its inner cell. We still find it appropriate to adhere to the definition (15.3) for the preprocessed recorded image. The contents of Ω_b are thus understood to be repeated endlessly throughout image space, thereby ensuring there is a replica of $a(\mathbf{x})$ in each basic cell. We call each replica of Ω_b an *original cell*. Since Ω_b is larger than Ω_f, the contents of each original cell spill over into neighbouring basic cells. The overlapping occurs only within cell borders. Consequently, $\text{pre}\{a(\mathbf{x})\}$ equals $a(\mathbf{x})$ within each inner cell, but it has an appropriately massaged form within each cell border. The massaging must, of course, arrange for $a(\mathbf{x})$ to be truly periodic, in the sense that its functional behaviour is duplicated in the neighbourhoods of *opposite points* (these are defined at the end of this paragraph). The right-handside of (15.35) still applies, except that the 'straight lines' now only traverse each cell border. Four constants are still needed since the amplitude of $\text{pre}\{a(\mathbf{x})\}$ at any point on the perimeter of a basic cell must mirror its behaviour at the opposite point, where the latter is defined as follows. Taking, merely for the purpose of this definition, the ξ-axis to pass right across the basic cell, a pair of opposite points is defined by where the ξ-axis intersects the perimeter of the basic cell.

It should be noted that the form of $\text{pre}\{a(\mathbf{x})\}$ provided by overlapped edge extension, within each basic cell, approximates the periodic overlapped ideal blurred image $p_b(\mathbf{x})$ defined by (12.6). The practical success of overlapped edge extension is critically dependent upon satisfying the following 'smoothness' constraint. The form of $\text{pre}\{a(\mathbf{x})\}$ within any cell border must be at least as smooth as $a(\mathbf{x})$ is within any inner cell. Visual inspection is adequate for assessing whether or not this constraint has actually been satisfied in any particular instance. The inclusion of spurious components of high spatial frequency in pre$\{a(\mathbf{x})\}$ is thereby prevented, which is conducive to the general stability of image restoration. Another advantage of the 'overlapped', as opposed to the 'simple' approach, is that its extension covers less than half the area and is more constrained. It involves the minimum amount of extrapolation required to fill each cell border, and it consequently tends to be significantly more accurate overall.

It is instructive to recall the discussion of the deconvolution problem in §14. The simple and overlapped edge-extension methods can be thought of as means for transforming the practical deconvolution problem into the idealized finite and periodic, respectively, deconvolution problems, but so as to minimize the deleterious effects of the extraneous imperfections which inevitably contaminate recorded images.

It is clear that overlapped edge extension can be implemented as straightforwardly as simple edge extension, but its practical performance is often markedly superior (see the introductory comments to this chapter). Simple edge extension has its uses nevertheless, especially when the overlapped method cannot be invoked for some technical reason. The techniques introduced in §§17 and 18 can benefit from the simple method, which makes all the difference to the important applications described in §30.

Loss of resolution during recording, due to deficiences of the imaging apparatus (e.g. aberrations which prevent attainment of the diffraction limit), can be considered as contributing to the overall contamination, because it causes the restored image to be unnecessarily degraded. It is therefore appropriate to treat super-resolution as a branch of preprocessing, since it can sometimes take advantage of the duality of the Fourier and image spaces to recover part of the lost resolution, without having to resort to deconvolution (the latter can of course be invoked afterwards in order to attempt to further reduce the resolution loss). There is also a significant psychological factor. Humankind is loth to accept limitations. So it is natural to try and beat the diffraction limit! Because a necesary corollary to the final paragraph of §3 is that any given $F(u)$ is effectively band-limited in its spatial frequencies, one is thus driven to enquire how to establish criteria for extending $F(\mathbf{u})$ beyond the implied band limit.

Suppose a spectrum is recorded within an interval of length 2ω of a straight line centred on the origin of Fourier space. It follows from the projection theorem (9.6) that the Fourier transform of the observed spectrum is a projection, resolved to the spatial frequency limit ω. Clearly, the projection can be super-resolved if a means is found for extending the spectrum outside the interval of length 2ω. Since the true image can be reconstructed from its projections (Chapter V), it is clear that it must be possible to super-resolve data in two dimensions if it is possible to do it for one-dimensional data. It is sufficient then to consider one-dimensional functions such as the g and G characterized by (13.1) through (13.5). Refer to the paragraph containing (10.3), which indicates why the integer M appearing in (13.3) effectively fixes the given band of spatial frequencies. Suppose the given data consists of $G(u)$, unavoidably contaminated of course, for $|u| < (M + \varepsilon)/L$, where $0 < \varepsilon < \frac{1}{2}$.

If $g(x)$ is of finite extent then $G(w)$ is an entire (integral) function (§13), implying that $G(w)$ can be analytically continued throughout all Fourier space, provided $G(u)$ is known exactly for a finite range of u. It is the 'exactly' which is the trouble. All the kinds of 'noise' that inescapably bedevil measurements usually restrict the range of 'continuable' spatial frequencies so severely that this idea is seldom of any practical use. It is worthwhile belabouring this point. When the quantity L introduced in (13.1) is palpably finite, the point raised in the previous paragraph of this section can be recast as: can any G_m, for $|m| > M$, be usefully estimated from $G(u)$ in the range $|u| < M/L$? Since (13.5) shows that $G(m/L) = G_m$, the error-sensitivity can be expected to be minimized if the $G((2l + 1)/2L)$, with l an integer and $|l| < M$, are used to estimate G_m for $M < |m| < M + N$, where $N \le M$. Note that $\mathrm{sinc}(l - m + \frac{1}{2}) = (-1)^{l-m}/(l - m + \frac{1}{2})\pi$, which when substituted into (13.5) gives

$$\sum_{|m|=M+1}^{M+N} (-1)^m G_m/(l - m + \tfrac{1}{2})$$

$$= (-1)^l \pi G((2l + 1)/2L) - \sum_{m=-M}^{M} (-1)^m G_m/(l - m + \tfrac{1}{2}), \quad (15.36)$$

the right side of which is 'given' because the G_m for $|m| \le M$ are found by inspection of the given data. Unless N is a small integer, in which case it is sometimes feasible to compensate for the contamination of the given data by a 'least squares' reduction, the matrix whose elements are $(-1)^m/(l - m + \frac{1}{2})$ tends to be too ill-conditioned to permit (15.36) to be usefully inverted. Some improvement can be realized by expanding $G(u)$ in spheroidal functions and taking advantage of their simultaneous orthogonality over a finite range, $|u| < (M + N)L$ say, and the infinite range $|u| < \infty$, but contamination again usually prevents this being of much use.

The main reason why the approach to super-resolution outlined in the previous paragraph is rarely successful (even when the initial resolution is comparatively low) is that no constraint is put on images to be real and non-negative (§3). Unfortunately, it is not clear how to incorporate this constraint into the above analytic continuation procedure. Other approaches are needed, of the kind outlined below.

Considerable claims (some of them almost semi-mystical!) have been made for the super-resolving powers of the *maximum entropy* method, which is most conveniently discussed in terms of the D.F.T. (§12). It is sufficient for our present purpose to consider just the one-dimensional form – e.g. (12.18). We define the entropy H by

$$\mathsf{H} = - \sum_{k=1}^{K} (f_k)^\nu \ln(f_k),$$
(15.37)

where the non-negative, real constant ν is conventionally taken to be either zero or unity; but there seems to be no very good reason why it should be thus constrained by any thermodynamic precedent – the only criterion for judging an image processing technique is the quality of the results; and so it may be worth experimenting with other values for ν. The method consists of continuing $F(\mathbf{u})$ so that H is maximized. Probably the crux of the method is that the logarithms in (15.37) inhibit any tendency for the real image to become negative. Impressive results are often obtained with the method, but it seems to be extremely sensitive to certain features in the true image – e.g. the form of $F(\mathbf{u})$ can change dramatically when a small background is added to $f(\mathbf{x})$. Not enough is presently known about maximum entropy to permit its worth to be judged really objectively. However, it is one of the few methods that inherently force the super-resolved image to be positive. We discuss maximum entropy further in §18.

A rather more rational super-resolution technique is Gerchberg's algorithm, which we have already introduced in §11. The algorithm behaves stably in a wide class of situations, although the contamination level of course limits the achievable degree of superresolution. Furthermore, the algorithm is versatile and suitable for efficient implementation in two dimensions. Several related algorithms, examples of which are discussed in §§23 and 24, are specializations of the following generalization of Gerchberg's algorithm.

Let $F_U(\mathbf{u})$ and $F_L(\mathbf{u})$ denote upper and lower bounds, respectively, for $|F(\mathbf{u})|$; and $f_U(\mathbf{x})$ and $f_L(\mathbf{x})$ denote corresponding bounds for $|f(\mathbf{x})|$:

$$F_L(\mathbf{u}) \le |F(\mathbf{u})| \le F_U(\mathbf{u}),$$
(15.38)

and

$$f_L(\mathbf{x}) \le |f(\mathbf{x})| \le f_U(\mathbf{x}).$$
(15.39)

Now assume there exists an $f(\mathbf{x})$ such that the constraints (15.38) and (15.39) are actually satisfied. This, of course, merely requires the constraints to be weak enough. We then state from practical experience that, when (15.39) and (15.38) are iteratively applied in turn, an initial estimate for $f(\mathbf{x})$ converges to a form, $\hat{f}(\mathbf{x})$ say, which satisfies (15.38) and (15.39). Furthermore, $|\hat{f}(\mathbf{x}) - f(\mathbf{x})|$ tends to decrease as the constraints are strengthened, provided such strengthening is done gradually as the iterations proceed. During the early iterations it is important to err on the weak side when choosing the bounds on the image and its spectrum. Bounds which are so strong that $f(\mathbf{x})$ does not actually satisfy (15.38) and (15.39) lead to spurious results.

The preceding paragraph gives valuable insight into the Fourier transform. We know from (6.12) and (6.13) that there is a duality between the Fourier and image spaces – i.e. knowing either $f(\mathbf{x})$ or $F(\mathbf{u})$ implies knowledge of the other. Our understanding of (15.38) and (15.39) is that there exists a complementary duality between the accuracies with which $f(\mathbf{x})$ and $F(\mathbf{u})$ can be recovered. Given $f_L(\mathbf{x})$ and $f_U(\mathbf{x})$, we can calculate valid bounds on $F(\mathbf{u})$, and vice versa. More importantly, knowing $f_L(\mathbf{x})$ and $f_U(\mathbf{x})$ in a limited region of image space, and knowing $F_L(\mathbf{u})$ and $F_U(\mathbf{u})$ in a limited region of Fourier space, we can generate a consistent $\hat{f}(\mathbf{x})$ throughout image space. In addition, we can extrapolate the bounds throughout the two spaces, the gaps between upper and lower bounds widening, of course, the further one extrapolates.

It follows that one may do well to constrain images with bounds as well as with positivity (as defined in §8) when attempting to extrapolate. Specific a priori information concerning $f(\mathbf{x})$ may be used directly when calculating $f_L(\mathbf{x})$ and $f_U(\mathbf{x})$. Super-resolution algorithms are most effective where the differences between $f_L(\mathbf{x})$ and $f_U(\mathbf{x})$ are least. It is therefore desirable that the part of Ω_f for which a priori information is available should be as large and as spread out as possible.

An interesting example of the application of these ideas is to images, gathered by space vehicles, of the mountainous surfaces of moons and planets, illuminated by the Sun at a low angle. Such images are essentially devoid of atmospheric effects, and include well spread regions of deep shadow. If the reasonable a priori assumption is made that the shadow regions are black and have sharp edges, we then have the basis of constraints suitable for superresolution. A recorded image is appropriately calibrated and enhanced, and is then interpolated to give more image samples (e.g. four times as many in each direction). The shadow regions are identified and the algorithm is applied. In one-dimensional simulations of this approach we have tripled the resolution over complete images.

Note that the difference between $f(\mathbf{x})$ and $\hat{f}(\mathbf{x})$ can be conveniently split into two parts:

$$f(\mathbf{x}) - \hat{f}(\mathbf{x}) = c_+(\mathbf{x}) + c_-(\mathbf{x}), \qquad (15.40)$$

where $c_+(\mathbf{x})$ and $c_-(\mathbf{x})$, which are here called the correctable and uncorrectable components of the contamination respectively, are defined as follows: $c_-(\mathbf{x})$ satisfies the constraints implying that it is impervious to the algorithm, whereas $c_+(\mathbf{x})$ does not satisfy the constraints which means that it is gradually reduced in amplitude as the iterations proceed. Note that we are here using the term 'contamination' in a somewhat different way from previously, but that does not materially change the argument. It is worth noting that super-resolution can only be effective if $c_+(\mathbf{x})$ and $c_-(\mathbf{x})$ do not fluctuate in opposition to each other, and if $|c_+(\mathbf{x})|$ generally exceeds $|c_-(\mathbf{x})|$. It seems these conditions often pertain in practice.

Any *a priori* information that can be incorporated into an extrapolation algorithm is valuable. An informative instance is our procedure for uncovering detail in a very poorly resolved image of the red-giant star Betelgeuse. There is usually little error in assuming a star is circular, which may seem obvious, but it is in fact a powerful piece of *a priori* information which should not be wasted. It implies that a circular frame, having a diameter somewhat greater than the star, may be introduced, with $\hat{f}(\mathbf{x})$ being set to zero outside it. This is a good example of the kind of simple but powerful constraint on which super-resolution algorithms can operate very effectively.

16   Multiplicative deconvolution

It is RHS (14.3) that gives rise to the idea of deconvolution being a *filtering* operation, as has already been pointed out in §14, in the paragraph containing (14.3). The OTF $H(\mathbf{u})$ can be regarded as performing a filtering action in (14.3) because it alters the spatial frequency content of $B(\mathbf{u})$ by division. Now, *division* is not a conventional filtering operation, as usually understood in signal processing contexts, but *multiplication* is. Situations characterized by (14.3) are elementary in that $B(\mathbf{u})$ is multiplied by what is here called the *simple inverse filter*, which is denoted by

$$\mathfrak{S}(\mathbf{u}) = 1/H(\mathbf{u}). \tag{16.1}$$

It follows that (14.3) can be rewritten as

$$f(\mathbf{x}) = \mathsf{F}\{B(\mathbf{u})\mathfrak{S}(\mathbf{u})\}. \tag{16.2}$$

The operation on RHS (16.2) takes no account of the inevitable contamination. We see from (14.1) and (14.2) that the Fourier transform of (14.9) is

$$A(\mathbf{u}) = B(\mathbf{u}) + C(\mathbf{u}) = F(\mathbf{u})H(\mathbf{u}) + C(\mathbf{u}), \tag{16.3}$$

where $A(\mathbf{u})$ and $C(\mathbf{u})$ are the spectra of $a(\mathbf{x})$ and $c(\mathbf{x})$ respectively. We call $C(\mathbf{u})$ the *spectral contamination*. It is clear that simple inverse filtering can only be adequate when an image needs to be enhanced (see §2), rather than restored or reconstructed. Such filtering tends to be very error-sensitive because of the large values assumed by $|A(\mathbf{u})\mathfrak{S}(\mathbf{u})|$ whenever $|B(\mathbf{u})/C(\mathbf{u})|$ is appreciably less than unity. To solve the practical deconvolution problem usefully, means have to be found for avoiding such 'noise amplification'.

It is convenient to identify two regions R_A and R_+ in Fourier space: R_A is the region where $|A(\mathbf{u})|$ has significant value; R_+ spans those points $\mathbf{u}$ for which $|H(\mathbf{u})| \gtrsim |C(\mathbf{u})|$. It is further convenient to denote by $\bar{R}_A$ the part of R_A not intersected by R_+.

Since, by definition, $(C(\mathbf{u}))/H(\mathbf{u})$ swamps $F(\mathbf{u})$ within $\bar{R}_A$, it is necessary to improve on simple inverse filtering when $\bar{R}_A$ is sizable (as it often is). What is needed is a 'modified inverse filter', denoted here by $\mathfrak{M} = \mathfrak{M}(\mathbf{u})$, which is in some useful sense computationally impotent when $\mathbf{u}$ lies in $\bar{R}_A$. An obvious from for $\mathfrak{M}(\mathbf{u})$ is

$$\mathfrak{M}(\mathbf{u}) = 1/H(\mathbf{u}) \qquad \text{for } \mathbf{u} \in R_+$$
$$= 0 \qquad \text{for } \mathbf{u} \in \bar{R}_A. \qquad (16.4)$$

Although such a modified inverse filter is a palpable improvement, it is not altogether satisfactory because the discontinuous nature of $\mathfrak{M}(\mathbf{u})$ on the boundary of $\bar{R}_A$ introduces artefacts into the restored image. For reasons of the kind discussed in detail in §§14 and 15 it is better if the form of $\mathfrak{M}(\mathbf{u})$ is smoother.

Consider another modified inverse filter:

$$W(\mathbf{u}) = (H^*(\mathbf{u}))/(|H(\mathbf{u})|^2 + \Phi(\mathbf{u})), \qquad (16.5)$$

which is known as the *Wiener filter*. The function $\Phi = \Phi(\mathbf{u})$ is a measure of the noise-to-signal ratio as a function of spatial frequency (the 'noise' here is equivalent to what is defined in §4 as 'contamination'), i.e. Φ is an estimate of $|C/F|^2$. Because the filter defined by (16.5) satisfies

$$W(\mathbf{u}) \simeq 1/H(\mathbf{u}) \qquad \text{when } |H(\mathbf{u})| \gg \Phi(\mathbf{u}),$$
$$\simeq (H^*(\mathbf{u}))/\Phi(\mathbf{u}) \qquad \text{when } |H(\mathbf{u})| \ll \Phi(\mathbf{u}), \qquad (16.6)$$

it has similar properties to those of $\mathfrak{M}(\mathbf{u})$, as defined by (16.4), but it is everywhere smooth. It is also known to be optimum in a mean-square sense, provided the contamination is Gaussianly distributed.

The presence of $c(\mathbf{x})$ on RHS (14.9) makes it impossible to obtain an 'exact' solution – such as (14.3) for the idealized case – to the practical deconvolution problem. However, one can postulate that some modified inverse filter is a useful approximation to the OTF of the modified psf introduced in §14. On choosing the Wiener filter (which is certainly the most popular multiplicative filter), this assumption becomes

$$W(\mathbf{u}) = \mathsf{F}\{\hat{h}^{-1}(\mathbf{x})\} \qquad \text{with } \hat{h}^{-1}(\mathbf{x}) \, \Theta \, \hat{h}(\mathbf{x}) = \delta(\mathbf{x}), \qquad (16.7)$$

where $\delta(\mathbf{x})$ is the two-dimensional delta function introduced in (6.18). Consequently, taking the Fourier transform of (14.10), and appealing to the convolution theorem (7.7), suggests that a useful version of the recoverable true image is obtained from

$$\hat{f}(\mathbf{x}) \simeq \mathsf{F}\{A(\mathbf{u})W(\mathbf{u})\}, \qquad (16.8)$$

which has the same functional form as (16.2) but it reflects, through the definition (16.5), the uncertainties due to the inescapable contamination of measured data. Although experience of the assessment of previously restored images of (apparently) similar kinds may help one to estimate the noise-to-signal ratio in any particular instance, it is only rarely that the dependence upon $\mathbf{u}$ of $\Phi(\mathbf{u})$ can be predicted with confidence. One is often reduced to taking $\Phi(\mathbf{u})$ to be a constant, here called the *filter constant* Φ. Its value can be chosen experimentally – i.e. different values of Φ are tried and the one giving the most 'satisfactory' results (there is no escaping subjectivity in the real-world!) is taken to be the 'correct' value.

There is an equivalent approach to deconvolution known as *homomorphic* or *cepstral* filtering. The *complex cepstrum* of a function $g = g(\mathbf{x})$ is here denoted by $\mathbf{C}_g = \mathbf{C}_g(\mathbf{u})$ and is defined, in terms of $G = G(\mathbf{u}) = \mathsf{F}\{g(\mathbf{x})\}$, by

$$\mathbf{C}_g(\mathbf{u}) = \ln(G(\mathbf{u})), \tag{16.9}$$

where ln is shorthand for $\log_e$. In the ideal case, characterized by (14.1) and (14.2), the complex cepstrum of $b(\mathbf{x})$ is given by

$$\mathbf{C}_b(\mathbf{u}) = \ln(F(\mathbf{u})) + \ln(H(\mathbf{u})) = \mathbf{C}_f(\mathbf{u}) + \mathbf{C}_h(\mathbf{u}) \tag{16.10}$$

so that the operation, here called *simple homomorphic filtering*, which is equivalent to simple inverse filtering as defined by (14.3), is

$$f(\mathbf{x}) = \mathsf{F}\{\exp(\mathbf{C}_b(\mathbf{u}) - \mathbf{C}_h(\mathbf{u}))\}. \tag{16.11}$$

Simple homomorphic filtering fails under the same sets of circumstances that lead to the failure of simple inverse filtering. In practice, when $a(\mathbf{x})$ is appreciably contaminated, it is convenient to introduce the *homomorphic Wiener filter* defined by

$$\mathbf{W}(\mathbf{u}) = \ln(W(\mathbf{u})), \tag{16.12}$$

where $W(\mathbf{u})$ itself is defined by (16.5). The recoverable true image is then given by

$$\hat{f}(\mathbf{x}) = \mathsf{F}\{\exp(\mathbf{C}_a(\mathbf{u}) - \mathbf{W}(\mathbf{u}))\}, \tag{16.13}$$

which is exactly equivalent to (16.8), whereas (16.11) is exactly equivalent to (16.2).

17   Subtractive deconvolution

Any function can be thought of as a distribution of impulses. In particular, the true image can be expressed as

$$f(\mathbf{x}) = \iint\limits_{\Omega_f} f(\mathbf{x}')\, \delta(\mathbf{x} - \mathbf{x}')\, d\sigma(\mathbf{x}') = \iint\limits_{\Omega_f} f(x', y')\, \delta(x - x')\, \delta(y - y')\, dx'\, dy',$$

$$\tag{17.1}$$

where Ω_f is the image frame, introduced in §7, and $\delta(\mathbf{x})$ is the two-dimensional delta function defined by (6.18) through (6.21). It is often useful, especially in computational contexts, to approximate $f(\mathbf{x})$ by its point samples, or pixels (see §5), at a rectangular grid of points $(l\Delta_1, m\Delta_2)$, i.e.

$$f(x) \simeq \sum_{l=-M_1}^{M_1} \sum_{m=-M_2}^{M_2} f_{l,m} \, \delta(x - l\Delta_1) \, \delta(y - m\Delta_2). \tag{17.2}$$

All quantities appearing on RHS (17.2) are introduced in the first paragraph of §12.

On replacing $\delta(\mathbf{x})$ by $h(\mathbf{x})$ in the middle expression in (17.1), the latter becomes the same as RHS (14.1). So, the approximate formula for $b(\mathbf{x})$, corresponding to (17.2) for $f(\mathbf{x})$, is

$$b(\mathbf{x}) \simeq \sum_{l=-M_1}^{M_1} \sum_{m=-M_2}^{M_2} f_{l,m} h(x - l\Delta_1, y - m\Delta_2). \tag{17.3}$$

Multiplying (17.3) through by $(\delta(x - l\Delta_1)\delta(y - m\Delta_2))$, and integrating over image space, gives

$$b_{l,m} = \sum_{l'=-\lambda}^{\lambda} \sum_{m'=-\mu}^{\mu} f_{l-l',m-m'} h_{l',m'}, \tag{17.4}$$

where $2\lambda\Delta_1$ and $2\mu\Delta_2$ are the effective x-extent and y-extent, respectively, of Ω_h, and where the pixels of $b(\mathbf{x})$ and $h(\mathbf{x})$ are defined by

$$z_{l,m} = \iint_{\Omega_z} z(x, y) \, \delta(x - l\Delta_1) \, \delta(y - m\Delta_2) \, dx \, dy \qquad \text{for } z = b \text{ or } h.$$

$$\tag{17.5}$$

When the contamination is included, as in (14.9), the formula equivalent to (17.4) is

$$a_{l,m} = \sum_{\Omega_h(l',m')} f_{l-l',m-m'} h_{l',m'} + c_{l,m}, \tag{17.6}$$

where the *sampled image frame* $\Omega_h(l, m)$ of the psf defines the ranges of the integers l and m that span Ω_h, i.e.

$$\sum_{\Omega_h(l, m)} = \sum_{l=-\lambda}^{\lambda} \sum_{m=-\mu}^{\mu}. \tag{17.7}$$

Each $f_{l,m}$ can be thought of as a 'true impulse' because it is a single pixel. Similarly, for l and m fixed, and for $l', m' \in \Omega_h(l', m')$, one can think of $(f_{l-l',m-m'} h_{l',m'})$ as a 'spread impulse' because it is an array of pixels. For the practical deconvolution problem then (see §14), the pixels representing the data – i.e. the $a_{l,m}$ – can be reduced down to 'contamination level' by successively 'subtracting out' each of the spread impulses.

This is how the concept of *subtractive deconvolution* arises. Because contamination is never negligible in practice, it is of course impossible to accurately cancel the whole of each spread impulse by a single subtraction. Useful results can often be obtained, however, by proceeding more gradually.

Subtractive deconvolution is particularly apposite when $f(\mathbf{x})$ is composed of isolated, unresolvable points and when the main effect of $h(\mathbf{x})$ is to reduce contrast rather than resolution. A typical example is a radio-astronomical synthesis telescope observing a region of the heavens containing bright radio stars. The latter often have small angular diameters and the telescope beam – equivalent to $h(\mathbf{x})$ here – , while narrow enough to separate the images of the stars, is usually afflicted with very pronounced side lobes. The latter tend to obscure fine detail in the actually recorded image $\mathfrak{a}(\mathbf{x})$. [Actually, synthesis telescopes make their measurements in Fourier space, because they are species of interferometers (refer to Chapter VI). Since Fourier transformation is a linear process, we can think of the actual measurements as being merely an alternative manifestation of $\mathfrak{a}(\mathbf{x})$, provided the data are Fourier-transformed straightforwardly without any preliminary massaging (which would, of course, constitute some form of preprocessing).] Rather than attempt any kind of preprocessing (using this term in the sense implied in §15), it is found to be effective to immediately invoke a subtractive deconvolutional technique. [There are several factors, which it would be inappropriate to discuss here, which militate against devising effective preprocessing (again in the sense of §15) for radio-astronomical interferometric data.] The archetype of subtractive techniques is Högbom's simple cleaning algorithm, which is an iterative procedure. We find that Högbom's delightfully graphic terminology helps to emphasize the essence of his algorithm.

Since there is no preprocessing, we take the preprocessed recorded image $a(\mathbf{x})$ to be identical to the actually recorded image $\mathfrak{a}(\mathbf{x})$. The pixels $a_{l,m}$ comprising $a(\mathbf{x})$ are said to belong to the *dirty image*, because $a(\mathbf{x})$ is blurred and contaminated. We wish to form a *cleaned image* corresponding to what is called the recoverable true image in §14. We denote the pixels of the cleaned image by $fc_{l,m}$. All of the $fc_{l,m}$ are set to zero before the algorithm is initiated.

The $fc_{l,m}$ gradually take on value as information is extracted from the dirty image. The iterations are characterized by a positive constant, here denoted by τ and called the *loop gain*, which is generally taken to be appreciably less than unity. An estimate of the average of the $|c_{l,m}|$, here denoted by $\bar{c}$, must be available ($\bar{c}$ is here called the *contamination level*). The algorithm consists of the following steps:

S1: The *brightest pixel*, denoted by $a_{j,k}$, in the dirty image is found, i.e.

$$a_{j,k} > a_{l,m} \qquad \text{for all } l \neq j \text{ and } m \neq k. \tag{17.8}$$

S2: A new cleaned image is formed by the *additive operation:*

$$fc_{l,m} + \tau a_{j,k}\, \delta_{l-j,m-k} \rightarrow fc_{l,m}. \tag{17.9}$$

S3: A new dirty image is formed by the *subtractive operation:*

$$a_{l,m} - \tau a_{j,k} \sum_{\Omega_h(l,m)} h_{l-j,m-k} \rightarrow a_{l,m} \quad \text{with } \tau < 1. \tag{17.10}$$

S4: Return to S1, unless

$$a_{l,m} < \bar{c} \quad \text{for all } l \text{ and } m. \tag{17.11}$$

S5: The cleaned image is convolved with a *cleaned psf,* which is $h(\mathbf{x})$ with its mainlobe smoothed and its side lobes heavily attenuated – this is an essentially 'cosmetic' operation (a species of 'enhancement' in fact) applying a final 'touching up' to the appearance of the cleaned image – to give what is considered the best recoverable true image $\hat{f}(\mathbf{x})$.

It would constitute too much of a diversion for us to go into the interpolative properties of Högbom's algorithm – refer to §11, in the paragraph following that containing (11.15) and (11.16). Interpolation by subtractive deconvolution is, however, central to the success of radio-astronomical map-making. The interested reader is urged to consult the relevant references quoted in the introductory comments to this chapter.

Simple cleaning tends to be inefficient when $f(\mathbf{x})$ does not have an impulsive character, i.e. when $f(\mathbf{x})$ is an ordinary sort of image! However, there is a related technique called *subtractive reblurring* which can successfully restore a very wide class of blurred images. It is convenient to re-adopt continuous representations of images when describing this technique. One could retain representations in terms of pixels easily enough, but the resulting explanation would be somewhat clumsier than the one offered below.

Sequences $\{a_j(\mathbf{x}); j = 0, 1, \ldots\}$ and $\{f_j(\mathbf{x}); j = 1, 2, \ldots\}$ of, respectively, *reblurred recorded images* and *estimated true images* are required. Each reblurred recorded image is defined in terms of its predecessor by

$$a_j(\mathbf{x}) = a_{j-1}(\mathbf{x}) - \tau \iint_{\Omega_a} a_{j-1}(\mathbf{x}')h(\mathbf{x}-\mathbf{x}')\, d\sigma(\mathbf{x}') \tag{17.12}$$

for $j \in \{1, 2, \ldots\}$, where

$$a_0(\mathbf{x}) = a(\mathbf{x}), \tag{17.13}$$

and Ω_a is the image frame of the preprocessed recorded image (see §14). The positive (less than unity) constant τ is again called the loop gain. Successive estimated true images are related by

$$f_j(\mathbf{x}) = f_{j-1}(\mathbf{x}) + \tau a_{j-1}(\mathbf{x}), \tag{17.14}$$

with $f_0(\mathbf{x}) = \tau a(\mathbf{x})$. The jth *difference image* is defined by

$$fd_j(\mathbf{x}) = |f_j(\mathbf{x}) - f_{j-1}(\mathbf{x})|. \qquad (17.15)$$

The iterations are stopped at that value J of j for which

$$fd_j(\mathbf{x}) < \bar{c} \qquad \text{for all } x \in \Omega_a. \qquad (17.16)$$

The recoverable true image is defined to be the estimated true image at the Jth iteration:

$$\hat{f}(\mathbf{x}) = f_J(\mathbf{x}). \qquad (17.17)$$

It is convenient to introduce the operator H which is defined by

$$\mathsf{H}\{a_j\} = \iint_{\Omega_a} a_j(\mathbf{x}')h(\mathbf{x} - \mathbf{x}')\, d\sigma(\mathbf{x}'). \qquad (17.18)$$

It follows that (17.12) can be rewritten with the aid of (17.18) as

$$a_j = a_{j-1} - \tau\mathsf{H}\{a_{j-1}\} = (1 - \tau\mathsf{H})^j\{a\}, \qquad (17.19)$$

so that the operator form of (17.14) is

$$f_j = \tau \sum_{k=0}^{j-1} (1 - \tau\mathsf{H})^k\{a\}. \qquad (17.20)$$

Provided the operator inverse of $(1 - \tau\mathsf{H})$ exists, RHS (17.20) can be regarded as a finite geometrical progression, implying that

$$f_j = ((1 - [1 - \tau\mathsf{H})^j)\{a\})\mathsf{H}^{-1}\{a\}. \qquad (17.21)$$

If

$$0 < \|\mathsf{H}\| < 1/\tau, \qquad (17.22)$$

where the *norm* of H is defined by

$$\|\mathsf{H}\| = |\mathsf{H}\{a\}|, \qquad (17.23)$$

which in turn implies that

$$\|1 - \tau\mathsf{H}\| < 1, \qquad (17.24)$$

then

$$\lim_{j \to \infty} \|(1 - \tau\mathsf{H})^j\| = 0, \qquad (17.25)$$

thereby indicating that

$$\lim_{j \to \infty} f_h = \mathsf{H}^{-1}\{a\}. \qquad (17.26)$$

Since there is no point in attempting to iterate to down below contamination level, it can be assumed that

$$f_J \simeq \lim_{j \to \infty} f_j, \tag{17.27}$$

where $\simeq$ here means 'is virtually equivalent to'. It follows from (14.9) and (17.18) that

$$\mathbf{H}^{-1}\{a\} = f(\mathbf{x}) + \mathbf{H}^{-1}\{c\}. \tag{17.28}$$

For the ideal case, characterized by negligible contamination (i.e. $\bar{c} = 0$), $a(\mathbf{x})$ reduces to $(f(\mathbf{x}) \odot h(\mathbf{x}))$ so that $\mathbf{H}^{-1}\{a\}$ is an operational representation of simple inverse filtering – refer to the first paragraph of §16. In practical situations, for which $c(\mathbf{x})$ is never negligible, $\mathbf{H}^{-1}\{a\}$ is an operational representation of Wiener filtering – refer to the paragraph containing (16.5) and (16.6) –, thereby implying that

$$\mathbf{H}^{-1}\{a\} \simeq \hat{f}(\mathbf{x}) \odot \hat{h}^{-1}(\mathbf{x}), \tag{17.29}$$

where $\hat{h}^{-1}(\mathbf{x})$ is the inverse to the modified psf $\hat{h}(\mathbf{x})$ introduced in §14. Since

$$\hat{h}^{-1}(\mathbf{x}) \simeq \mathbf{F}\{W(\mathbf{u})\} \tag{17.30}$$

it follows that

$$\hat{h}(\mathbf{x}) \simeq \mathbf{F}\{1/W(\mathbf{u})\}. \tag{17.31}$$

Subtractive reblurring is found to be generally successful in computational practice, but it is computationally expensive. By comparison, simple cleaning is computationally undemanding, but it is only successful for a very limited class of images. In the presence of appreciable contamination, simple inverse filtering must be replaced by Wiener filtering. Both simple cleaning and simple inverse filtering are based on the given psf $h(\mathbf{x})$ or, equivalently, its inverse $h^{-1}(\mathbf{x})$. Wiener filtering is equivalent to convolving $a(\mathbf{x})$ with $\hat{h}^{-1}(\mathbf{x})$. This suggests that simple cleaning might be improved by making the substitution

$$h_{l-j,m-k} \to \hat{h}_{l-j,m-k} \tag{17.32}$$

in (17.10).

Inspection of (16.5) and (17.31) reveals that

$$\hat{h}(\mathbf{x}) = h(\mathbf{x}) + q(\mathbf{x}), \tag{17.33}$$

where

$$q(\mathbf{x}) = \mathbf{F}\{Q(\mathbf{u}) = (\Phi(\mathbf{u}))/H^*(\mathbf{u})\}. \tag{17.34}$$

In many instances, $Q(\mathbf{u})$ occupies an effectively larger region of Fourier space than does $H(\mathbf{u})$. Furthermore, $|Q(\mathbf{u})|$ is usually smaller near the origin of Fourier space than further out (it, drops off, of course, beyond this). Consequently, $q(\mathbf{x})$ tends to have a narrower main lobe than does $h(\mathbf{x})$, but with a few pronounced nearby side lobes. So, $\hat{h}(\mathbf{x})$ often takes the form of $h(\mathbf{x})$ with an oscillatory 'spike' superimposed upon it. This spike tends to stabilize the cleaning operation even when the true image is a smoothly varying function of $\mathbf{x}$ (i.e. it has no impulsive character whatsoever). One can therefore replace simple cleaning with *modified cleaning*, which is still characterized by the above steps S1 through S5, but with the incorporation of (17.32).

18   Direct deconvolution

There are available a wide variety of deconvolutional techniques that can be applied directly in image space. These include the subtractive methods already discussed in §17. Other direct techniques are discussed here and are compared in particular with the inverse filtering approaches described in §16. We include few analytical details in this section and avoid derivations altogether, for the reasons given in the introductory comments to this chapter. We concentrate on reviewing the main conceptual and practical aspects of the techniques.

We start by considering recursive filters which, according to the definition given in §14, generate output pixels as weighted sums of given pixels and previously calculated output pixels. Such filters are widely used in 1-D signal processing applications where sampling is done in time rather than over distance. They are easily implemented. Furthermore, the recursion can allow a relatively short filter to have a comparatively long effective extent.

A number of approaches have been taken to the problem of extending recursive digital filters to 2-D, and there have been attempts to apply these filters to image restoration. Their usefulness is limited for several reasons. First, because of the lack of a 2-D factorization theorem (i.e. it is impossible in general to write a 2-D filter as a sequence of 1-D recursive filters – it is worth noting that this bears on the uniqueness question for the Fourier phase problem – see §22), many results valid for 1-D filters do not hold in 2-D. This makes it difficult to approximate a desired frequency response with a stable 2-D filter. Second, to be effectively implemented, a general 2-D recursive filter needs to be representable as a finite sum of 2-D separable recursive filters. Third, there is the difficulty, in the case of the practical deconvolution problem, of choosing boundary values with which to begin the recursion. Although the latter can be obtained by simple edge extension (see §15), the effect of errors in the chosen boundary values may spread throughout the whole image, especially since the number of sample points is often far from large (e.g. 128×128 pixels). For all of these reasons, recursive filters are seldom used for image restoration. They are best avoided altogether in 2-D, and they are most appropriate for 1-D when correcting for an extended causal psf (in which case a recursive filter can be much shorter – i.e. have far fewer elements – than a corresponding nonrecursive filter). The reader interested in design details of recursive filters can examine the references quoted in the introductory comments to this chapter.

We next discuss non-recursive image restoration using a finite filter array (as defined in §14) in image space. Note that the filters introduced in §16 are non-recursive, but are not finite in the sense meant here because they are periodic (or circular).

The use of direct non-recursive filters for image restoration is widespread. The outputs of such filters depend only upon the given pixels, so that there is no feedback from previously calculated output pixels (unlike with recursive filters). One consequence of this is that the border between the frame Γ_o, within which faithful restoration can be achieved without preprocessing, and the frame Γ, which encloses the actually recorded image, is at least as narrow as for any other proven approach to deconvolution. The width of the said border can, of course, be further substantially reduced by employing suitable preprocessing (see §15). Non-recursive filters have the further advantages of being computationally stable and easily implementable. Their noise analysis happens to be relatively simple also.

Non-recursive filters can be applied either by convolution in image space or by multiplication in Fourier space. Digital computational folk lore nowadays has it that the F.F.T. permits convolutions to be evaluated with spectacularly increased efficiency by multiplication in Fourier space (even though the multiplication must be both preceded and followed by Fourier transformation). This is almost always true if both of the quantities to be convolved contain comparable numbers of pixels. If the array characterizing a non-recursive filter is small enough, however, it may be more efficient to apply it directly in image space (various points pertaining to this are discussed in detail in §46). To put this into perspective, we think it worth mentioning that, for a 512×512 pixel image, the breakeven point is with a 6×6 array. Something that should be kept in mind, though, is that it can be easier to handle the effects of contamination in Fourier space. For instance, the quantity $\Phi(\mathbf{u})$ introduced in (16.5) is a very convenient device for taking account of the statistics of the comtamination. When applying a non-recursive filter directly (i.e. in image space), on the other hand, it transpires that one must iteratively vary a 'noise control parameter' until the appearance of the restored image is optimized (the reader interested in details should refer to the sources quoted in the introductory comments to this chapter) – it does not seem to be possible to determine this parameter straightforwardly from the contamination's statistics.

Direct non-recursive filters can be implemented either purely digitally;

or purely optically, using incoherent light to superimpose images of displaced transparencies; or electro-optically, by scanning a transparency of the blurred image with a specially designed aperture and writing the filtered result directly on to film.

There are a number of popular signal processing techniques for designing 1-D filters having effectively finite impulse responses. Their usual purpose, however, is to approximate some desired frequency response (e.g. a band-pass filter). In image restoration the aim is to make the resultant psf as close as possible, in an appropriate sense, to a delta function. It follows that it is preferable to optimize direct recursive filters in image space rather than in Fourier space.

We now consider the design of direct non-recursive filters in 1-D. It is inappropriate here to go in to great detail. We concentrate instead on the crucial design choices that have to be made. The 1-D treatment can be straightforwardly extended to 2-D, in the manner outlined later.

The problem is to design a filter $w(x)$ that, when convoluted with the psf $h(x)$, produces a *resultant psf* $\bar{h}(x)$ which is optimum in some sense, without unduly increasing the contamination level. Some authors assume an explicitly continuous $h(x)$ and try to synthesize an optimum continuous $\bar{h}(x)$. Here, however, we take the much simpler approach of operating on arrays of sampled versions of $w(x)$, $h(x)$, and $\bar{h}(x)$, i.e. these quantities become, respectively, **w**, **h**, and **h̄**, which are to be thought of as vectors. Their respective components are w_n with n running from $-N$ to N, h_n with n running from $-J$ to J and $\bar{h}_n$ with n running from $-(N+J)$ to $(N+J)$, so that

$$\bar{h}_j = \sum_{n=-N}^{N} w_n h_{j-n} \qquad \text{with } h_j = 0 \text{ for } |j| > J. \qquad (18.1)$$

It is understood that the samples are equispaced, in the sense that $t_n = t(n\Delta)$ where t represents either w or h or $\bar{h}$, and Δ is the distance between adjacent samples (Δ is also here called the spacing of the filter elements).

Various criteria have been proposed for deriving **w**. The more effective of these attempt to optimize $\bar{h}$ in some way. Ideally, $\bar{h}$ should effectively be a Kronecker delta, but this is obviously impossible when the number of elements of h is finite. A popular approach is to minimize the radius of gyration of $\bar{h}$, i.e. to minimize the quantity

$$\sum_{n=-(N+J)}^{N+J} n^2 (\bar{h}_n)^2$$

under a constraint. Suitable constraints are either

$$\sum_{n=-(N+J)}^{N+J} (\bar{h}_n)^2 = 1, \qquad (18.2)$$

or

$$\sum_{n=-(N+J)}^{N+J} \tilde{h}_n = 1. \tag{18.3}$$

Provided one invokes appropriate matrix techniques, the latter constraint is easier to implement than the former. It should be noted that the factor n^2 in the above expression for the radius of gyration tends to ensure the stability of **w**. This is because, for a large enough N, $|\tilde{h}_n|$ becomes proportional to $1/|n|$. This prevents $\tilde{h}_n$, and hence w_n, varying with n in a markedly oscillatory manner, as $n \to N$.

The idea behind minimizing the radius of gyration is to ensure that $|\tilde{h}_n|$ exhibits a tendency to decrease with increasing $|n|$. However, we have not found this to be generally successful because it often overemphasizes some $\tilde{h}_n$, which are, in fact, quite small and for each of which $|n|$ is only moderately large. We prefer to maximize $\tilde{h}_n$ under the constraint (18.2), which in this case produces more stable results than (18.3). The method of Lagrangean multipliers can then be invoked to derive **w** straightforwardly. The numerical accuracy has to be watched because an $N \times N$ matrix must be inverted. An important property of our criterion is that it forces $\tilde{h}_0$ to approach unity as $N \to \infty$, which implies, of course, that $\tilde{\mathbf{h}}$ really does tend to a Kronecker delta as $N \to \infty$. So, provided N is small enough for our criterion to be conveniently applied, it performs comparably to the multiplicative deconvolution techniques described in §16.

It is interesting to note that maximizing $\tilde{h}_0$ is equivalent to minimizing

$$\sum_{j=1}^{N+J} (\tilde{h}_j)^2,$$

because of the normalization implied by (18.2). This is another useful aspect of our approach.

It is worth remarking that it is usually best to maximize $\tilde{h}_0$ when $h(x)$ is an even function. When this is not the case, it may be preferable to maximize some other $\tilde{h}_n$. Another possibility is to increase N, because it transpires that a filter array containing N elements which maximizes $\tilde{h}_0$ must perform at least as well as a filter array of $(N-2)$ elements which maximizes $\tilde{h}_1$.

The ratio $\tilde{h}_0/h_0$ is a measure of the restorative power of the filter **w**. However, convolving with a filter usually amplifies the inevitable contamination present in the data. So, the improvement represented by (a large value of) $\tilde{h}_0/h_0$ has to be weighed against the contamination-amplification produced by the filter. It must also be kept in mind that filtering tends to alter the statistics of the contamination. Uncorrelated noise becomes correlated over regions of extent $2N\Delta$. The effect of noise can be taken into account when deriving **w** by placing an upper limit on $\tilde{h}_0/h_0$, thereby obtaining the best **w** for a given value of N consistent with a particular contamination level.

Our experience suggests that our criterion is significantly superior to criteria based on minimizing the radius of gyration. The rate at which $\tilde{h}_0$ approaches unity with increasing N varies considerably with the type of psf. For a Gaussian blur, N can be less than J, whereas for uniform linear blur, N is often much larger than J. The noise amplification tends to increase rapidly with $\tilde{h}_0$ but it can be controlled effectively by adopting the strategy outlined in the previous paragraph. It must be emphasized, however, that direct filter arrays are most useful for psfs whose extents are comparatively small.

In practice, one often wants a constant function to remain constant after convolution with $w(x)$. This usually requires that $\tilde{\mathbf{h}}$ be normalized according to (18.3), rather than (18.2) as demanded by our criterion. This tends to destabilize the numerical derivation of $\mathbf{w}$. Nevertheless, it is sometimes possible to ameliorate this difficulty by replacing w_n in (18.1) by Cw_n, where C is an appropriate constant – but there is a consequent disadvantage: $\tilde{h}_0$ need no longer tend to unity monotonically as N increases.

It is straightforward to extend the above argument to derive an optimum 2-D filter array, because a 2-D convolution can always be written as a 1-D convolution. The size of the matrix which has to be inverted increases markedly of course, because in order to derive an optimum $M \times M$ array a $M^2 \times M^2$ matrix must be inverted.

In general, image restoration results obtained using overlapped edge extension (see §15) followed by Wiener filtering (see §16) are superior to those obtained with a non-recursive filter array derived on the basis of our criterion. This is because the inverse psf for multiplicative deconvolution is effectively of infinite extent because of the periodic (or circular) property of the F.F.T. However, when the contamination is slight, the difference between the results given by the two approaches can be negligible. While non-recursive filter arrays are efficient and easily implemented (for N small enough), they are not as versatile as Wiener filters for handling contamination.

There is a direct method which can be invoked to solve the idealized finite deconvolution problem in the case of uniform linear blur. Although this is a very restricted application, it gives further useful insight into simple inverse filtering (see §16). Let the extent of the psf be σ as in (15.32). Then if L, the extent of Ω_b (in the x-direction), is chosen sufficiently large to ensure that L/σ is an irrational number, the problem can be solved immediately by Wiener filtering (see §16). This is because none of the zeros of $H(\mathbf{u})$ can lie at sample points – refer to the discussion following (14.8). However, if L/σ happens to be an integer, J say, the zeros of $H(\mathbf{u})$ lie along the straight lines in Fourier space defined by

$$u = \zeta_m = m/\sigma \qquad \text{for all integers } m \neq 0. \qquad (18.4)$$

There is a zero of $H(\mathbf{u})$ at every Jth sampling point in the u-direction. Numerous books and papers state that Wiener filtering necessarily loses information if $H(\mathbf{u})$ is zero at any sampling point. However, this is not true, at least for the idealized finite deconvolution problem. The restoration can be performed in Ω_b, treating it as periodic, thereby obtaining an estimate $\hat{f}(\mathbf{x})$ of the true image. In general, $\hat{f}(\mathbf{x})$ has value throughout Ω_b, whereas we know that $f(\mathbf{x})$ is confined to Ω_f. The zeros ζ_m defined by (18.4) are the cause of the spreading of $\hat{f}(\mathbf{x})$ outside of Ω_f, because every Jth sample (in the u-direction) of $\hat{F}(\mathbf{u})$ is missing. The erroneous component $\hat{e}(\mathbf{x})$ of $\hat{f}(\mathbf{x})$ is of period σ in the x-direction and is repeated J times within Ω_b. In fact, $\hat{e}(\mathbf{x})$ is the only part of $\hat{f}(\mathbf{x})$ which is non-zero outside Ω_f. This means, in principle, that $\hat{e}(\mathbf{x})$ can be unambiguously determined by inspection of $\hat{f}(\mathbf{x})$, provided σ is less than the sum of the widths, on the left-hand and right-hand sides, of the border (using this term as defined in §15) between Ω_f and Ω_b. So, $f(\mathbf{x})$ is obtained by subtracting $\hat{e}(\mathbf{x})$ from $\hat{f}(\mathbf{x})$ throughout Ω_f. Thus, there is no irretrievable loss of information at the zeros of $H(\mathbf{u})$. In fact, for any kind of linear blur, knowledge of $\hat{f}(\mathbf{x})$ within the border between Ω_f and Ω_b can be sufficient to permit reconstruction of those samples of $F(\mathbf{u})$ which lie at the zeros of $H(\mathbf{u})$.

We conclude this section by briefly comparing *direct matrix methods* with multiplicative deconvolution. What we mean by a direct matrix method is to rewrite (14.9) as a matrix equation and then invoke matrix methods to obtain an estimate of the true image.

One advantage of matrix methods (shared, or course, by the subtractive techniques discussed in §17) is that they can solve practical deconvolution problems involving spatially varying psfs. For a 1-D image consisting of N contiguous pixels, a solution requires the inversion of an $N \times N$ matrix. In the 2-D case, with which we are primarily concerned, the problem is reformulated as a 1-D matrix deconvolution problem. Each $N \times N$ image becomes a vector possessing N^2 components, so that an $N^2 \times N^2$ matrix must be inverted. Even for only a moderate value of N, 256 say, the matrices are very large, so that their inversion tends to be computationally expensive. Such a matrix is often nearly singular and so its conventional inverse is very sensitive to errors in the matrix elements. The singularity difficulty can be ameliorated by invoking singular value decomposition to generate a pseudo-inverse which can be expected to be much more stable. This approach is only practicable for small images, however, so that to be feasible it must be combined with some sort of 'sectioning' technique such as that described in §15.

Once formulated in terms of matrices, a deconvolution problem is solved with the aid of least squares methods and Lagrangian multipliers, thereby minimizing (or maximizing) a chosen image characteristic subject to one or more selected constraints. This approach makes it possible to obtain direct matrix analogues of multiplicative techniques. One can, for example, minimize the mean square restoration error given sufficient information concerning the statistics of the image and the contamination (refer to the related discussion in §16). The entropy as defined by (15.37), or any appropriately defined 'smoothness', of the image can be maximized. Alternatively one can employ the maximum likelihood method in which the most likely restored image is generated consistent with both the given blurred image and whatever information is available concerning the statistics of the original and blurred images. By introducing a constraint on the allowed resultant contamination level, one has a control parameter which can be iteratively adjusted at each restoration step in order to obtain as satisfactory results as possible.

The maximum entropy and maximum likelihood methods have the advantage that the restored image is necessarily positive. It must be emphasized that direct matrix methods of this type are usually iterative, and therefore time-consuming, in practice. If the psf is spatially invariant, any such technique is nevertheless practicable, even for quite large images. This is because inversion of large matrices can be avoided with the aid of the F.F.T. In fact, results generated by direct matrix methods can be comparable to those obtained with the overlapped edge-extension technique (see §15) followed by Wiener filtering (see §16). We favour multiplicative deconvolutional approaches on the whole, however, because of their ease of implementation and their computational simplicity.

19   Consistent deconvolution

One of the points made in §14 is that periodic convolutions suffer from far fewer difficulties relating to the mutual consistency of $a(\mathbf{x})$ and $h(\mathbf{x})$ than do finite convolutions. It is shown here that this implied affliction of finite convolutions can be put to good use, nevertheless, when the preprocessed image is thought to be less contaminated than the available version $\mathfrak{h}(\mathbf{x})$ of the psf. The argument is essentially one-dimensional, so that it has wider application in signal processing than in image processing, although it can be invoked for restoring blurred projections of two-dimensional images (recall from §9 that an individual projection is characterized by a particular value of a parameter – e.g. the angle ϕ – while being a function of a single variable – e.g. the Cartesian coordinate ξ).

The one-dimensional version of (14.9) is

$$a(x) = f(x) \ominus h(x) + c(x), \tag{19.1}$$

and, invoking notation introduced in §§6, 8, and 14, its Fourier transform is

$$A(w) = F(w)H(w) + C(w), \tag{19.2}$$

where the real variable u has been generalized to the complex variable w, as in (14.4). It is to be understood that all the images (i.e. functions of x) appearing in (19.1) are of finite length.

It is convenient to invoke the notation, introduced in §13 and recalled in (14.5), for the sets of (complex) zeros of spectra of images of finite length. Since the available data are represented by $a(x)$ and $\mathfrak{h}(x)$, they can be transformed computationally into the sets of zeros $\mathbf{Z}_a$ and $\mathbf{Z}_b$. Now refer back to the paragraph containing (14.4). Note that, even if a faithful version of $h(x)$ was available, so that $\mathbf{Z}_h$ could be computed, it would not be possible to express $\mathbf{Z}_a$ as simply as $\mathbf{Z}_b$ is in (14.5). The presence of $C(w)$ on RHS (19.2) means that

$$\mathbf{Z}_a \neq \mathbf{Z}_f \cup \mathbf{Z}_h. \tag{19.3}$$

However, $\mathbf{Z}_a$ *can* be expressed as

$$\mathbf{Z}_a = \mathbf{Z}_{\hat{f}} \cup \mathbf{Z}_{\hat{h}} \tag{19.4}$$

where $\hat{f} = \hat{f}(x)$ and $\hat{h} = \hat{h}(x)$ are defined in §14.

Suppose now that the differences between $\mathfrak{h}(x)$ and $h(x)$ are small enough that those members of $\mathbf{Z}_a$ which belong to $\mathbf{Z}_{\hat{h}}$ are sufficiently close to the members of $\mathbf{Z}_h$ that they can be recognized by comparison of the computed sets $\mathbf{Z}_a$ and $\mathbf{Z}_b$. This recognition procedure can be carried out adequately entirely in the computer when $\mathfrak{h}(x)$ is only a moderately distorted version of $h(x)$. When the distortions are appreciably severe (but still small enough for the aforementioned recognition procedure to be successful, in the majority of instances at least), visual recognition is mandatory (the pattern recognition capabilities of the human brain are as yet unrivalled!). The 'guessed' members of $\mathbf{Z}_{\hat{h}}$ are then removed from $\mathbf{Z}_a$, leaving only what are assumed to be the members of $\mathbf{Z}_{\hat{f}}$, from which $\hat{f}(x)$ is reconstructed as indicated in §13. This algorithm is called *consistent deconvolution*. Its advantage is that it can restore an image to the accuracy with which the blurred image is recorded, even when the available estimate of the psf is none too faithful.

Some of the recognitions can be erroneous in practice, of course, implying that some members of the 'guessed' $\mathbf{Z}_{\hat{f}}$ should have been allotted to $\mathbf{Z}_h$. The reconstructed $\hat{f}(x)$ is then not as faithful as it might have been, but its imperfections are usually only objectionable if the number of 'false recognitions' is an appreciable fraction of the number of members of $\mathbf{Z}_b$. This 'uncertainty' over the allotment of zeros to $\mathbf{Z}_{\hat{f}}$ or $\mathbf{Z}_{\hat{h}}$ degrades reconstructions in much the same way as does the presence of the quantity $\Phi = \Phi(\mathbf{u})$ in the Wiener filter (see §16). The basic cause of the degradations is contamination of the data. The more severe the contamination, the larger must $|\Phi|$ be, thereby reducing the resolution of the Wiener filter. Similarly, the number of wrongly recognized zeros increases with the contamination level.

Example III – Wiener filtering

It is our opinion that the multiplicative technique called Wiener filtering (see §16) is, in practice, the most widely applicable deconvolutional method. This example illustrates various practical aspects of the Wiener filter, invoking some of the preprocessing techniques described in §15.

Figure IIIa is a sub-scene of the image shown in Fig. Ia of Example I. We take Fig. IIIa to be the true image (refer to first paragraph of §4) for this example. We now suppose that the sub-scene is photographed with a camera which is out of focus, so that a blurred image (Fig. IIIb) is obtained. The psf with which Fig. IIIa was convolved to produce Fig. IIIb was a disc (refer to the out-of-focus psf shown in Table I.1) of diameter 15 pixels, which corresponds to the image shown in Fig. IIa of Example II. We imagine that, apart from being incompetent as regards lens settings, our photography is of such high quality that the only contamination present in the blurred image shown in Fig. IIIb is quantization noise associated with digitizing to 8-bit accuracy. We show later why starting with such a virtually-uncontaminated ideal blurred image – refer to the definitions (3.4) and (14.1) – does not significantly detract from the realism of our illustrations of Wiener filtering.

Figure IIIb shows the whole of the blurred version of Fig. IIIa. The latter consists of 242×242 pixels, whereas Fig. IIIb consists of 256×256 pixels. Since the diameter of the psf is 15 pixels, the maximum horizontal (or vertical) spreading of each point in Fig. IIIa is by 7 pixels, left and right (or up and down), which explains why Fig. IIIb has 14 more pixels than Fig. IIIa in both the horizontal and vertical directions. Because Fig. IIIb is 'uncontaminated', there is no difficulty in restoring Fig. IIIa virtually perfectly by Wiener filtering. Even more significant than the lack of contamination is the 'completeness' of Fig. IIIb. The kinds of blurred image which figure in practical applications are always recorded in finite frames, so that we can usually expect the information in any 'given' blurred image to be 'cut-off' round the edges. As soon as we try to recover the true image from a truncated blurred image, we tend to run into appreciable difficulties.

We now suppose that, instead of being presented with Fig. IIIb, we are given Fig. IIIc which consists of the central 228×228 pixels of Fig. IIIb. The sharp truncation along the perimeter of Fig. IIIc plays havoc with restoration procedures unless appropriate precautions are taken. In fact, these edge effects tend to outweigh contamination in practice. Consequently, the quality of the following restorations of Fig. IIIc would be little affected if we had added to Fig. IIIc a realistic level of contamination. [It would, of course, have been easy enough to contaminate Fig. IIIc, but we would have then felt constrained to show the effects of varying the contamination level, which would have inordinately lengthened this example. The reader in need of reassurance, that it is indeed justifiable to claim that truncation can be expected to swamp contamination in the real-world of blurred images, can examine specific instances referenced in the introductory comments to this chapter.]

Figures IIId and IIIe show what happens when one applies Wiener filters without taking any of the aforesaid precautions (i.e. without performing any of the preprocessing discussed in §15). The quantity $\Phi(\mathbf{u})$ introduced in (16.5) was assumed constant – i.e. it was taken to be the filter constant Φ introduced after (16.8). Values for Φ of 0.0009 and 0.01 were used when generating Figs. IIId and

IIIe respectively. The 'ripply' artefacts and poor quality of the 'restored' detail are typical for blurred images subjected to Wiener filtering without preprocessing. these two figures emphasize another important practical point, which is that the value chosen for Φ somewhat affects the quality of the restored image.

There is a practical computational point associated with Figs. IIId and IIIe which is worth explaining in a little detail. Since Wiener filtering involves Fourier transformations, it is mandatory for computational economy to invoke the F.F.T. (refer to §12 and Example II). This implies that images to be transformed should consist of $2^n \times 2^n$ pixels, with n being a positive integer. We could have 'packed' Fig. IIIc 'with zeros' – refer to paragraph in §12 preceding that containing (12.14) through (12.17). We preferred, however, to scale Fig. IIIc using bilinear interpolation – refer to §45 and to paragraph containing (47.1) – to produce a 256×256 pixel image. After completing the Wiener filtering, each restored image (i.e. Figs. IIId and IIIe) was rescaled to 228×228 pixels.

Before we can demonstrate the virtues of preprocessing we must consider the three frames Γ, Ω and Ω_b, which are defined in §15 and are depicted in Fig. IIIn. The frame Γ is the recording frame, within which the actually recorded image (see §4) exists. For this example, Γ is the frame, consisting of 228×228 pixels, which encloses the image shown in Fig. IIIc. It is sometimes appropriate to think of an 'original image', of which the true image (i.e. Fig. IIIa, in this example) is only a small part. However, if the actually recorded image represents all the data available to us then the only part of the original image that we can gain any knowledge of is that part which directly affects the actually recorded image. The action of the psf is to spread each point in the original image over an region of the image plane, centred on the said point, of area equal to that of the point spread frame. The latter is the rectangular frame which just encloses all points $\mathbf{x}$ for which $|h(\mathbf{x})|$ is significant – i.e. a 15×15 pixel square for this example. In general, the psf spreads differently to the left (by m_1 pixels, say), to the right (by m_2 pixels), up (by n_1 pixels) and down (by n_2 pixels) – note that the size of the point spread frame is then $(m_1 + m_2 + 1) \times (n_1 + n_2 + 1)$ pixels because zero blurring is characterized by a point spread frame possessing one pixel (and not zero pixels!) – note that $m_1 = m_2 = n_1 = n_2 = 7$ for this example. So, the psf spreads the part of the original image lying within Γ throughout a frame (which we denote by Ω – see Fig. IIIn) whose dimensions are those of Γ plus $(m_1 + m_2)$ horizontally and $(n_1 + n_2)$ vertically – for this example, Ω contains 242×242 pixels. Conversely, the psf spreads into Γ at least part of every point in the original image which lies within Ω. Therefore, the part of the original image existing within Ω is the part which directly affects the actually recorded image. We call this part of the original image the true image (which is represented by Fig. IIIa for this example). This is not yet the whole story, however, because the psf spreads the true image into a larger frame (which we denote by Ω_b – see Fig. IIIn) whose dimensions are those of Ω plus $(m_1 + m_2)$ horizontally and $(n_1 + n_2)$ vertically – for this example, Ω_b contains 256×256 pixels (it is the frame enclosing Fig. IIIb).

It is appropriate to adopt an objective criterion for the 'goodness' of a restoration. We have here chosen what we call the mean absolute error

$$\mathsf{E} = \sum_{l \in \Gamma} |f_l - \hat{f}_l|$$

between the pixels f_l of the true image and the pixels $\hat{f}_l$ of a restored image; the notation $l \in \Gamma$ identifies all the pixels lying within Γ. The quantity E is plotted

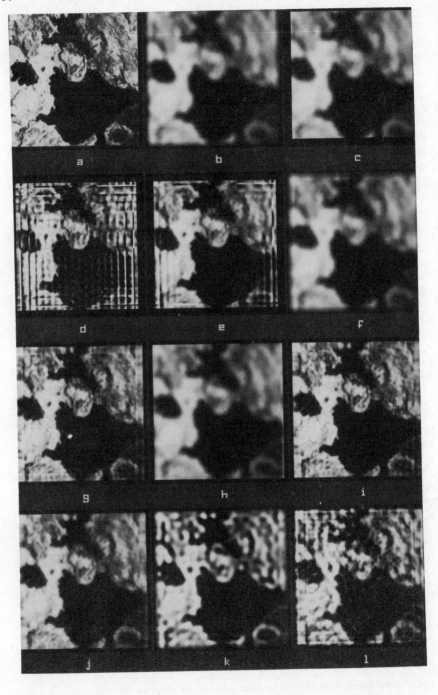

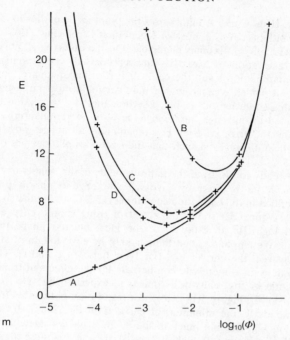

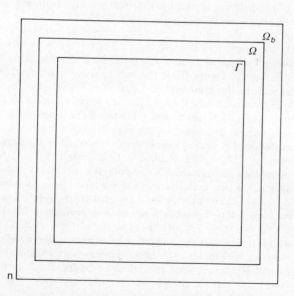

Fig. III. IIIm shows the dependence upon filter constant Φ of mean absolute error E, when attempting to restore the true image from the following blurred images: A, ideal blurred image; B, unpreprocessed truncated blurred image; C, simple-edge-extended blurred image; D, overlapped-edge-extended blurred image. IIIn shows the image frames.

versus Φ in Fig. IIIm. Curve A emphasizes the point made earlier in this example that Fig. IIIa can be recovered more or less perfectly from Fig. IIIb – provided Φ is taken small enough, E becomes negligible. On the other hand, curve B shows that the best restoration of Fig. IIIc (*un*preprocessed) corresponds to a very appreciable contamination level (the equivalent root-mean-square 'noise level' is some 20 per cent, which is much higher than occurs in most important practical applications – this confirms our earlier assertion that the truncation of Fig. IIIc could be expected to outweigh any realistic level of contamination).

To ensure that the error criterion E is meaningful, all images presented in this example have been normalized such that their pixel amplitudes are quantized in the range 0–255.

We are now ready to illustrate the improvements in the quality of the restored image which can be realized by invoking the edge-extension preprocessing technique described in the paragraph containing (15.34), and in the six paragraphs following that. Figure IIIf shows the result of using simple edge extension to generate from Fig. IIIc an estimate of the ideal blurred image (which exists throughout Ω_b). We emphasize that there cannot be any thought of attempting to faithfully reconstruct the part of Fig. IIIb lying between Γ and Ω. The rationale behind edge extension is merely to compensate for the effects of truncation – it is completely idiotic in any real-world context to expect to be able to restore the true image outside Γ. Figure IIIg is the best restoration obtainable from Fig. IIIf (i.e. it corresponds to the minimum of curve C in Fig. IIIm). It is a palpable improvement on Fig. IIIe but much of the detail remains somewhat blurred and there is residual 'rippling', which is caused by the inability of simple edge extension to generate a blurred image that is a close approximation to the convolution of the true image and the psf – this is a manifestation of the inherent inconsistency of finite convolutions (as noted in §14).

Figure IIIh shows the result of using overlapped edge extension to generate from Fig. IIIc an estimate of the overlapped blurred image (which repeats itself throughout image space in contiguous cells, each congruent to Ω – Fig. IIIh occupies one such cell). Figure IIIi is the best restoration obtainable for Fig. IIIh (i.e. it corresponds to the minimum of curve D in Fig. IIIm). The improvement over Fig. IIIg is marked. While some of the fine detail apparent in Fig. IIIa is not clearly restored in Fig. IIIi, much less remains of the 'rippling' which disfigures Figs. IIId and IIIe, and to a lesser extent Fig. IIIg. This is a manifestation of the inherent consistency of periodic convolutions – remarked in the paragraph following that containing (14.8). We emphasize how important it is, from a purely practical point of view, to make one's computations as consistent as possible with the mathematical physics underlying whatever processing is being attempted, provided, of course, that this can be done without having to introduce great complications. Figure IIIj shows the result of Wiener filtering Fig. IIIh when Φ is set to 0.01. The fact that Fig. IIIj is only slightly less faithful than Fig. IIIi confirms our experience that the quality of a restored image is relatively insensitive to the value of Φ, provided the latter is reasonably close to optimum.

It is instructive to demonstrate the effect of basing the Wiener filter on an incorrect psf. Figure IIIk is the restoration obtained from Fig. IIIh when a Gaussian psf (see Table I.1, p. 16) is assumed instead of an out-of-focus psf. The effective diameter of the Gaussian was the same as that of the disc for the out-of-focus psf (while the disc is shown in Fig. IIa, the Gaussian is shown in Fig.

If). Figure IIIl is the restoration obtained when an out-of-focus psf is assumed but its disc is taken to be of diameter 19, rather than 15, pixels. The value of Φ chosen for Figs. IIIk and IIIl was the same as that for Fig. IIIi. Figures IIIk and IIIl are both noticeably inferior to Fig. IIIi, which confirms our experience that estimating the form of the psf correctly is more important than adopting the optimum value for Φ.

IV

PHASE RECOVERY

Perhaps the dominant characteristic of our 'technical age' is the host of measurement apparatus, diagnostic devices, and data transmission and reception systems which permeate most facets of human existence nowadays. Much of the data gathered by such systems can be regarded as possessing intensities and phases. The relative accuracy, with which intensity and phase can be sensed, depends upon the range of frequencies (or equivalently wavelengths) spanned by the emanations carrying the information.

For frequencies of a few MHz or less, phase can often be measured more precisely than intensity. For higher frequencies, however, the accurate determination of phase becomes progressively more difficult. Phase can be successfully measured at several tens of GHz, even if a great deal of complication is involved [Thompson, 1]. When one reaches the optical bands (infra-red and visible light) phase can no longer be directly recorded; it has to be inferred from interference patterns, as in holography [Cathey, B1; Smith, B1] and its adaptations to quantitative sensing of surface shape and displacement [Ennos, 1; Erf, B1]. In the ultraviolet and beyond, phases can only be recovered by performing lengthy digital computations, because it is virtually impossible to observe the kinds of interference from which phase can be deduced with any sort of immediacy.

This chapter describes the underlying theory, and the algorithms proven in computational practice, for recovering the phases of spectra from records of their intensities. We are typically concerned with a spectral intensity that is continuous throughout the (convex) region of Fourier space within which it has significant value (we of course examine the consequences of the important practical consideration that only discrete samples of such a Fourier intensity are often available). Measurement science boasts two very different classes of phase problem. The first relates to periodic images, such as occur in crystallography (especially of the X-ray variety, but also in electron microscopy and neutron diffraction). An enormous effort has been expended on estimating crystallographic phases, and the techniques which have been developed [Beurger, B1; Ramachandran, B1; Sedláček, B1; Woolfson, B1] bear very little on the methods described in this chapter, because crystallographic spectra exist only at discrete points in Fourier space. The second class of phase problem concerns images of finite size, such as are of interest in astronomy, radio-engineering, often in microscopy, and throughout most of image science in fact.

The emphasis in this chapter is on the second class of phase problems. The methods are so recently developed, they have yet to percolate around the engineering science community. The results are likely to be useful in many branches of measurement science, because they may permit clear inferences to be made with far fewer qualifications than would have been thought mandatory only a few years ago. Once it is widely appreciated that the forms of two-dimensional, positive (i.e. real and non-negative) images can be recovered unambiguously from the intensities of their Fourier transforms, a steady stream of important applications can be expected.

A branch of science that is likely to benefit spectacularly from improved methods of phase recovery is optical astronomical interferometry, for which it is improbable that reliable phase data will ever become available. To form useful images then, it will be necessary to develop versatile computational procedures for inferring phases from (sufficiently closely spaced samples of) measured intensities. Furthermore, the new phase recovery techniques can be expected to lead to substantial improvements in the quality of images reconstructed from data gathered using speckle interferometry and its various extensions (see Chapter VI), which presently represent the most successful means of forming high resolution images in optical astronomy.

The Fourier phase problem is introduced in §20, which also deals with other necessary preliminaries [Bates, 7, 10, 11; Fienup, 1].

Fourier holography is interpreted in §21 as a straightforward solution to the phase problem [Bates, 4, 5; Fienup, 4, 6]. Some of the phase recovery techniques developed by electron microscopists [Hawkes, B1: Misell, B1, 1] and X-ray crystallographers [Beurger, B1; Ramachandran, B1], can be usefully interpreted in holographic terms.

The remarkable difference between one-dimensional and multidimensional phase problems – i.e. that solutions to the former are almost never unique whereas solutions to the latter are almost always unique – is established in §22 [Bates, 14; Bruck, 1; Fienup, 2; Hayes, 1; Taylor, 1]. It is appropriate to note here that throughout the past decade there have been hints of 'reduced non-uniqueness' for two-dimensional, as opposed to one-dimensional, phase problems. These are due mainly to the proponents of 'maximum entropy' (see §18) [Ables, 1; Bryan, 1; Gull, 1; Nityananda, 1; van Schooneveld, B1], although an extension of one-dimensional Fourier processing has been relevant [Napier, 1]. Neither of these techniques, nor any others of similar genre, are examined in this chapter because they do not seem to be either as powerful or as widely applicable as the methods that are treated here.

Algorithms that have proved their worth in the recovery of two-dimensional phase distributions are presented in §23. The main theme is Fienup's adaptations of Gerchberg's algorithm [Fienup, 1, 2, 3; Gerchberg, 2]. Unadulterated, this theme can be ineffective, however, especially when played with measured data contaminated with appreciable amounts of noise [Bates, 15]. Appropriately orchestrated themes and variations show promise of successful phase recovery in widening ranges of situations of practical interest [Bates, 13, 15; Fiddy, 1; Fienup, 4, 5].

There are many classes of problem that can either be transformed so as to be cast as pure phase problems or are to be thought of as 'restricted phase problems' in that extra *a priori* information is available. A selection of these are treated in §24 [Bates, 6; Gerchberg, 1; Nawab, 1; Saxton, B1; Schiske, 1]. An important related problem which is only mentioned here is 'signal reconstruction from Fourier amplitudes' [Van Hove, 1], where Fourier amplitude means 'magnitude plus the sign of the phase'. It is not included in §24 because it is similar in many ways to KT-processing, which is one of the techniques examined in §37.

While Example IV relates mainly to §23, it also illustrates aspects of §§21 and 22. The images which are presented all have the character of celestial objects; groups of 'stars' either isolated or superimposed upon 'nebulosities'. Our reason for choosing such images is that the need for phase restoration is very likely to prove most pressing in the near future in optical astronomical contexts. We have designed this example to persuade the reader that phase retrieval is fast becoming a practicable proposition.

20   The Fourier phase problem

In many practical applications, measurements have to be made in Fourier space. So, the true image has to be inferred from its spectrum. Even in those situations where phase$\{F(\mathbf{u})\}$ can be estimated directly from recorded data, it usually transpires that the magnitude of $F(\mathbf{u})$ can be measured more accurately than its phase. There is considerable interest, therefore, in algorithms that attempt to recover as much information as possible about $f(\mathbf{x})$ from $|F(\mathbf{u})|$.

Reference to (6.13) shows that

$$f(\mathbf{x}-\boldsymbol{\xi}) = \int\limits_{-\infty}^{\infty}\!\!\int F(\mathbf{u})\exp(-\mathrm{i}\,2\pi[\mathbf{x}-\boldsymbol{\xi}]\cdot\mathbf{u})\,d\Sigma(\mathbf{u}) = \mathsf{F}\{F(\mathbf{u})\exp(\mathrm{i}\,2\pi\boldsymbol{\xi}\cdot\mathbf{u})\},$$

$$(20.1)$$

where the notation introduced in (6.16) has been invoked. Furthermore,

$$(f(-\mathbf{x}))^* = \left(\int\!\!\int_{-\infty}^{\infty} F(\mathbf{u}) \exp(\mathrm{i}\, 2\pi \mathbf{x} \cdot \mathbf{u})\, d\Sigma(\mathbf{u}) \right)^*$$

$$= \int\!\!\int_{-\infty}^{\infty} F^*(\mathbf{u}) \exp(-\mathrm{i}\, 2\pi \mathbf{x} \cdot \mathbf{u})\, d\Sigma(\mathbf{u}) = \mathsf{F}\{F^*(\mathbf{u})\}. \qquad (20.2)$$

Remember that

$$(f(\mathbf{x}))^* = f^*(\mathbf{x}) = f(\mathbf{x}), \qquad (20.3)$$

because the true image is assumed real (and non-negative) in this book, except where explicitly stated otherwise. Since

$$|F(\mathbf{u})\exp(\mathrm{i}\, 2\pi\boldsymbol{\xi}\cdot\mathbf{u})| = |F(\mathbf{u})| = |F^*(\mathbf{u})|, \qquad (20.4)$$

it is clearly impossible to determine the location of Ω_f (the image frame – see §7), or to be able to distinguish $f(\mathbf{x})$ from its *mirror image* $f(-\mathbf{x})$, when only $|F(\mathbf{u})|$ is given.

Because of the points raised in the previous paragraph, it is convenient to introduce the concept of the *form* of an image. The appearance or shape of something is not changed merely by moving it or by looking at it in a mirror. So it makes sense to say that $f(\mathbf{x}-\boldsymbol{\xi}_1)$ and $f(-\mathbf{x}+\boldsymbol{\xi}_2)$ have the same *image-form* as $f(\mathbf{x})$, where $\boldsymbol{\xi}_1$ and $\boldsymbol{\xi}_2$ are arbitrary constant position vectors.

Only in special cases (e.g. simple holography – see §21) can phase $\{F(\mathbf{u})\}$ be recovered from $|F(\mathbf{u})|$ by analogue devices. It is usually necessary to get the data into a digital computer so that operations of the kind described in §§21 through 24 can be carried out. This means that the data must be sampled, even when an effectively continuous recording medium (such as photographic film or magnetic tape) is used. The spacing of the sample points in Fourier space is a critical consideration because the techniques discussed in this chapter are predicated upon being able to immediately generate the autocorrelation of $f(\mathbf{x})$ from the given $|F(\mathbf{u})|$.

It is assumed throughout this chapter that the autocorrelation of $f(\mathbf{x})$ is of finite size, which effectively accords with most practical applications. The x-extent and the y-extent of $(f(\mathbf{x})\divideontimes f(\mathbf{x}))$ are here denoted by $2L_1$ and $2L_2$ respectively. It is further assumed that $f(\mathbf{x})$ is positive – see (20.3) and refer to the reasoning presented in §§3 and 8.

The spacings of the given samples of $|F(\mathbf{u})|$ are crucial for the Fourier phase problem. We call these samples the *primitive samples* of $|F(\mathbf{u})|$ and we suppose that they lie on a rectangular grid (called the *primitive sampling grid*) in Fourier space. This grid is aligned with the u- and v-axes and its spacings parallel to them are $1/2\mathfrak{L}_1$ and $1/2\mathfrak{L}_2$ respectively.

Except in very special cases, solutions to the phase problem are only possible when $\mathcal{L}_1 \geq L_1$ and $\mathcal{L}_2 \geq L_2$, as is made clear later in this section.

The *Fourier phase problem* is here posed as: given the primitive samples of $|F(\mathbf{u})|$, reconstruct the image-form of $f(\mathbf{x})$.

It is useful to introduce the set of images whose spectra all have the same magnitude $|F(\mathbf{u})|$. A typical member of this set is defined by

$$\mathfrak{f}(\mathbf{x}) = \mathsf{F}\{F(\mathbf{u}) \exp(i\,\Psi(\mathbf{u}))\} = \mathsf{F}\{\mathfrak{F}(\mathbf{u})\}, \qquad (20.5)$$

where $\Psi(\mathbf{u})$, called the *phase function*, is an arbitrary real function of $\mathbf{u}$. Note that

$$|\mathfrak{F}(\mathbf{u})| = |F(\mathbf{u})|. \qquad (20.6)$$

The central question for the phase problem is thus: does there exist more than one image $\mathfrak{f}(\mathbf{x})$ belonging to the set? – remember that images are only considered to be different if they have different image-forms. It has already been noted that phase functions of the type

$$\Psi(\mathbf{u}) = \exp(\pm i\, 2\pi \boldsymbol{\xi} \cdot \mathbf{u}), \qquad (20.7)$$

where $\boldsymbol{\xi}$ is an arbitrary constant position vector, do not change the image-form. So, no generality is lost by centring each image frame $\Omega_{\mathfrak{f}}$ at the origin of image space.

Only positive images $\mathfrak{f}(\mathbf{x})$ are to be considered. So, the extent–constraint theorem (8.15) ensures that the extents of all $\mathfrak{f}(\mathbf{x})$ must necessarily be half of those of $(f(\mathbf{x}) \divideontimes f(\mathbf{x}))$, indicating that

$$\Omega_{\mathfrak{f}} \equiv \Omega_f. \qquad (20.8)$$

It follows that $f(\mathbf{x})$ and $F(\mathbf{u})$ can be characterized as in the first two paragraphs of §10. Furthermore, any of the images $\mathfrak{f}(\mathbf{x})$ can be written as

$$\mathfrak{f}(\mathbf{x}) = (1/L_1 L_2) \sum_{l=-M_1}^{M_1} \sum_{m=-M_2}^{M_2} \mathfrak{F}_{l,m} \exp(-i\,2\pi[lx/L_1 + my/L_2]) \quad (20.9)$$

for $\mathbf{x} \in \Omega_f$, so that its spectrum is

$$\mathfrak{F}(\mathbf{u}) = \sum_{l=-M_1}^{M_1} \sum_{m=-M_2}^{M_2} \mathfrak{F}_{l,m} \operatorname{sinc}(L_1 u - l) \operatorname{sinc}(L_2 v - m), \qquad (20.10)$$

where the $\mathfrak{F}_{-l,-m} = \mathfrak{F}_{l,m}^*$ are the Fourier coefficients of $\mathfrak{f}(\mathbf{x})$.

A further consequence of the positivity of $f(\mathbf{x})$ is that the auto-correlation theorem (7.10) reduces to

$$f(\mathbf{x}) \divideontimes f(\mathbf{x}) = \mathsf{F}\{|F(\mathbf{u})|^2\}. \qquad (20.11)$$

It follows from the sampling theorem (see §10) that the autocorrelation of $f(\mathbf{x})$ can only be immediately computed from the given primitive samples of $|F(\mathbf{u})|$ if the spacings of the primitive sampling grid are sufficiently small, i.e. if

$$1/\mathfrak{L}_1 \leq 1/L_1 \quad \text{and} \quad 1/\mathfrak{L}_2 \leq 1/L_2, \qquad (20.12)$$

in accord with the final sentence of the paragraph which precedes the statement of the phase problem. It is worth noting that we can always tell, in practice, whether $1/\mathfrak{L}_1$ and $1/\mathfrak{L}_2$ are small enough, because $\mathsf{F}\{|F(\mathbf{u})|^2\}$ should be effectively zero outside the rectangular frame defined by $|x| < 2L_1, |y| < 2L_2$. If $\mathsf{F}\{|F(\mathbf{u})|^2\}$ is appreciable at the edge of the frame defined by $|x| < 2\mathfrak{L}_1, |y| < 2\mathfrak{L}_2$ then either $\mathfrak{L}_1$ and/or $\mathfrak{L}_2$ must be too small. This is discussed further in §23.

It is interesting to compare the phase problem, as posed in this section, with the oldest and most famous of phase problems, that which arises in X-ray crystallography. The goal of the crystallographer is to recover the image-form of the unit cell – denoted by $f(\mathbf{x})$ here, for convenience, although real-world crystals are three-dimensional (but it makes no difference to the present discussion) – from the measured magnitudes of the *structure factors* (these are what crystallographers call the primitive samples). However, the X-ray diffraction measurements are made on the whole crystal, represented here by the periodic image $\mathrm{p}(\mathbf{x})$ introduced in the final paragraph of §10 (but with N being necessarily unity), and not merely the unit cell. So, it is $|\mathfrak{P}(\mathbf{u})|$ rather than $|F(\mathbf{u})|$ which is measured; but $|\mathfrak{P}(\mathbf{u})|$ exists effectively only at points spaced by $1/L_1$ and $1/L_2$ in the u and v directions respectively. This, of course, means that

$$\mathfrak{L}_1 = L_1/2 \quad \text{and} \quad \mathfrak{L}_2 = L_2/2, \qquad (20.13)$$

which is not consistent with (20.12). So, the methods described in this chapter cannot be directly applied to crystallographic structure determination. The reason that crystallographers can actually solve their phase problem is that they know their unit cells are made up of particular atoms, whose individual diffraction characteristics are well understood. These conditions are very specialized from the viewpoint of the approach adopted in this chapter, which makes no *a priori* assumptions concerning the configurations of individual parts of the true image.

The remainder of this chapter is concerned with, on the one hand, how to recover the image-forms compatible with the given primitive samples of $|F(\mathbf{u})|$, and on the other, to reveal under what conditions only one such image-form can be expected to exist.

21 Fourier holography

Sufficient two-dimensional phase problems have by now been solved successfully (and not only for ideal, computer-generated data – many laboratory and other experimental measurements have also been processed) that one can feel confident of being able to recover the image-form from only the magnitude of the spectrum in virtually any situation of practical importance. [The measurements must, of course, be accurate enough that the data actually correspond to the image-form which one hopes to reconstruct!] However, the method of solution (see §23) demands considerable computational effort, which it is worth avoiding, if possible. This section is concerned with a class of Fourier phase problems that can be solved (comparatively) straightforwardly.

The essence of *Fourier holography*, as here defined, is that the image contains a known *reference*. The latter is either specified explicitly or is implied by an ancillary condition. The unknown part of the image (here simply called the *unknown*) is denoted by $g(\mathbf{x})$. Since the reference plays a part not unlike that of the psf in other chapters of this book, it is conveniently denoted by $h(\mathbf{x})$. So, the complete true image is given by

$$f(\mathbf{x}) = g(\mathbf{x}) + h(\mathbf{x}). \tag{21.1}$$

It follows from the autocorrelation and correlation theorems (see §7) that the Fourier transform of the given $|F(\mathbf{u})|^2$ is

$$\mathsf{F}\{|F(\mathbf{u})|^2\} = f(\mathbf{x}) \divideontimes f(\mathbf{x}) = g(\mathbf{x}) \divideontimes g(\mathbf{x}) + h(\mathbf{x}) \divideontimes h(\mathbf{x})$$
$$+ g(\mathbf{x}) \divideontimes h(\mathbf{x}) + h(\mathbf{x}) \divideontimes g(\mathbf{x}). \tag{21.2}$$

The Fourier phase problem is simplified significantly if the final two terms on RHS (21.2) can be isolated. This can often be achieved when the reference is *offset*, which implies that non-overlapping frames can enclose the two regions of image space occupied by $g(\mathbf{x})$ and $h(\mathbf{x})$. The x-axis is chosen to lie along the shortest line (of length s, which is called the *separation*) joining the peripheries of these two regions. The *offset condition*

$$s > 0 \tag{21.3}$$

ensures the image frames (see §7) of the reference and the unknown do not intersect:

$$\Omega_g \cap \Omega_h = \varnothing. \tag{21.4}$$

It is convenient to denote by $\mathfrak{w}$, here called the *width*, the larger of the x-extents of Ω_g and Ω_h, i.e. (invoking notation introduced in §7)

$$\mathfrak{w} = \text{larger of } L_1^{(g)} \text{ and } L_1^{(h)}. \tag{21.5}$$

The origin of image space is chosen for convenience half-way along the x-extent of Ω_h.

Before analysing the consequences of (21.3), it is convenient to examine the tighter *offset separation condition*:

$$s > \mathfrak{w}. \tag{21.6}$$

Reference to (21.2) and (21.5) then confirms that $\Omega_{h * g}$ does not intersect the image frames of the other three terms on RHS (21.2). The frames Ω_{g*g} and Ω_{h*h} overlap because they are both centred at the origin of image space, while Ω_{g*h} is separated from them on the other side from Ω_{h*g}. So, $(h(\mathbf{x}) \divideontimes g(\mathbf{x}))$ can be isolated by inspection, and removed from the rest of $(f(\mathbf{x}) \divideontimes f(\mathbf{x}))$ and stored for further processing.

The most straightforward of the species of holography discussed in this section, and probably the one closest to what technically educated people think of as 'true holography', is what is here called *elementary offset holography*. This is characterized by the offset separation condition (21.6) and by the reference being unresolvable by the imaging system which measures $|F(\mathbf{u})|$. On normalizing the integrated brightness (or intensity) of the reference to unity, for convenience, its unresolvableness is formalized by writing

$$h(\mathbf{x}) = \delta(\mathbf{x}). \tag{21.7}$$

The definitions introduced in the final paragraph of §6 then indicate that

$$h(\mathbf{x}) \divideontimes g(\mathbf{x}) = g(\mathbf{x}), \tag{21.8}$$

confirming that the Fourier phase problem has been solved (virtually trivially).

It is worth emphasizing that only the image-form of $g(\mathbf{x})$ can be recovered because no part of phase$\{F(\mathbf{u})\}$ figures amongst the given data. So, it would be just as acceptable to isolate and store $(g(\mathbf{x}) \divideontimes h(\mathbf{x}))$, in which case it would be the mirror image of $g(\mathbf{x})$ that would be reconstructed; but remember that there is no reason to prefer $g(\mathbf{x})$ to $g(-\mathbf{x})$ or vice versa. This consideration applies of course to any holographic solution to the phase problem.

The conditions for what is here called *simple offset holography* are that the form of $h(\mathbf{x})$ is given and that (21.6) holds. Since inspecting (7.5) and (7.6) shows that

$$h(\mathbf{x}) \divideontimes g(\mathbf{x}) = h(-\mathbf{x}) \circledcirc g(\mathbf{x}), \tag{21.9}$$

it follows that $g(\mathbf{x})$ can be recovered from $(h(\mathbf{x}) \divideontimes g(\mathbf{x}))$ with the aid of an appropriate deconvolutional technique chosen from amongst those described in Chapter III. Note that we cannot be certain *a priori* that $(h(\mathbf{x}) \divideontimes g(\mathbf{x}))$ is not, in actual fact, $(g(\mathbf{x}) \divideontimes h(\mathbf{x}))$. So, we must first deconvolve with $h(\mathbf{x})$ and then with $h(-\mathbf{x})$. The more compact of the resulting images corresponds necessarily to the image-form of $g(\mathbf{x})$, because the less compact one is the Fourier transform of either $(G(\mathbf{u}) \exp(-\mathrm{i}\, 2 \,\mathrm{phase}\{H(\mathbf{u})\}))$ or $(G^*(\mathbf{u}) \exp(\mathrm{i}\, 2 \,\mathrm{phase}\{H(\mathbf{u})\}))$. Therefore, the phase problem has again been solved relatively straightforwardly.

Offset holography, as defined here, is characterized by (21.3) and by the form of $h(\mathbf{x})$ being given. The image-form of the unknown can still be reconstructed comparatively straightforwardly, provided $g(\mathbf{x})$ is sufficiently small compared with $h(\mathbf{x})$ that

$$f(\mathbf{x}) \circledast f(\mathbf{x}) - h(\mathbf{x}) \circledast h(\mathbf{x}) \simeq g(\mathbf{x}) \circledast h(\mathbf{x}) + h(\mathbf{x}) \circledast g(\mathbf{x}) \qquad (21.10)$$

is a *useful* first approximation (in the sense that the following iterative procedure converges). It follows from (21.3) that $\Omega_{g * h}$ and $\Omega_{h * g}$ do not intersect so that an estimate of either $(h(\mathbf{x}) \circledast g(\mathbf{x}))$ or $(g(\mathbf{x}) \circledast h(\mathbf{x}))$ – it does not matter which because it is only the image-form that is being sought – can be isolated and stored for further processing. Referring to (21.9) it is again seen that an estimate $\hat{g}(\mathbf{x})$ of $g(\mathbf{x})$ can be reconstructed, by deconvolution, from the stored estimate of, say, $(h(\mathbf{x}) \circledast g(\mathbf{x}))$ – (21.2) then shows that the approximate formula (21.10) can be improved:

$$f(\mathbf{x}) \circledast f(\mathbf{x}) - h(\mathbf{x}) \circledast h(\mathbf{x}) - \hat{g}(\mathbf{x}) \circledast \hat{g}(\mathbf{x}) \simeq g(\mathbf{x}) \circledast h(\mathbf{x}) + h(\mathbf{x}) \circledast g(\mathbf{x}).$$
$$(21.11)$$

A new estimate of $(h(\mathbf{x}) \circledast g(\mathbf{x}))$ is thus obtained, so that a new estimate $\hat{g}(\mathbf{x})$ of the unknown can be generated by deconvolution and substituted back into LHS (21.11), etc., etc. The iterations are continued until the differences between successive versions of $\hat{g}(\mathbf{x})$ are below some 'uncertainty' threshold set by the estimated 'contamination level'. It may be helpful to think of this solution as a simple analogue of the 'heavy atom method' employed by X-ray crystallographers – the reference $h(\mathbf{x})$ corresponds to the 'heavy atom' because it dominates $(f(\mathbf{x}) \circledast f(\mathbf{x}))$ in the sense that $(g(\mathbf{x}) \circledast g(\mathbf{x}))$ can be neglected to a first approximation.

There is an instructive variant of offset holography (here called *modified offset holography*) which is characterized by merely $|H(\mathbf{u})|$, rather than the whole of $H(\mathbf{u})$ or equivalently $h(\mathbf{x})$, being given. Invoking techniques described in §§22 and 23, the image-form of $h(\mathbf{x})$ is reconstructed from $|H(\mathbf{u})|$. Since it is not known whether this image-form corresponds to $h(\mathbf{x})$ or $h(-\mathbf{x})$, the contents of $\Omega_{g * h}$ and $\Omega_{h * g}$ are separately stored and subjected to the abovementioned deconvolutional procedure for reconstructing the image-form of $g(\mathbf{x})$. When the iterations have been completed, that version of $\hat{g}(\mathbf{x})$ which is the more compact (see §7) is taken to be the desired image-form. The image-form can even be recovered in one dimension, by an adaptation of consistent deconvolution (see §§13 and 19 for details of notation). From the stored contents of $\Omega_{h * g}$ and from the given $|H(\mathbf{u})|$ the sets of zeros $\mathbf{Z}_{h * g}$ and $\mathbf{Z}_{h * h}$ are computed. Since

$$\mathbf{Z}_{h * g} = \mathbf{Z}_h^* \cup \mathbf{Z}_g, \qquad (21.12)$$

and

$$\mathbf{Z}_{h * h} = \mathbf{Z}_h^* \cup \mathbf{Z}_h, \qquad (21.13)$$

those members of $\mathbf{Z}_{h * g}$ that correspond to the members of $\mathbf{Z}_h^*$ can be recognized by inspection, so that the members of $\mathbf{Z}_g$ can be isolated and

$g(\mathbf{x})$ can be reconstructed. This procedure can be extended to two dimensions, although it may seem hardly worthwhile since solutions to two-dimensional phase problems (see §23) can now be obtained so readily. One begins by computing the projections (see §9) of the contents of Ω_{h*g}. On denoting the projection at angle ϕ by $pc(\xi; \phi)$ then, for any ϕ, the subscript g is replaced in (21.12) by pc. It is convenient to replace the subscript h appearing in both (21.12) and (21.13) by $h\phi$, where $\mathbf{Z}_{h\phi}$ is the set of zeros which is computed from the given $|H(\mathbf{u})|$ for values of $\mathbf{u}$ corresponding to the straight line passing through the origin of Fourier space at an angle ϕ to the u-axis. This permits one to reconstruct the set $\mathbf{Z}_{pg}$ of zeros corresponding to the projection at angle ϕ of (the image-form of) $g(\mathbf{x})$. Then the projections $pg(\xi; \phi)$ are recovered for as many values of ϕ as are deemed necessary, thereby allowing the reconstruction of the image-form of $g(\mathbf{x})$ by conventional means (see §§25 and 33).

When Ω_g and the image frame(s) of any reference(s) overlap, so that (21.3) no longer holds, it is still possible to solve the phase problem comparatively straightforwardly, provided *triple-reference holography* (as it is here called) can be invoked. Three separate references – i.e. $h_l(\mathbf{x})$ for $l = 1$, 2, and 3 – are needed and their forms must be given. Besides the three $h_l(\mathbf{x})$, the three $|F_l(\mathbf{u})|$ are also required where each $F_l(\mathbf{u})$ is the Fourier transform of

$$f_l(\mathbf{x}) = g(\mathbf{x}) + h_l(\mathbf{x}). \qquad (21.14)$$

It follows that F, f, and h in (21.2) can be replaced by F_l, f_l, and h_l. On defining

$$q_{mn}(\mathbf{x}) = f_m(\mathbf{x}) * f_m(\mathbf{x}) - h_m(\mathbf{x}) * h_m(\mathbf{x}) - f_n(\mathbf{x}) * f_n(\mathbf{x}) + h_n(\mathbf{x}) * h_n(\mathbf{x}),$$
$$(21.15)$$

with m and n, both of which are different, being either 1, 2, or 3. However, in order to be able to recover $g(\mathbf{x})$ only two different pairs of values of m and n need be invoked. Remembering that $g(\mathbf{x})$ is real, it is seen from (8.2) and the correlation theorem (7.8) that

$$\mathfrak{Q}_{mn}(\mathbf{u}) = \mathsf{F}\{q_{mn}(\mathbf{x})\} = G^*(\mathbf{u})H_{mn}(\mathbf{u}) + G(\mathbf{u})H_{mn}^*(\mathbf{u}), \qquad (21.16)$$

where

$$H_{mn}(\mathbf{u}) = \mathsf{F}\{h_m(\mathbf{x}) - h_n(\mathbf{x})\}. \qquad (21.17)$$

For each point $\mathbf{u}$ in Fourier space it is convenient to define

$$\mathfrak{Q}_{mn}(\mathbf{u}) = \mathfrak{X}_{mn} + i\,\mathfrak{Y}_{mn}, \qquad G(\mathbf{u}) = \mathfrak{A} + i\,\mathfrak{B}, \qquad H_{mn}(\mathbf{u}) = \mathfrak{x}_{mn} + i\,\mathfrak{y}_{mn},$$
$$(21.18)$$

where $\mathfrak{X}_{mn}$, $\mathfrak{Y}_{mn}$, $\mathfrak{A}$, $\mathfrak{B}$, $\mathfrak{x}_{mn}$, and $\mathfrak{y}_{mn}$ are all real. Note that the $\mathfrak{X}_{mn}$, $\mathfrak{Y}_{mn}$, $\mathfrak{x}_{mn}$, and $\mathfrak{y}_{mn}$ can all be computed directly from the given data. Substituting (21.18) into (21.16), and remembering that $|F_l(\mathbf{u})|$ and $h_l(\mathbf{x})$ are given for three values of l, shows that there are only two unknowns (i.e. $\mathfrak{A}$ and $\mathfrak{B}$) in the two linear algebraic equations

$$\mathfrak{X}_{lm}/2 = \mathfrak{A}x_{lm} + \mathfrak{B}\mathfrak{y}_{lm} \quad \text{and} \quad \mathfrak{X}_{mn}/2 = \mathfrak{A}x_{mn} + \mathfrak{B}\mathfrak{y}_{mn}. \qquad (21.19)$$

It follows that the phase problem is solved because $\mathfrak{A}$ and $\mathfrak{B}$ are recovered for all $\mathbf{u}$, thereby permitting the image-form of $g(\mathbf{x})$ to be reconstructed by Fourier transformation. This solution is a simple analogue of the 'isomorphous replacement' technique employed by crystallographers.

Triple-autocorrelation holography, which is the last of the 'holographic' techniques discussed in this section, applies when the true image consists of isolated, unresolvable 'points', i.e.

$$f(\mathbf{x}) = \sum_{j=1}^{J} b_j \,\delta(\mathbf{x} - \mathbf{x}_j), \qquad (21.20)$$

where the b_j are positive constants and the $\mathbf{x}_j$ are constant position vectors. Instead of an explicit reference, one specifies that all $|\mathbf{x}_{mn}| = |\mathbf{x}_{nm}|$, where

$$\mathbf{x}_{mn} = \mathbf{x}_m - \mathbf{x}_n, \qquad (21.21)$$

are different for different pairs of the integers m and n, each running from 1 to J. It is convenient to introduce two different pairs, μ, ν and μ', ν', of fixed integers (each chosen from the set $\{1, 2, \ldots, J\}$ arbitrarily except that $\mu \neq \nu \neq \nu'$ and $\nu' \neq \mu' \neq \mu$) and also the shorthand notation

$$A(\mathbf{x}) = f(\mathbf{x}) \ast f(\mathbf{x}) \qquad (21.22)$$

for the autocorrelation (which, as for all methods discussed in this chapter, is obtained from the given $|F(\mathbf{u})|^2$ by Fourier transformation). Inspection of (21.20) confirms that the triple product

$$AAA(\mathbf{x}) = A(\mathbf{x})A(\mathbf{x} - \mathbf{x}_{\mu\nu})A(\mathbf{x} - \mathbf{x}_{\mu'\nu'}) \qquad (21.23)$$

is zero except at a set of 'points' having the same juxtapositions and relative brightnesses as either $f(-\mathbf{x})$ or $f(\mathbf{x})$, as defined by (21.20). So, the phase problem is solved straightforwardly; but it must be emphasized that (21.23) only holds in general if all of the $\mathbf{x}_{mn}$ differ by more than the diffraction limit of the imaging system. This solution is analogous to the 'product' and 'superposition' techniques used in X-ray crystallography. If the reader finds this argument hard to follow analytically, it will become immediately clear if the procedure is checked by example with the aid of three identical versions of $A(\mathbf{x})$ drawn on transparent paper (so that they can be slid over each other with all three being simultaneously visible).

22 Uniqueness questions

It is understood throughout this section that the spacings of the primitive samples of $|F(\mathbf{u})|$ satisfy (20.12), so that $(f(\mathbf{x}) \ast f(\mathbf{x}))$ can be reconstructed and the amplitude of $|F(\mathbf{u})|$ itself can be computed for any $\mathbf{u}$. This is, of course, equivalent to stating that the autocorrelation is of finite size. The extent–constraint theorem (8.15) then ensures that $f(\mathbf{x})$, being positive, is itself of finite size.

It is instructive to start with the one-dimensional case. Suppose that $f(x)$ is of finite length and of extent L (see §7). Since the primitive samples of $|F(u)|$ are spaced close enough to permit $(f(x) \ast f(x))$ to be reconstructed, the set $\mathbf{Z}_{f \ast f}$ of zeros can be computed (see §13). It is convenient to partition this set in the same way that $\mathbf{Z}_g$ is partitioned in (13.12), i.e.

$$\mathbf{Z}_{f \ast f} = \mathbf{Z}^r_{f \ast f} \cup \mathbf{Z}^c_{f \ast f}. \tag{22.1}$$

It follows from the final paragraph of §13 – refer to (13.24) in particular – that $\mathbf{Z}^c_{f \ast f}$ must have an even number, here denoted by $2M$, of members which are grouped in complex conjugate pairs, i.e. if

$$w_m = \rho_m \exp(\mathrm{i}\, \phi_m), \tag{22.2}$$

with ϕ_m real and ρ_m positive, is found to belong to $\mathbf{Z}^c_{f \ast f}$ then so must w^*_m;

$$\mathbf{Z}^c_{f \ast f} = \{w_m, w^*_m;\, m = 1, 2, \ldots, M\}. \tag{22.3}$$

Because

$$\mathbf{Z}^c_{f \ast f} = (\mathbf{Z}^c_f)^* \cup \mathbf{Z}^c_f \tag{22.4}$$

it is seen that $\mathbf{Z}^c_f$ has M members. However, for each m, there is no way of ascertaining whether w_m or w^*_m belongs to $\mathbf{Z}^c_f$, i.e.

$$\mathbf{Z}^c_f = \{\rho_m \exp(\tau_m \mathrm{i}\, \phi_m);\, m = 1, 2, \ldots, M\}, \tag{22.5}$$

where each τ_m can be either $+1$ or -1. Note that there are 2^M distinct sets $\{\tau_m;\, m = 1, 2, \ldots, M\}$. The reasoning leading up to (13.20) confirms that each of these sets, when taken in conjunction with (22.5), corresponds to an image of length L. If $\tilde{\tau}_m$ indicates a particular choice of either $+1$ or -1, for each m, then the sets $\{\tilde{\tau}_m;\, m = 1, 2, \ldots, M\}$ and $\{-\tilde{\tau}_m;\, m = 1, 2, \ldots, M\}$ correspond to images having the same image-form because each of the images is the other's mirror image, as can be seen for instance from (13.7) when $G(w)$ is replaced by $G^*(w^*)$. So, corresponding to the set $\mathbf{Z}^c_{f \ast f}$ defined by (22.3) there are in general 2^{M-1} different image-forms, provided none of the ϕ_m are either 0 or π. While all these image-forms must be real, not all of them need be positive. However, it can be expected in general that more than one of them will be positive.

Consequently, the one-dimensional Fourier phase problem does *not* in general possess a unique solution.

On replacing the $G(w)$ of §13 by $F(w)$, the polynomial $P(w, M)$ introduced in (13.11) can be regarded as the seat of the ambiguity between the several image-forms compatible with a given $|F(u)|$. The *fundamental theorem of algebra* ensures that $P(u, M)$ can be decomposed into M factors of the type $((u - w_m)(u + w^*_m))$. The magnitude of each such factor, considered as a function of u, is unchanged when w_m is replaced by w_m^*. This aspect of the phase problem is completely altered in the two-dimensional case, even though one can express a two-dimensional spectrum – i.e. $F(w', w^`)$ where $w'$ and $w^`$ are two different complex variables – similarly to the scheme set out in (13.9)–(13.11), but with necessary changes such as replacing $\sin(\pi Lw)$ by $(\sin(\pi L_1w') \sin(\pi L_2w))$, etc. Furthermore, $P(w, M)$ must be replaced by $P(w', w', M)$. Now, there is *no* fundamental theorem of algebra in two, or more, dimensions. In fact, it is impossible to factorize almost all multidimensional polynomials. This suggests that solutions to multidimensional Fourier phase problems are unique except in cases so special that they have no practical significance. This is a heartening conclusion, comforting to those concerned with developing iterative phase restoration schemes of the kind described in §23, but it gives no inkling as to the connection between phase$\{F(u)\}$ and $|F(u)|$. The required insight can, however, be acquired through the line of reasoning presented below.

Before examining the two-dimensional phase problem, as posed in §20, it is convenient to look into the *ancillary phase problem* which is here posed as: given samples of $|\mathfrak{E}(\mathbf{u})|$ at the points $(l/2NL_1, m/2NL_2)$ in Fourier space, where N is an arbitrary positive integer and l and m are integers running (in principle) from $-\infty$ to ∞, reconstruct the image-form of $e(\mathbf{x})$; on the understanding that the sample points are closely enough spaced that $|\mathfrak{E}(\mathbf{u})|$ could be accurately reconstructed from them for any $\mathbf{u}$, and that $e(\mathbf{x})$ is defined by

$$e(\mathbf{x}) = \mathsf{F}\{\mathfrak{E}(\mathbf{u})\} = p(\mathbf{x})w(x/NL_1)w(y/NL_2), \qquad (22.6)$$

where $p(\mathbf{x})$ is introduced in (10.7) and the *window function* $w(x)$ is defined by

$$\begin{aligned} w(x) &= [1 + \cos(\pi x)]/2 &&\text{for } |x| < 1, \\ &= 0 &&\text{for } |x| > 1. \end{aligned} \qquad (22.7)$$

Reference to the final paragraph of §10 then indicates that

$$e(\mathbf{x}) = 0 \qquad \text{for } \mathbf{x} \in \Omega_{p,l,m} \text{ with } |l| > 1 \text{ and/or } |m| > 1. \qquad (22.8)$$

Furthermore, $e(\mathbf{x})$ is seen to have value in only one-half of each of the four frames $\Omega_{p,\pm 1,0}$ and $\Omega_{p,0,\pm 1}$ and only one-quarter in each of the four frames $\Omega_{p,\pm 1,\pm 1}$. Also note that

$$w(0) = 1 \quad \text{and} \quad dw(0)/dx = w(\pm 1) = dw(\pm 1)/dx = 0. \qquad (22.9)$$

Since $p(\mathbf{x})$ is defined, by (10.7), to be zero throughout each $\bar{\Omega}_{f,l,m}$, it follows that

$$\begin{aligned} \lim_{N \to \infty} e(\mathbf{x}) &= f(\mathbf{x}) &&\text{for } \mathbf{x} \in \Omega_{p,0,0} \\ &= 0 &&\text{for } \mathbf{x} \in \Omega_{p,l,m} \text{ with } l \neq 0 \text{ and/or } m \neq 0, \quad (22.10) \end{aligned}$$

implying that $e(\mathbf{x}) = 0$ outside $\Omega_{p,0,0}$, in the limit as $N \to \infty$, so that

$$\underset{N \to \infty}{\mathrm{Lim}}\, e(\mathbf{x}) = f(\mathbf{x}). \tag{22.11}$$

On defining

$$s(u, L) = (1/NL)\mathbf{F}_{(1u)}\{w_{(x)}(x/NL)\}, \tag{22.12}$$

it is seen from (22.7), with the aid of (6.3) and (6.4) that

$$s(u, L) = \mathrm{sinc}(2NLu) + [\mathrm{sinc}(2NLu + 1) + \mathrm{sinc}(2NLu - 1)]/2 \tag{22.13}$$

The notation

$$\alpha_1 = 1/NL_1 \quad \text{and} \quad \alpha_2 = 1/NL_2 \tag{22.14}$$

is introduced to simplify subsequent expressions. It follows from (22.6), (22.12) through (22.14), the convolution theorem (7.3) and from (10.8) through (10.10) that

$$\begin{aligned}
\mathfrak{E}(\mathbf{u}) &= \sum_{l=-M_1'}^{M_1'} \sum_{m=-M_2'}^{M_2'} \mathfrak{P}_{l,m}(\delta(u - l\alpha_1)\,\delta(v - m\alpha_2))\, \Theta\,(s(u, L_1)s(v, L_2)) \\
&= \sum_{l=-M_1'}^{M_1'} \sum_{m=-M_2'}^{M_2'} \mathfrak{P}_{l,m}s(u - l\alpha_1, L_1)s(v - m\alpha_2, L_2),
\end{aligned} \tag{22.15}$$

because of the simple form to which any one-dimensional convolution involving a delta function reduces:

$$g(u) \Theta \delta(u - \alpha) = \int_{-\infty}^{\infty} g(\beta)\,\delta(u - \alpha - \beta)\,d\beta = g(u - \alpha). \tag{22.16}$$

It is seen from (22.13) through (22.15), with the aid of (6.4), that

$$\mathfrak{E}((2l + 1)\alpha_1/2, m\alpha_2) = (\mathfrak{P}_{l,m} + \mathfrak{P}_{l+1,m})/2, \tag{22.17}$$

$$\mathfrak{E}(l\alpha_1, (2m + 1)\alpha_2/2) = (\mathfrak{P}_{l,m} + \mathfrak{P}_{l,m+1})/2, \tag{22.18}$$

and

$$\mathfrak{E}(l\alpha_1, m\alpha_2) = \mathfrak{P}_{l,m} = a_{l,m}\exp(i\,\theta_{l,m}), \tag{22.19}$$

where the positive constants $a_{l,m}$ and the real constants $\theta_{l,m}$, here called the *actual sample magnitudes* and the *actual sample phases*, are introduced for later convenience.

Returning to the ancillary phase problem, as posed above, it is seen from (22.15) that the magnitudes of LHS (22.17) through (22.19) correspond to the given data. It is convenient to divide these samples into three sets whose members are called the *actual samples* $a_{l,m}$, the *in-between along-rows samples* $b_{l,m}^r$ and the *in-between along-columns samples* $b_{l,m}^c$, where

$$b_{l,m}^r = |\mathfrak{E}((2l + 1)\alpha_1/2, m\alpha_2)| \quad \text{and} \quad b_{l,m}^c = |\mathfrak{E}(l\alpha_1, (2m + 1)\alpha_2/2)|. \tag{22.20}$$

Note that the $a_{l,m}$ are defined by (22.19). The actual samples are so-called because they would be the only ones needed to reconstruct $\mathfrak{E}(\mathbf{u})$ for any $\mathbf{u}$, if the actual sample phases were available. Each in-between sample lies midway between two actual samples, some along rows and some along columns (hence the respective superscripts r and c) of the *actual sampling grid*, which is the array of *actual sample points* $(l\alpha_1, m\alpha_2)$.

Inspection of (22.17) through (22.20) reveals the existence of the simple formulas

$$\cos(\theta_{l+1,m} - \theta_{l,m}) = (4(b^r_{l,m})^2 - (a_{l,m})^2 - (a_{l+1,m})^2)/2a_{l,m}a_{l+1,m}, \quad (22.21)$$

and

$$\cos(\theta_{l,m+1} - \theta_{l,m}) = (4(b^c_{l,m})^2 - (a_{l,m})^2 - (a_{l,m+1})^2)/2a_{l,m}a_{l,m+1}. \quad (22.22)$$

Now, $e(\mathbf{x})$ is positive, as follows from (10.7) and the positivity of both $f(\mathbf{x})$ and $\mathfrak{w}(\mathbf{x})$ – refer to (22.7). The definition (6.12) of the Fourier transform then confirms that $F(0, 0)$ is positive, implying that

$$\theta_{0,0} = 0. \quad (22.23)$$

Recall that there is a twofold ambiguity associated with the argument of any trigonometric quantity – e.g. if the value of $\cos(\beta)$ is given, the value of β itself is uncertain unless other information is available, because both $+\beta$ and $-\beta$ satisfy what is given. Substituting (22.23) into (22.21) therefore gives

$$\theta_{1,0} = \pm\Gamma_{1,0}, \quad (22.24)$$

where $\Gamma_{1,0}$ is seen to be the arc-cosine of a real number which can itself be immediately calculated from data given for the ancillary phase problem (it is worth emphasizing here that all the $\Gamma_{l,m}$ and $\Gamma_{l,m,n}$ introduced below can be regarded as given data). Since only the image-form of $e(\mathbf{x})$ is required, it is acceptable to choose the, say, positive value of $\Gamma_{1,0}$, because this is equivalent to choosing either $e(\mathbf{x})$ or $e(-\mathbf{x})$ – it does not matter that there is no way of deciding which is which because they both have the same image-form (see §20). It is then assumed that

$$\theta_{1,0} = \Gamma_{1,0}, \quad \text{with } \Gamma_{1,0} \text{ defined to be positive.} \quad (22.25)$$

It is now appropriate to introduce a shorthand notation for sample points – i.e. (l, m) identifies the actual sample point $(l\alpha_1, m\alpha_2)$, while $(l+\frac{1}{2}, m)$ and $(l, m+\frac{1}{2})$ identify, respectively, the in-between sample points: $((2l+1)\alpha_1/2, m\alpha_2)$ along rows, and $(l\alpha_1, (2m+1)\alpha_2/2)$ along columns. Consider what is here called a *Fourier space unit cell* consisting of the sample points $(0, 0)$, $(\frac{1}{2}, 0)$, $(1, 0)$, $(1, \frac{1}{2})$, $(1, 1)$, $(\frac{1}{2}, 1)$, $(0, 1)$, and $(0, \frac{1}{2})$. The formula (22.24) expresses the connection between (the given samples at) $(0, 0)$, $(\frac{1}{2}, 0)$, and $(1, 0)$. The three further connections – i.e. between

$(1, 0)$, $(1, \frac{1}{2})$, and $(1, 1)$, between $(1, 1)$, $(1, \frac{1}{2})$, and $(0, 1)$ and between $(0, 1)$, $(0, \frac{1}{2})$, and $(0, 0)$ – are seen from (22.21) and (22.22) to be

$$\theta_{1,1} - \theta_{1,0} = \pm\Gamma_{1,1,0}, \qquad \theta_{1,1} - \theta_{0,1} = \pm\Gamma_{0,1,1} \quad \text{and} \quad \theta_{0,1} = \pm\Gamma_{0,1},$$

$$(22.26)$$

where the three RHSs equal the corresponding RHSs of (22.21) and (22.22) which as noted above consist entirely of given data. Invoking (22.25) and eliminating $\theta_{0,1}$ from (22.26) gives

$$\theta_{1,1} = \Gamma_{1,0} \pm \Gamma_{1,1,0}, \qquad\qquad (22.27)$$

and

$$\theta_{1,1} = \text{either } (\Gamma_{0,1} \pm \Gamma_{0,1,1}) \text{ or } (-\Gamma_{0,1} \pm \Gamma_{0,1,1}). \qquad (22.28)$$

Provided

$$\Gamma_{0,1} \neq \Gamma_{1,0} \quad \text{and} \quad \Gamma_{0,1} \neq \Gamma_{0,1,1}, \qquad\qquad (22.29)$$

there can only be one 'match' between the quantities on the RHSs of (22.27) and (22.28) – e.g. if $\theta_{1,1} = (\Gamma_{1,0} - \Gamma_{1,1,0}) = (-\Gamma_{0,1} + \Gamma_{0,1,1})$ then $(\Gamma_{1,0} - \Gamma_{1,1,0})$ cannot equal any of the other three quantities on RHS (22.28). By drawing a phasor diagram representation of (22.26), the reader will find this argument transparent.

Exact equalities never occur in the real-world, where all measurements are inescapably contaminated. Consequently, neither of the equalities postulated in (22.28) can hold precisely in practice. However, one of the two 'equalities' is always more nearly satisfied than the other. Consequently, while $\theta_{1,1}$ can never be found 'exactly', it can always be estimated unambiguously.

Another practical point worth emphasizing is that contaminated data can cause the magnitudes of RHS (22.21) and/or RHS (22.22) to exceed unity on occasion. We are not then forced to introduce complex phase angles; we merely take $(\theta_{l+1,m} - \theta_{l,m})$ and/or $(\theta_{l,m+1} - \theta_{l,m})$ to be 0 or π according as to whether the respective RHSs are greater than 1 or less than -1.

It follows from the previous three paragraphs that one can expect to find ('estimate' would, of course, be a more precise term) $\theta_{1,1}$ uniquely. Substituting from (22.28) into (22.26) gives

$$\theta_{0,1} = \theta_{1,1} - \pm\Gamma_{0,1,1} \quad \text{and} \quad \theta_{0,1} = \pm\Gamma_{0,1}. \qquad (22.30)$$

There can again be effectively only one 'match', so that $\theta_{0,1}$ is found uniquely.

What the above reasoning has shown is that, given the actual sample phases at two adjacent corners of a unit cell, the actual sample phases at the other two corners can be determined straightforwardly. Now consider the unit cell consisting of $(0, 1)$, $(\frac{1}{2}, 1)$, $(1, 1)$, $(1, \frac{3}{2})$, $(1, 2)$, $(\frac{1}{2}, 2)$, $(0, 2)$,

$(0, \frac{3}{2})$. The phases are known at the corners $(0, 1)$ and $(1, 1)$, so that the phases at the corners $(1, 2)$ and $(0, 2)$ can be found by repeating the previous reasoning. Consequently, all actual sample phases can be evaluated recursively in pairs. So, the ancillary Fourier phase problem is solved uniquely.

In principle, there is no limit to how small α_1 and α_2 can be, or equivalently how large N can be. In the limit as $N \to \infty$, it is seen from (22.11) that $e(\mathbf{x})$ reduces to $f(\mathbf{x})$, so that $\mathfrak{E}(\mathbf{u})$ reduces to $F(\mathbf{u})$. It follows therefore that, as N increases, the given data for the ancillary phase problem tend towards the given data for the Fourier phase problem. The phases of the actual samples of $F(\mathbf{u})$ can thus be obtained by the recursive procedure described above.

There is thus only one image-form compatible with a given $|F(\mathbf{u})|$, at least if the latter has been measured. The inevitable contamination must, of course, distort the reconstructed image-form from what it would ideally be in the absence of any corruption, but the reconstruction can be expected to be unique.

There are two further important points worth making about the above uniqueness argument. First, it breaks down if the image can be factored into a function of x multiplied by a function of y, because $|F(\mathbf{u})|$ then separates into the product of two one-dimensional spectral intensities, e.g.

$$|F(\mathbf{u})| = |F_1(u)| \, |F_2(v)|. \tag{22.31}$$

Perfect factorizations of this kind never occur in the real-world and are therefore only of interest in ideal situations postulated purely for theoretical convenience. The second point is that the argument can be immediately extended to any number of dimensions, so that solutions to K-dimensional Fourier phase problems can *almost always* be expected to be unique, provided $K > 1$.

Since it is not feasible to take N arbitrarily large in computational practice, the simple recursive algorithm introduced in this section cannot by itself be expected to recover real-world image-forms all that faithfully. It is explained in §23 that this algorithm can, however, play a central part in a widely applicable composite phase restoration scheme.

23   Iterative phase restoration

A powerful conclusion to be drawn from §22 is that, if a viable version of phase$\{F(\mathbf{u})\}$ can be constructed, it must correspond to the correct image-form for $f(\mathbf{x})$. This section describes what may at present be the most effective approach to estimating phase$\{F(\mathbf{u})\}$. As with many of the more interesting applications of mathematics to physical science, there is no known way of rigorously 'proving' that this approach is 'correct'. 'Science' is based on experiment, however, and cannot be successfully practised from the depths of an armchair with one's eyes permanently closed. Experiments can be carried out in the computational laboratory just as they can in the physical laboratory. If it is 'discovered' that a computational procedure performs adequately under particular conditions, and continues to do so as the conditions are widened, one should assume it to be generally valid until contrary evidence appears. While theoretical justification should be sought, there is no need to feel upset if one cannot find it.

The first step in generating a 'phase' to go with the given primitive samples of $|F(\mathbf{u})|$ is to estimate the extents of Ω_f. Recall from §12 that the Fourier transform of a quantity that is only given at regularly spaced sample points is necessarily periodic. With the aid of the autocorrelation theorem (7.10), it is then seen that

$$\mathsf{F}\{(primitive\ samples\ of\ |F(\mathbf{u})|)^2\} = \mathfrak{A}(\mathbf{x}) \simeq f(\mathbf{x}) \circledast f(\mathbf{x})$$

$$\text{for } \mathbf{x} \in \Omega_{\mathfrak{A},0,0}, \quad (23.1)$$

where $\Omega_{\mathfrak{A},l,m}$ is the l, mth frame in image space within which the estimate of $(f(\mathbf{x}) \circledast f(\mathbf{x}))$ is repeated (this notation is consistent with that introduced in the final paragraph of §10). A threshold ε is chosen (on the basis of the estimated contamination level of the data) such that $\mathfrak{A}(\mathbf{x})$ can be considered negligible wherever

$$\mathfrak{A}(\mathbf{x}) < \varepsilon. \quad (23.2)$$

Even though $\mathfrak{A}(\mathbf{x})$ is computed directly from the primitive samples of $|F(\mathbf{u})|$, its form reveals whether the primitive samples do in fact satisfy (20.12). The point is that $\mathfrak{A}(\mathbf{x})$ should satisfy (23.2) all along the perimeter of $\Omega_{\mathfrak{A},0,0}$. One hopes, of course, that $\mathfrak{L}_1$ and $\mathfrak{L}_2$ are somewhat larger than L_1 and L_2, respectively, so that (23.2) can be seen to be satisfied everywhere within $\Omega_{\mathfrak{A},0,0}$ outside a rectangular frame which is noticeably smaller than $\Omega_{\mathfrak{A},0,0}$. The smaller frame is the best available estimate of $\Omega_{f \circledast f}$. A preprocessed autocorrelation $\hat{\mathfrak{A}}(\mathbf{x})$ is defined as the positive part of $\mathfrak{A}(\mathbf{x})$ existing within $\Omega_{f \circledast f}$ (it makes sense to remove any negative parts because they can only be due to some sort of contamination, since $f(\mathbf{x})$ is positive by definition).

The extents of Ω_f are necessarily half of those of $\Omega_{f \circledast f}$, as follows from the extent–constraint theorem (8.15). The values of L_1 and L_2 can thus be deduced directly from the primitive samples of $|F(\mathbf{u})|$.

It is worth mentioning what can be done when $\mathfrak{A}(\mathbf{x}) > \varepsilon$ on parts of the perimeter of $\Omega_{\mathfrak{A},0,0}$, implying that (20.12) does not hold. Further preprocessing of $\mathfrak{A}(\mathbf{x})$ is then needed. It is multiplied by a window (see §15) which is unity throughout most of the interior of $\Omega_{\mathfrak{A},0,0}$ but falls to zero with zero slope on the perimeter. The idea is to make the contents of $\Omega_{\mathfrak{A},0,0}$ look as much like an autocorrelation as possible, while destroying as little as possible of the information content of $\mathfrak{A}(\mathbf{x})$. Consequently, one proceeds as described in §15. The positive part of this windowed version of $\mathfrak{A}(\mathbf{x})$ becomes $\hat{\mathfrak{A}}(\mathbf{x})$, so that Ω_{f*f} and $\Omega_{\mathfrak{A},0,0}$ must be assumed equal. While $\hat{\mathfrak{A}}(\mathbf{x})$ is likely to be significantly different from $(f(\mathbf{x})\circledast f(\mathbf{x}))$, it is probably the best available estimate of the autocorrelation of the true image.

A *preprocessed intensity spectrum* $\mathfrak{I}(\mathbf{u})$ is now defined by

$$\mathfrak{I}(\mathbf{u}) = \mathsf{F}\{\hat{\mathfrak{A}}(\mathbf{u})\}. \tag{23.3}$$

Further appropriate definitions are

$$(\hat{a}_{\lambda,\mu})^2 = \mathfrak{I}_{\lambda,\mu} = \mathfrak{I}(\lambda/2L_1, \mu/2L_2), \tag{23.4}$$

$$\hat{F}(\mathbf{u}) = (\mathfrak{I}(\mathbf{u}))^{\frac{1}{2}} \exp(\mathrm{i}\,\mathrm{phase}\{F(\mathbf{u})\}) = \mathsf{F}\{\hat{f}(\mathbf{x})\}, \tag{23.5}$$

and

$$\hat{a}_{\lambda,\mu} \exp(\mathrm{i}\,\hat{\theta}_{\lambda,\mu}) = \hat{F}_{\lambda,\mu} = \hat{F}(\lambda/2L_1, \mu/2L_2), \tag{23.6}$$

where λ and μ are arbitrary integers and $\hat{f}(\mathbf{x})$ is the version of $f(\mathbf{x})$ that is recoverable from $\hat{\mathfrak{A}}(\mathbf{x})$ (observe that this is consistent with the notation introduced in §14). Note that the quantity phase$\{F(\mathbf{u})\}$ appearing in (23.5) is actually the version of that quantity which is recoverable from $\hat{\mathfrak{A}}(\mathbf{x})$. The $\hat{a}_{\lambda,\mu}$, $\mathfrak{I}_{\lambda,\mu}$ and $\hat{F}_{\lambda,\mu}$ are here called the *modified primitive samples* of, respectively, $|F(\mathbf{u})|$, $|F(\mathbf{u})|^2$ and $F(\mathbf{u})$. Estimation of the $\hat{\theta}_{\lambda,\mu}$, which are here called the *modified primitive sample phases*, is of course the purpose of the approach to phase restoration described in this section. Note that the $\hat{a}_{\lambda,\mu}$ now effectively constitute the given data for the Fourier phase problem.

To generate accurate versions of the $\hat{\theta}_{\lambda,\mu}$ one has to resort to iterative techniques. Algorithms, such as Gerchberg's (see §11) or any of the variants of maximum-entropy (see §15) can be adapted more or less directly. The reasons for the efficacy or otherwise of such algorithms are little understood, so that practical experience is the only reasonable criterion for preferring one to another. Anyway, whichever is chosen, one has to begin with initial estimates $\tilde{\theta}_{\lambda,\mu}$ for the $\hat{\theta}_{\lambda,\mu}$. What are here called the *estimated primitive samples* of $F(\mathbf{u})$ are defined by

$$\tilde{F}_{\lambda,\mu} = \hat{a}_{\lambda,\mu} \exp(\mathrm{i}\,\tilde{\theta}_{\lambda,\mu}). \tag{23.7}$$

Means of generating the $\tilde{\theta}_{\lambda,\mu}$ are discussed below. It is here important to emphasize that $\mathsf{F}\{estimated\ primitive\ samples\ of\ F(\mathbf{u})\}$ can be expected to have value throughout Ω_{f*f}, except in the unlikely circumstance that the $\tilde{\theta}_{\lambda,\mu}$ constitute a set of phases corresponding precisely to the image-form of $\hat{f}(\mathbf{x})$.

When the image-form of $f(\mathbf{x})$ is 'clumpy', in the sense that it possesses well separated and clearly defined detail, there are several iteratively-based phase restoration procedures that function quite satisfactorily. The essential point seems to be that clumpy images have spectra that exhibit 'strong interference', i.e. deep fringes criss-cross most of the part of Fourier space wherein $|F(\mathbf{u})|$ has appreciable value. This is reminiscent of elementary and simple offset holography (see §21) for which the wide separation of the reference and the unknown ensures that there are pronounced, closely spaced fringes in the spectrum.

The image-forms that really challenge phase restoration algorithms consist of faint detail superimposed upon comparatively bright and more or less uniform backgrounds. There is then only weak interference apparent in Fourier space. The spectra tend to consist of large central lobes surrounded by small ripples. To make any headway with phase restoration in such cases, it is first necessary to carry out what is here called *defogging* (it is given this name because the aforementioned background is like a dense fog through which the faint detail fitfully shines). The 'skirt' of the central lobe (i.e. the closed curve in Fourier space, surrounding the central lobe, which appears to be where the central lobe merges into the rest of $|\hat{F}(\mathbf{u})|$) is identified, either by 'eye' (which is usually most effective because of the superior pattern-recognition facility of the human brain) or by application of a threshold. A smoothed version, here denoted by $S(\mathbf{u})$, of the central lobe of $|\hat{F}(\mathbf{u})|$ is then formed by removing any small ripples from this lobe. The defogged version $|F_{df}(\mathbf{u})|$ of $|\hat{F}(\mathbf{u})|$ is defined by

$$|F_{df}(\mathbf{u})| = |\hat{F}(\mathbf{u})| - kS(\mathbf{u}),\tag{23.8}$$

where k is chosen such that $|F_{df}(0)|$ is about 1.5 times the next largest peak in $|\hat{F}(\mathbf{u})|$. This has the effect of transforming the essentially interferenceless $|\hat{F}(\mathbf{u})|$ into a spectral magnitude exhibiting appreciable interference, in the sense that the magnitudes of the 'hills and ridges' of $|F_{df}(\mathbf{u})|$ do not vary all that much throughout the part of Fourier space wherein $|F(\mathbf{u})|$ is significant.

It is emphasized that there is no detailed theoretical justification for defogging. It just works! It is also fairly insensitive to the value chosen for k. The point of introducing $S(\mathbf{u})$ is to ensure that any ripples present in the central lobe of $|\hat{F}(\mathbf{u})|$ are preserved in $|F_{df}(\mathbf{u})|$. In order to prevent the notation becoming unduly complicated, it is convenient to redefine

$$\hat{F}(\mathbf{u}) = |F_{df}(\mathbf{u})| \exp(\mathrm{i}\,\mathrm{phase}\{\hat{F}(\mathbf{u})\}).\tag{23.9}$$

The recoverable image $\hat{f}(\mathbf{x})$ then approximates the aforementioned faint detail. The image-form of the fog is reconstructed from $((1-k^2)^{\frac{1}{2}}S(\mathbf{u}))$, employing the same approach as that (described below) for reconstructing $\hat{f}(\mathbf{x})$ from $|\hat{F}(\mathbf{u})|$. The faint detail – i.e. $\hat{f}(\mathbf{x})$ – can then be superimposed upon the fog.

The next step is to calculate the quantities

$$a_{l,m} = |\hat{F}(l\alpha_1, m\alpha_2)|, \qquad b_{l,m}^r = |\hat{F}((2l+1)\alpha_1/2, m\alpha_2)|,$$

$$\text{and} \quad b_{l,m}^c = |\hat{F}(l\alpha_1, (2m+1)\alpha_2/2)|, \tag{23.10}$$

where α_1 and α_2 are defined by (22.14). Note that the symbols originally introduced in (22.19) and (22.20) have been redefined, for two reasons: to avoid having to invent new notation, and because it is helpful later. The quantities on LHSs of equations (23.10) are now to be thought of as the actual and in-between samples of $|\hat{F}(\mathbf{u})|$, as opposed to $|\mathfrak{E}(\mathbf{u})|$. The purpose of all this is to be able to generate the $\tilde{\theta}_{\lambda,\mu}$ with the aid of the simple recursive algorithm described in the paragraphs of §22 which contain (22.21) through (22.30). Before this algorithm can be invoked, one must choose a (finite) value for the positive integer N introduced in §22 at the start of the paragraph containing (22.6). It turns out to be quite unnecessary to attempt to use a large value. In fact, computational experience suggests that it is usually best to choose $N = 2$.

When $N = 2$, the actual samples of $|\hat{F}(\mathbf{u})|$ are identical to its modified primitive samples, which is computationally convenient, of course.

The quantities $\theta_{l,m}$, originally introduced in (22.19), are now defined to be the phase estimates obtained, after replacing $|\mathfrak{E}(\mathbf{u})|$ by $|\hat{F}(\mathbf{u})|$, from the recursive algorithm described in §22. It is seen, when $N = 2$, that

$$\tilde{\theta}_{\lambda,\mu} = \theta_{\lambda,\mu}. \tag{23.11}$$

This method of generating initial phase estimates is here called *crude phase estimation*. Computational experience suggests that it can make the difference between success and failure to the iterative approach described below when the given data are derived from significantly contaminated measurements.

The iterative procedure that has proved most successful for restoring phases of two-dimensional spectra is the adaptation of Gerchberg's algorithm (see §11) known as *Fienup's algorithm*. Two versions of it are invoked here.

The starkest form of Fienup's algorithm is the *error-reduction*, or *error-correction algorithm*. To start it one needs an initial phase function or image. Fienup himself uses pseudo-random initial quantities. An advantage of beginning with a pseudo-random image is that its frame can be made the same size as Ω_f, whose extents are obtained from inspection of $\mathfrak{A}(\mathbf{x})$, as explained earlier in this section. Our experience suggests that it is equally satisfactory to start with an initial phase function, denoted here by $\Psi(\mathbf{u})$. In fact, the following description of the error-correction algorithm assumes that the initial quantity is a phase function.

A pseudo-random $\Psi(\mathbf{u})$ tends to be perfectly satisfactory for computer-generated data or for measured data when the images are clumpy and the contamination level is low. However, crude phase estimation seems to be more effective in general. Furthermore, the iterations sometimes fail to converge (or converge so slowly as to be virtually useless) when a pseudo-random $\Psi(\mathbf{u})$ is used. Whatever type of $\Psi(\mathbf{u})$ is employed, a first estimate of the image is defined by

$$\tilde{f}(\mathbf{x}) = \mathbf{F}\{|F(\mathbf{u})| \exp(i\,\Psi(\mathbf{u}))\}. \tag{23.12}$$

We want $\tilde{f}(\mathbf{x})$ to exist within a frame $\Omega_{\tilde{f}}$ having the same extents as Ω_f, but we cannot tell *a priori* where to place it in image space. We estimate the position of the centre of $\Omega_{\tilde{f}}$ as that point (a, b) for which $\Re$ is a maximum, where

$$\Re = \int_{b-L_2/2}^{b+L_2/2} \int_{a-L_1/2}^{a+L_1/2} \tilde{f}(\mathbf{x})\,d\sigma(\mathbf{x}). \tag{23.13}$$

The positive part of $\tilde{f}(\mathbf{x})$ within $\Omega_{\tilde{f}}$ is then taken as the current estimate of $\hat{f}(\mathbf{x})$. It is here said that $\tilde{f}(\mathbf{x})$ has been *boxed* with a variable frame. The frame is variable because a and b are varied until $\Re$ is maximized. A new estimate of the phase function is then formed from the Fourier transform of $\hat{f}(\mathbf{x})$, i.e.

$$\Psi(\mathbf{u}) = \text{phase}\{\mathbf{F}\{\hat{f}(\mathbf{x})\}\}, \tag{23.14}$$

which is substituted back into (23.12) to give another version of $\tilde{f}(\mathbf{x})$, thereby setting up an iterative loop. Note that $\Omega_{\tilde{f}}$ is positioned differently, in general, at each iteration.

An objective *stopping criterion* is needed so one knows when to cease iterating. We choose a threshold $\tilde{\varepsilon}$, which is based, like the threshold ε introduced in (23.2), on the estimated contamination level. After each iteration we take the given actual sample magnitudes and the latest estimates of their phases, and calculate the corresponding magnitudes of the in-between samples by sinc-interpolation (refer to §§10 and 11), writing these calculated magnitudes as $\tilde{a}_{\lambda,\mu}$ for the set, denoted by $\mathfrak{S}$(in-between), of values of λ and μ identifying the in-between samples. The iterations are stopped when

$$\sum_{\lambda,\mu \in \mathfrak{S}(\text{in-between})} |a_{\lambda,\mu} - \tilde{a}_{\lambda,\mu}|^2 < \tilde{\varepsilon}. \tag{23.15}$$

Remember that the actual sample magnitudes never change, because they represent the given data, and so it would be pointless (and computationally wasteful) to include them in (23.15).

While the error-correction algorithm is ideal for the early iterations, it is usually unsuitable for refining $\hat{f}(\mathbf{x})$ down to the ultimate image-form, which is reached, of course, when the stopping criterion (23.15) can be applied.

In the *hybrid input–output algorithm*, the previous version $\hat{f}_-(\mathbf{x})$ of the recoverable image acts as what is called a *driver*, in that the current version of $\hat{f}(\mathbf{x})$ is formed by modifying $\hat{f}_-(\mathbf{x})$. To explain how this is effected, it is convenient to denote by Ω_+ the set of points belonging to

Ω_{f*f} that lie outside Ω_f, and to denote by Ω_{-+} and Ω_{--} the sets of points belonging to Ω_f where $\bar{f}(\mathbf{x})$ is, respectively, positive and negative. The current version of $\hat{f}(\mathbf{x})$ is defined by

$$\hat{f}(\mathbf{x}) = \bar{f}(\mathbf{x}) \qquad \text{for } \mathbf{x} \in \Omega_{-+}$$
$$= \hat{f}_-(\mathbf{x}) - l\bar{f}(\mathbf{x}) \qquad \text{for } \mathbf{x} \in \Omega_{--} \cup \Omega_+, \qquad (23.16)$$

where the positive constant l is called the *gain constant*.

Towards the end of the iterations, a *fixed* and *enlarged frame* $\bar{\Omega}_f$, occupying the domain $0 < |x| < L_1'/2$, $0 < |y| < L_2'/2$, is used where L_1' and L_2' are typically 10 per cent larger than L_1 and L_2 respectively.

It is advantageous to carry out the iterations in what are here called *Fienup cycles*. Each of these consists of ν_1 error-correction iterations followed, or preceded, by ν_2 hybrid input/output iterations, where $(\nu_1 + \nu_2)$ is typically 50, but can, of course, be significantly more or less. The choice of the parameters k, l, L_1'/L_1, L_2'/L_2, ν_1 and ν_2 is determined by computational experience (see the quoted references), as is the decision to follow or precede error-correction with hybrid input/output in each Fienup cycle.

24 Related problems

Perhaps the most striking feature of the Fourier phase problem is the effect the number of dimensions has on the uniqueness question (see §22). In two or more dimensions, solutions to phase problems can be expected to be unique in practice. In one dimension, the uniqueness evaporates. It is worthwhile, therefore, examining whether related phase problems, which at first sight appear one-dimensional, can be manipulated into two-dimensional forms.

An important seemingly-one-dimensional problem is the *moving-window phase problem* which is here posed as: given $|\mathfrak{T}(u, \tau)|$, for $-\infty < u < \infty$ and $-\infty < \tau < \infty$, reconstruct $f(x)$; on the understanding that $\mathfrak{T}(u, \tau)$ is defined by

$$\mathfrak{T}(u, \tau) = \mathbf{F}_{(1u)}\{f_{(x)}(x, \tau)\}, \qquad (24.1)$$

where $f(x, \tau)$ is defined by

$$f(x, \tau) = f(x)\,\text{window}(x - \tau), \qquad (24.2)$$

so that, regarding x as 'time', the 'window' function can be thought of as selecting a segment of the 'signal' $f(x)$ centred at the 'instant' τ. As τ changes, so the window 'moves'.

It is convenient to define

$$t(y, v) = \int\!\!\!\int_{-\infty}^{\infty} \mathfrak{T}(u, \tau) \exp(i\,2\pi[v\tau - yu])\, du\, d\tau, \qquad (24.3)$$

which is the two-dimensional Fourier transform of $\mathfrak{T}(u, \tau)$. It would not make sense to employ the $\mathbf{F}\{\cdot\}$ notation because the factors $(v\tau)$ and (yu) have opposite signs allotted to them in the exponent – this is due to $t(y, v)$ being an 'image' in one of its dimensions and a 'spectrum' in the other. Substituting (24.1) and (24.2) into (24.3) gives

$$t(y, v) = \int\!\!\int\!\!\int_{-\infty}^{\infty} f(x)\,\text{window}(x - \tau)\,\exp(\mathrm{i}\,2\pi[(x - y)u + \tau v])\,du\,d\tau\,dx,$$

$$= \int\!\!\int_{-\infty}^{\infty} f(x)\,\text{window}(x - \tau)\,\exp(\mathrm{i}\,2\pi v\tau)\,\delta(x - y)\,d\tau\,dx, \quad \text{from (6.3)},$$

$$= f(y)\int_{-\infty}^{\infty} \text{window}(y - \tau)\,\exp(\mathrm{i}\,2\pi v\tau)\,d\tau, \quad \text{from (6.8)},$$

$$= f(y)\,\exp(\mathrm{i}\,2\pi vy)\int_{-\infty}^{\infty} \text{window}(-\tau)\,\exp(\mathrm{i}\,2\pi v\tau)\,d\tau,$$

$$= f(y)\mathfrak{S}(v)\,\exp(\mathrm{i}\,2\pi vy), \tag{24.4}$$

where

$$\mathfrak{S}(v) = \mathbf{F}_{(1v)}\{\text{window}_{(x)}(-x)\}. \tag{24.5}$$

Note, first, that the moving-window problem does *not* reduce to a pair of one-dimensional phase problems, because the y and v dependencies of RHS (24.4) do not 'separate' – refer to the penultimate paragraph of §22. Note next that the forms of $f(y)$ and $\mathfrak{S}(v)$ are given by $(t(y, v)\,\exp(-\mathrm{i}\,2\pi vy))$ for any fixed values of v and y respectively. So, the form of window(x), as well as the form of $f(x)$, can be recovered from $|\mathfrak{T}(u, \tau)|$. Finally, note that, if $f(y)$ and $\mathfrak{S}(v)$ are respectively replaced on RHS (24.4) by $\mathbf{F}_{(1y)}\{F(u)\,\exp(\mathrm{i}\,\Psi(u))\}$ and $\mathbf{F}_{(1v)}\{\text{window}(\tau)\,\exp(\mathrm{i}\,\Phi(\tau))\}$, where $\Psi(u)$ and $\Phi(\tau)$ are any real functions, the magnitude of the two-dimensional Fourier transform of RHS (24.4) depends upon the forms of $\Psi(u)$ and $\Phi(\tau)$. Remarkably, therefore, the solution to the moving-window phase problem appears to remain unique even when either $f(x)$ and/or $\mathfrak{S}(u)$ is *not* positive.

It is worthwhile giving some consideration to the class of *restricted Fourier phase problems* which can be typically posed as: given $|F(\mathbf{u})|$ throughout Fourier space, together with extra *a priori* information, reconstruct $f(\mathbf{x})$. Even for two-dimensional spectra, this problem remains important, first because phase recovery is necessarily adversely affected by any contamination present in the data (the extra information may ameliorate this), and second because there are certain images – e.g. those that can be factored into functions of x multiplied by functions of y, or those that are not positive – whose image-forms can only be recovered uniquely if extra information is provided. There is the further point that the one-dimensional Fourier phase problem (as posed in §34) does not in general possess a unique solution.

As an example of an instructive one-dimensional restricted problem, suppose that a palpably contaminated estimate $\tilde{F}(u)$ of the spectrum is given as well as a version of $|F(u)|$ which has been measured much more accurately. This can occur, for instance, with a multiple interferometer formed from a linear antenna array, it being understood that for the frequency band being received, the visibility phase is much more difficult to measure accurately than the visibility intensity, where *visibility* is astronomical terminology for (effectively) the Fourier transform of the image. Provided the errors present in $\tilde{F}(u)$ are not excessive, the problem can be solved by a variant of consistent deconvolution (see §19). In order to take advantage of the notation introduced in §13, it is convenient to define

$$\tilde{f}(x) = \mathsf{F}\{\tilde{F}(u)\}. \tag{24.6}$$

From the given $|F(u)|^2$ and $\tilde{F}(u)$ one calculates the respective sets $\mathbf{Z}_{f*f}$ and $\mathbf{Z}_{\tilde{f}}$ of zeros. Provided the members of $\mathbf{Z}_{\tilde{f}}$ are not too far removed from those of $\mathbf{Z}_f$, one can recognize by inspection which members of $\mathbf{Z}_{f*f}$ belong to $\mathbf{Z}_f^*$ and which belong to $\mathbf{Z}_f$, thereby permitting the reconstruction of $f(x)$ to the accuracy characterizing the measurement of $|F(u)|$. It is worth remarking that, in practice, this recognition procedure can usually be carried out significantly more effectively by 'eye' even though it is, in principle, easy enough to write a computer programme to do it.

There is an important restricted problem that first arose in the context of electron microscopy. It involves an image $\mathfrak{f}(\mathbf{x})$ that can be complex. The problem is posed as: given $|\mathfrak{F}(\mathbf{u})|$ and $|\mathfrak{f}(\mathbf{x})|$, throughout Fourier space and image spaces respectively, reconstruct $\mathfrak{f}(\mathbf{x})$; on the understanding that

$$\mathfrak{f}(\mathbf{x}) = \mathsf{F}\{\mathfrak{F}(\mathbf{u})\}. \tag{24.7}$$

This problem is conveniently solved with the aid of the *Gerchberg–Saxton algorithm*, which was actually developed before either Gerchberg's algorithm (see §11) or Fienup's algorithm (see §23). Because both $|\mathfrak{F}(\mathbf{u})|$ and $|\mathfrak{f}(\mathbf{x})|$ are given, it is unnecessary to estimate the size of $\Omega_{\mathfrak{f}}$ before setting up an iterative loop. This is just as well because the extent-constraint theorem (8.15) cannot be invoked, on account of $\mathfrak{f}(\mathbf{x})$ not being necessarily positive. One begins by introducing a pseudo-random phase function $\Psi(\mathbf{u})$ and computing

$$\hat{\mathfrak{f}}(\mathbf{x}) = \mathsf{F}\{|\mathfrak{F}(\mathbf{u})| \exp(i\,\Psi(\mathbf{u}))\} \tag{24.8}$$

The next step is to compute

$$\hat{\mathfrak{F}}(\mathbf{u}) = \mathsf{F}\{|\mathfrak{f}(\mathbf{x})| \exp(i\,\text{phase}\{\hat{\mathfrak{f}}(\mathbf{x})\})\} \tag{24.9}$$

The iterative loop is set up by re-defining

$$\Psi(\mathbf{u}) = \text{phase}\{\hat{\hat{\mathfrak{F}}}(\mathbf{u})\} \qquad (24.10)$$

so that $\hat{\hat{\mathfrak{F}}}(\mathbf{u})$ and $\hat{f}(\mathbf{x})$ can be repeatedly computed in turn. Only the actual samples of $|F(\mathbf{u})|$ and $|f(\mathbf{x})|$ are needed to implement this algorithm. In-between samples are unnecessary (recall the terminology introduced in §22). Consequently, we cannot invoke a stopping criterion equivalent to (23.15). What we can do is calculate the mean square difference between the actual samples of $|F(\mathbf{u})|$ and $|\hat{F}(\mathbf{u})|$. We can also do the same for the actual samples of $|f(\mathbf{x})|$ and $|\hat{f}(\mathbf{x})|$. It is appropriate, in fact, to evaluate these mean square differences alternatively, in Fourier space and image space respectively. One ceases iterating when both stopping criteria are satisfied during the same iteration.

Although the two-dimensional uniqueness argument presented in §22 cannot be applied directly to the Gerchberg–Saxton algorithm, because of the absence of any size constraint in the iterative loop, similar reasoning can be developed (see the quoted references).

Example IV – Illustrations of phase retrieval

Figure IVa shows a simple image having the character of an isolated group of stars. When the magnitude of the spectrum of this image is combined with its crude phase estimate (for $N = 2$) – see paragraph containing (23.11) and preceding paragraph – the image shown in Fig. IVb is obtained by Fourier transformation. Even though Fig. IVb is a far from faithful restoration of Fig. IVa, it is a useful basis for further processing.

Figures IVc and IVd are reconstructions of the original image (shown in Fig. IVa) obtained after single Fienup cycles – see final paragraph of §23 – for which $\nu_1 = 1$ and $\nu_2 = 4$. The initial phase function $\Psi(\mathbf{u})$ – see paragraph containing (23.12) and preceding paragraph – is pseudo-random for Fig. IVc and is the crude phase estimate for Fig. IVd. Both reconstructions can be greatly improved, of course, by further Fienup cycling. However, comparison of Figs. IVc and IVd indicates the advantage of starting with the crude phase estimate rather than a pseudo-random initial phase function.

The image shown in Fig. IVe is of the kind that can be reconstructed immediately by offset Fourier holography – see paragraph containing (21.9). Figure IVf is the magnitude of the spectrum of Fig. IVe. Since each image in this Example is real, the magnitude of its spectrum is symmetrical about the origin of Fourier space – see (8.2). It is therefore only necessary to display half of each spectral magnitude. So, in each of Figs. IVf, IVi and IVk, the origin of Fourier space lies at the middle of the left hand edge of the display. Note that the 'central lobe' of each spectrum is itself centred at the origin. The square of Fig. IVf can be thought of as the hologram, because its Fourier transform is the autocorrelation of the image shown in Fig. IVe. The most prominent fringe pattern in Fig. IVf corresponds to the 'interference' between the two separated parts of Fig. IVe. The other, less pronounced fringes are due to the detail existing within the larger of the two separated parts of Fig. IVe. The 'fog' on which the three 'stars' in this

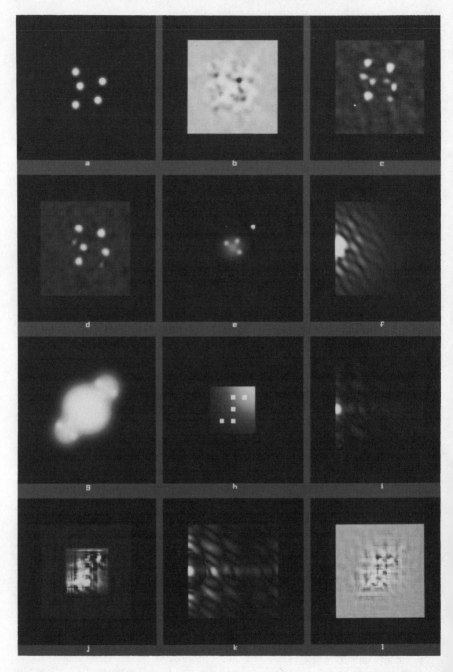

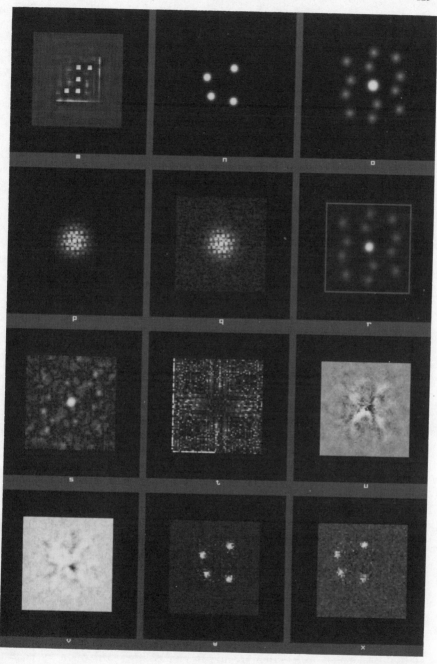

Fig. IV.

larger part are superimposed gives rise to little in the way of a fringe pattern, although its presence is, of course, implicit in the detailed distribution of brightness across Fig. IVf. Figure IVg, which is the autocorrelation of Fig. IVe, consists of the (at least) three separated parts characteristic of Fourier holography. The central part consists of the autocorrelations of the two separated parts of Fig. IVe. The two outer parts of Fig. IVg are cross-correlations of the two parts of Fig. IVe. The similarity of the left-hand outer part to the larger part of Fig. IVe is clear. Suppose the form of the smaller part of Fig. IVe is known *a priori* – i.e. it is a given reference (see second paragraph of §21). By deconvolving (multiplicatively, e.g. by Wiener filtering – see §16 and Example III) this reference from Fig. IVg, we can obtain a faithful version of the larger part of Fig. IVe.

Note that the larger part of Fig. IVe consists of detail superimposed upon a background 'fog' – i.e. the type of image which is difficult to reconstruct faithfully by immediate application of Fienup cycling (see §23). This is of no concern (unless the available dynamic range is exceeded) when the image consists of two parts satisfying the separation condition (21.6). The parts in Fig. IVe satisfy this condition. In a more general case, such as occurs when Fig. IVh is the image which we wish to recover, it is advisable to resort to defogging – see paragraphs containing (23.8) and (23.9).

Figure IVi is the magnitude of the spectrum of the image shown in Fig. IVh. Note that the central lobe is very much brighter than the rest of the spectrum. Figure IVj is the image obtained from the spectral magnitude shown in Fig. IVi after crude phase estimation and a single Fienup cycle ($\nu_1 = 76$, $\nu_2 = 80$). Note that little of the detail seen in Fig. IVh is apparent in Fig. IVj, even though such a large Fienup cycle is employed. The detail can, in fact, be readily recovered if the spectral magnitude is defogged before we attempt to retrieve the phase. This is demonstrated in the following paragraph.

Figure IVk is the magnitude of the defogged spectrum ($k = 0.15$) of the image shown in Fig. IVh. Note that the outer fringes of the spectrum are now much more visible (the dynamic range of the display is the same as for Fig. IVi). Figure IVl is the image obtained (by crude phase estimation) from the spectrum shown in Fig. IVk. There is negligible resemblance to Fig. IVh but the image is restricted to roughly the same area of image space as Fig. IVj and is of comparable quality. Starting with the spectral phase corresponding to Fig. IVl and the spectral magnitude shown in Fig. IVk, the image shown in Fig. IVm is generated by a single Fienup cycle having the same parameters as that which produced Fig. IVj. Note that the detail shown superimposed on the 'fog' in Fig. IVh is apparent in Fig. IVm. The improvement of Fig. IVm over IVj is dramatic and it emphasizes the restorative powers of defogging. A more faithful restored image can be obtained by refogging and further Fienup cycling.

A final set of images demonstrates how the preprocessing techniques described in the second to fifth paragraphs (inclusive) of §23 can ameliorate the effects of noise on a given spectral magnitude. The original image shown in Fig. IVn is sufficiently simple that it can illustrate the benefits of preprocessing, without it being necessary to complicate the discussion by having to invoke defogging. Figures IVo and IVp, respectively, are the autocorrelation and spectral magnitude of Fig. IVn. Figure IVq is the spectral magnitude after corruption by noise (generated from a pseudo-random number routine) having the very appreciable r.m.s. value of 10 per cent of the spectral magnitude at the origin of Fourier space. Note that the whole of the spectral magnitude is displayed in Figs. IVp and

IVq (unlike Figs. IVf, IVi, and IVk which each display only half of the spectral magnitude). Figure IVr is the Fourier transform of the square of the spectral magnitude shown in Fig. IVq. The rectangular frame in Fig. IVr is our estimate of Ω_{f*f}. The effect of the noise on the spectral magnitude can be gauged from Fig. IVs, which is the (magnified by a factor of 4) magnitude of the difference between Fig. IVo and the part of Fig. IVr inside the rectangular frame.

Figure IVt is the magnitude of the difference (magnified by a factor of 4) between Fig. IVq and the square root of the spectrum of Fig. IVr. So, if Fig. IVq is thought of as data (i.e. it represents what might actually be measured) then Fig. IVt shows the level of modification of the data produced by the preprocessing techniques described in §23.

Figure IVt represents all of the compensation which we can make for the noise present in the given spectral magnitude. Figure IVu is the image obtained, with the aid of crude phase estimation, from the magnitude of the Fourier transform of Fig. IVr. It is by no means a recognizable version of Fig. IVn, but it is concentrated within a region whose area is much the same as that of the image frame Ω_f. It is noticeably more concentrated than the image (shown in Fig. IVv) obtained by crude phase estimation from the *un*preprocessed spectral magnitude (Fig. IVq). Figures IVw and IVx are the results obtained by applying single Fienup cycles ($\nu_1 = 1$ and $\nu_2 = 9$) to the images shown in Figs. IVu and IVv respectively. The fact that Fig. IVw is a more faithful version of Fig. IVn than is Fig. IVx illustrates the usefulness of our preprocessing techniques.

We emphasize to the reader that the images used to illustrate this Example are somewhat rudimentary compared with those published in journals while this book was being printed. We felt it was more important to get our messages across clearly than to try and demonstrate our algorithms' most impressive performances (which are, of course, continually improving).

V

RECONSTRUCTION FROM PROJECTIONS

Most of the methods discussed in this chapter were originally inspired by the desire to contribute to the advancement of comput(eriz)ed (or computer-assisted) tomography, now generally referred to as CT [Hounsfield, 1]. This combination of engineering, mathematical physics and computational technique has had almost as big an impact on medical diagnostics as Roentgen's discovery of X-rays in 1895. It is becoming recognized, however, that the theoretical approaches and algorithms introduced in §§25 through 32 below can be usefully applied in many branches of physical science [Kak, 1; Bates, 16]. While the basic mathematical physics dates from World War I [Radon, 1], the modern theory of CT is usually acknowledged to have appeared in the early 1960s [Cormack, 1]. The earlier efforts of electron microscopists [as chronicled, for instance, by Misell, 1, B1], radio-astronomers [see various papers quoted in Part I of Van Schooneveld, B1], and a group working in Kiev in the middle 1950s [Barrett, 1; Tetelbaum, 1] should also be recognized.

Five useful general references for this chapter are four books [Herman, B1, B2, B3; Rosenfeld, B2] and a special issue of a journal [*Proc. IEEE*, S2].

The ideal CT situation [Lewitt, 1] and its radio-astronomical analogue [Napier, 2] are introduced in §25. This has its roots in the *Radon Transform* [Radon, 1], which is seldom invoked explicitly nowadays because of the power of modern Fourier techniques. Various departures from the 'ideal' are discussed in §§26 through 33, and are referenced in the remaining paragraphs of these introductory comments, but it seems worthwhile at this point to quote two particularly pertinent references [Macovski, 1; Part I of van Schooneveld, B1].

Problems associated with imaging distributions of radiating sources embedded within objects are studied in §26 [Gullberg, T1; Knoll, 1]. The connection between CT and radiometry does not seem to have received much explicit attention [Bates, 16] although it is implicit in the theoretical science of radiative transfer [Chandrasekhar, B1].

The imaging of scatterers within objects is discussed in §27, in the contexts of medical Compton-scatter systems [Battista, 1; Garden, T1] and conventional echo-location (i.e. sonar and radar [Bates, 16]).

Further aspects of echo-location problems are of interest in §28, which also treats the theory of image reconstruction for positron-emission tomography [Ter-Pogossian, 1; Thompson, 2].

The number of projections available in any practical application is inevitably finite. The consequences of this are examined in §29 [Lewitt, 1, 3; Smith, 2]. The 'limited-angle problem', for which the given projections cluster within a restricted segment of the complete 180° range of angles, is neglected here because it only seems to have useful solutions in very special cases, as has recently been clearly established [Davison, 1].

It can be either inconvenient or impossible to irradiate the whole of the cross-section of an object, in which case the measured projections are incomplete (in the sense that they have 'gaps' in them). Ways of compensating for such incompleteness are described in §30 [Lewitt, 2].

The concern in §31 is with situations for which the simplified assumptions underlying most of the practical manifestations of CT are markedly inadequate. The problems introduced in this section have only recently been systematically attacked [Bates, 16; Garden, 1; Greenleaf, 1; McKinnon, 1].

It is quite impossible to devise straightforward image reconstruction algorithms when attempting to take strict account of the many 'non-ideal' factors that bedevil practical systems. Luckily, of course, it transpires that spectacularly useful results can be obtained in many CT applications without bothering with these niceties. If we do wish to bother with them, however, it seems that iterative approaches are forced upon us [Bates, 16], as is explained in §32.

The reconstruction of images from their projections is one of the great triumphs of applied Fourier theory. Paradoxically, perhaps, images of the highest quality are not obtained in standard CT practice by invoking this theory in what might be regarded, at first sight at least, as the obvious way. Straightforward Fourier reconstruction tends to produce images contaminated with 'speckly' artefacts (which provides a conceptually attractive bridge between this chapter and the next). The details are explored in §33, wherein it is explained why practical reconstruction of CT images is effected by what is called modified, or filtered, or convolutional, back-projection [Lewitt, 1, 3], which it is worth noting was originally put forward by electrical engineers who had strong astronomical leanings [Bracewell, 2]. As is noted in §25, the radio astronomical (interferometrical) image reconstruction problem is closely related to its CT counterpart. It is however rarely solved by modified back-projection, for the reasons touched on in the concluding paragraph of §33 [Part I of Van Schooneveld, B1; see also a number of papers in Roberts, B1].

Example V illustrates various aspects of transmission CT, which is the basis for the above-mentioned spectacular advances in X-ray diagnosis. This example highlights how powerful the CT concept is in situations for which the simple formula (9.4), for a projection, accurately characterizes the given data. The images presented in the example relate to §§25, 29, 30, 31, and 33.

25 The transmission CT problem – the interferometric problem

In conventional *computed tomography* (CT), or 'computer assisted tomography (CAT)' as it is sometimes called, an image of a cross-section through a body is inferred from measurements of the attenuation of X-ray beams *transmitted* through the body – hence the 'transmission CT problem'.

It is customary to think of a CT image as representing the distribution of 'density' throughout the cross-section. Actually, of course, it is the X-ray 'absorptivity' that is imaged, but since this quantity is roughly proportional to the density for most materials, it is conceptually useful to think of image 'brightness' or 'intensity' as being proportional to a 'density'.

Regarding $f(x, y)$ as the X-ray absorptivity (per unit length) of the cross-section in question, it is clear that the attenuation of a typical beam transmitted through the cross-section can be written as $\exp(-\kappa p(\xi; \phi))$, where $p(\xi; \phi)$ is defined by (9.4) – the infinite limits on the integral merely imply that the beam has traversed the whole cross-section; the latter is always finite in practical applications. Throughout this chapter the image is confined to a circular region of image space, centred on the origin, of radius $\hat{r}$, i.e.

$$f(x, y) = f(r, \theta) = 0 \qquad \text{for } r \geq \hat{r}. \tag{25.1}$$

Knowing the value of the constant κ, the projection $p(\xi; \phi)$ is recovered from the logarithm of the measured attenuation.

It is worth remembering that the 'absorption' of the transmitted beam is only apparent in the sense that most of the attenuation is due to scattering, rather than actual capture, of photons by atoms within the body. This is unimportant from a strictly image-processing point of view, but it introduces a significant practical complication. Collimators must be placed in front of the detectors of the X-rays to ensure that all the photons collected by any particular detector travelled along the straight line from the source to that detector (i.e. along the line parallel to the η-axis and distant ξ from it). This tends to limit the achievable resolution, although in this regard the discussion in (ii) and (iii) below is more apposite.

Note that (9.4) describes a 'beam' made up of 'parallel rays', the straight lines parallel to the η-axis. In CT practice, measurements characterized by (9.4) are known as 'parallel-beam projections'. It is more usual nowadays to employ 'fan-beam projections' with rays emanating from an effective 'point source' and collected by detectors positioned on an arc centred on the source. This need not affect the theory underlying the image reconstruction, although it is often found convenient to introduce certain modifications. The latter are not discussed here because they are very specialized and can only be properly appreciated if studied in detail (which can be done by perusing references quoted in the introductory comments to this chapter).

The *ideal transmission CT problem* is here posed as: given $p(\xi; \phi)$ for $-\hat{r} < \xi < \hat{r}$ and $0 \leq \phi < \pi$, reconstruct $f(x, y)$; on the understanding that $p(\xi; \phi)$ is defined by (9.4).

There are three important provisos:

(i) The projection $p(\xi; \phi)$ relates realistically to what can be measured in conventional CT if the X-ray beam is mono-energetic – i.e. monochromatic, such as is emitted by a single-isotope source. Such a source is impracticable for routine clinical CT practice, and so a standard X-ray tube is generally used. The X-rays then suffer from *beam-hardening* (higher attenuation of lower energy – i.e. lower frequency – photons), which is discussed further in §32.

(ii) It is, of course, impossible to measure $p(\xi; \phi)$ for a continuous range of ϕ. Only a finite number of discrete projections can actually be measured (this is discussed further in §29). In practice, however, one only wishes to reconstruct $f(x, y)$ to some finite resolution, Δx say. If $\Delta \phi$ is the angle between successive projections, their 'separation' on the circumference of the circle of radius $\hat{r}$ which encloses the image – refer to (25.1) – is $\hat{r} \Delta \phi$. The latter is seen to be roughly equal to Δx. This is a manifestation of the sampling theorem (§ 10). The finiteness of both the image and the resolution to which it is reconstructed therefore imply that a finite number of projections are in practice equivalent to projections given throughout the continuous interval $0 \leq \phi < \pi$ – all this is made more precise in §29.

(iii) X-ray detectors are necessarily of finite size, thereby setting another limit on the resolution of CT measurements. The 'density' varies in general with z (the coordinate perpendicular to the x, y-plane) as well as with x and y, and so should strictly be written as $f(x, y, z)$. Cross-quently, the 'density' appearing within the integral in (9.4) is in fact 'sampled', in the sense introduced in §11. The projection itself is, of course, sampled because the X-ray detector has a finite area, not merely a finite width. If $rp(\xi; \phi)$ denotes the response of an X-ray CT detector whose beam axis lies in the x, y-plane and is directed parallel to the η-axis at a distance ξ from it then the above discussion indicates that

$$rp(\xi; \phi) = \int\!\!\int\!\!\int_{-\infty}^{\infty} f(x, y, z) \, \text{inst}(\xi - \xi', z) \, d\xi' \, d\eta \, dz, \qquad (25.2)$$

where $\text{inst}(\xi, z)$ is the 'response function', or instrument function in the terminology of §11, of the detector. In those situations for which

$\partial f(x, y, z)/\partial z$ is generally smaller than $\partial f(x, y, z)/\partial x$ and $\partial f(x, y, z)/\partial y$ it is justifiable to re-express (25.2) as

$$rp(\xi; \phi) = \int\!\!\int_{-\infty}^{\infty} f(x', y') \operatorname{inst1}(\xi - \xi') \, d\xi' \, d\eta, \qquad (25.3)$$

where $\operatorname{inst1}(\xi)$ is the effective one-dimensional response function of the detector in the ξ-direction,

and
$$f(x, y) = f(x, y, 0), \qquad (25.4)$$

$$x' = \xi' \cos(\phi) - \eta \sin(\phi), \qquad y' = \xi' \sin(\phi) + \eta \cos(\phi). \qquad (25.5)$$

Provided $rp(\xi; \phi)$ is measured for values of ξ spaced much closer than the effective width of $\operatorname{inst1}(\xi)$ then $p(\xi; \phi)$, as defined by (9.4), can be recovered to (in principle) arbitrarily high resolution. This argument has involved a fair amount of 'hand-waving' (but it does make good physical sense!). It has been included to emphasize that the specification (in the ideal transmission CT problem, as posed above) of sample points spaced arbitrarily closely in the ξ-direction is not physically ridiculous. In practice, of course, sample points along the ξ-axis are spaced not all that much closer than the resolution in the z-direction. Anyway, resolution is usually limited more by the necessary finiteness of the number of projections – see (ii) above – than by the finite size of the detectors.

Even with the above provisos, (9.4) usefully models measured data for conventional CT (as confirmed by its spectacular successes; and note that (9.4) is the theoretical model adopted for the basic design of practical systems). Of course, it also models any physical system which measures the integrals of a two-dimensional quantity along *straight rays*. Examples are an optical telescope with a pupil that is 'long and thin', or a radio-telescope consisting of a linear array of individual small antennas all of which are fed 'in-phase'. Both telescopes have beams that are narrow in the directions parallel to the long dimensions (themselves taken for convenience here to be parallel to the ξ-axis) of the pupil and the array and are wide in the perpendicular directions (here taken parallel to the η-axis). When the pupil, or the array, is oriented appropriately, the total radiation intensity received by the telescope is proportional to the line integral of the brightness across that part of the heavens intersected by the beam. If the celestial object being observed is of effectively finite size, and if its brightness distribution is denoted by $f(x, y)$, then $p(\xi; \phi)$ as defined by (9.4) can be proportional to the received radiation intensity.

Reference to (9.6) shows that the one-dimensional Fourier transform of $p(\xi; \phi)$ with respect to ξ gives $F(\rho; \phi)$ along a single line passing through the origin of Fourier space and inclined at the same angle to the u-axis as the projection is to the x-axis in image space. The one-dimensional transforms of all projections, for the angular interval $0 \leq \phi < \pi$, consequently fill up Fourier space, at least out to a radius set by the resolution of the given projections – refer to (ii) and (iii) above. The two-dimensional Fourier transform of $F(\rho; \phi)$ then gives the image.

Although the *Fourier reconstruction* theory outlined in the previous paragraph completely characterizes the solution to the ideal transmission CT problem, it gives little hint of the actual reconstruction procedure that is almost universally adopted nowadays. The reasons for this are explained in §33. All that is noted here is that the procedure implied by the previous paragraph leads to an estimate of the spectrum on a *radial grid* – i.e. on straight lines emanating from the origin of Fourier space like the spokes of a wheel –, whereas one would like to be presented with samples of $F(u, v)$ on a *rectangular grid* – i.e. at sample points $(l\alpha, m\beta)$ where l and m are integers, and α and β are constants – so that $f(x, y)$ can be efficiently computed using the F.F.T. (see §12).

It is worth emphasizing that certain variants of *nuclear magnetic resonance* (NMR) imaging provide estimates of the spectrum on a rectangular grid (and in three dimensions) so that straightforward Fourier reconstruction is useful in NMR imaging practice.

Fourier reconstruction theory also completely characterizes image processing for radio-astronomical synthesis telescopes, which are in effect gigantic phase-sensitive (or Michelson) interferometers. They make their measurements in Fourier space, however. Consequently, one can pose an ideal *interferometric problem* as: given $F(\mathbf{u})$ inside the circle of radius $\hat{\rho}$ in Fourier space, reconstruct $f(\mathbf{x})$ to the resolution permitted by the value of $\hat{\rho}$. If, in fact, the given data are as perfect as is implied in this ideal form of the problem then the solution is trivial – i.e.: after appropriate preprocessing of the kind described in §15, reconstruct $f(\mathbf{x})$ using the F.F.T. In practice, naturally, there are difficulties, as is explained in §33.

26 The emission CT problem – the radiometric imaging problem

Consider the problem of imaging a localized distribution of spatially incoherent sources of radiation, whose density is here denoted by $f(x, y, z)$. Since three-dimensional images can be built up in parallel planes stacked one upon the other, it is adequate to discuss two-dimensional imaging in a single plane, say the one defined by $z = 0$. Appealing to the notation introduced in (25.4), the goal is then to reconstruct $f(x, y)$ from measurements of the radiation. Because $\hat{r}$ can have any finite value, the localization of the image can be specified by (25.1).

If the source distribution is *disembodied* in the sense that it hangs in free space unsupported and not embedded in any material medium, the radiation intensity $d\mathrm{q}$ due to sources within the volume element $(dx\,dy\,dz)$ weakens as $1/\Re^2$ with distance $\Re$ from the element, and so

$$d\mathrm{q} \propto f(x, y, z)\,dx\,dy\,dz/\Re^2. \tag{26.1}$$

the presence of the factor $1/\Re^2$ emphasizes that the position of the radiation detector must enter into any expression characterizing measured data, unlike in the transmission CT problem analysed in §25.

Measurements must be made from all around the source distribution, if enough data are to be gathered to permit a faithful image to be reconstructed. It is convenient, therefore, to envisage a linear array of detectors, with the centre of the array describing a circle of radius $\hat{r}$ as it rotates around the sources. In many practical applications, the sources are comparatively weak, so that it is not very meaningful to postulate (as does make sense for transmission CT – see §25) arbitrarily small fields of view for individual detectors. The latter must be endowed with appreciable collecting areas and must be sensitive to incoming radiation over quite a wide range of angles, in order to achieve a useful signal-to-noise ratio. The axes of the detectors' receiving beams must all lie in the same plane (i.e. that coinciding with the cross-section being imaged) and there seems to be no good reason why they should not be parallel and spaced at equal intervals from each other. In keeping with the notation introduced in §§9 and 25, we take the x, y-coordinates to be fixed in the cross-section (i.e. these coordinates are fixed in the source distribution) and the ξ, η-coordinates to rotate with the array (the latter being parallel to the ξ-axis). We envisage the beam axes to be parallel to the negative η-axis, with the latter always coinciding with the axis of the central detector's beam. Cartesian coordinates $z, \tilde{\xi}$ identify points on the face of the array, with the $\tilde{\xi}$-axis being parallel to the ξ-axis.

At each point on the face of the array it is convenient to introduce the polar angle ϑ, with its pole parallel to the negative η-axis, and the azimuthal angle φ, with $\varphi = 0$ corresponding to the $\tilde{\xi}$-axis. Note that $\vartheta = \vartheta(\tilde{\xi}, z)$ and $\varphi = \varphi(\tilde{\xi}, z)$ for each point on the face of the array.

Consider a radiation detector whose beam axis intersects the face of the array at the particular point $(\xi, 0)$. Its response Φ to radiation incident upon any point $(\tilde{\xi}, z)$ within its collecting area, from the direction identified by the angles ϑ and φ, must in general exhibit a complicated functional dependence, i.e.

$$\Phi = \Phi(\tilde{\xi}, z; \vartheta(\tilde{\xi}, z); \varphi(\tilde{\xi}, z)), \qquad (26.2)$$

which falls to zero for values of $|\tilde{\xi} - \xi|$ and $|z|$ that lie outside the collecting area. The total radiation intensity received by the detector is

$$q = \iiint_{\mathfrak{B}} \Phi \, dx \, dy \, dz, \qquad (26.3)$$

where $\mathfrak{B}$ is the volume enclosing the source distribution; and where the coordinates x, y, z and the distance $\Re$ appearing in (26.1) are all seen to be functions of $\tilde{\xi}, z, \vartheta$, and φ.

The integral in (26.3) is so complicated that nobody has yet found a way of reconstructing f from measured values of q, without introducing extensive approximations. It should be recognized, however, that there is little practical point trying to recover f very accurately because measurements of the sort implied above are usually poorly resolved, mainly for technical reasons.

It is both technically reasonable and analytically necessary to accept the resolution limitation set by the finite area of a typical radiation detector, but to postulate that this area is sufficiently small that the response of the detector can be characterized by a simple beam, whose axis intersects the face of the array at the point $(\xi, 0)$. The functional dependencies of the angles ϑ and φ then simplify to $\vartheta = \vartheta(\xi)$ and $\varphi = \varphi(\xi)$. For the imaging system to have a resolution which is useful in practice, the beamwidth must be narrow enough that $\Re$ can be replaced in (26.1) by distance along the beam axis – i.e. for all points in the ξ, z-plane, $\Re$ can be approximated by

$$\Re \simeq \hat{r} - \eta \tag{26.4}$$

This states that, for any point (ξ', η, z) within the sensible part of the beam, $((\xi' - \xi)^2 + z^2)$ is negligible compared with $(\hat{r} - \eta)^2$. It follows that $\tan(\vartheta) \simeq \vartheta$. Keeping within the spirit of these approximations, the beam shape is taken to be 'elliptical' and characterized by effective beamwidths Θ_1 and Θ_2 in the ξ- and z-directions respectively. It follows that the functional dependence of the beam simply reduces to

$$\Phi = \Phi(\vartheta, \varphi) \simeq \Phi([(\xi' - \xi)^2/\Theta_1^2 + z^2/\Theta_2^2]^{\frac{1}{2}}/(\hat{r} - \eta)). \tag{26.5}$$

There is no reason to employ detectors that are not symmetrical about their beam axes. This implies that $\Phi(\vartheta, \varphi)$ is unchanged when $\vartheta \to -\vartheta$ and/or $\varphi \to -\varphi$, so that the functional dependence of Φ can be further simplified:

$$\Phi \simeq \Phi([(\xi' - \xi)^2/\Theta_1^2 + z^2/\Theta_2^2]/(\eta - \hat{r})^2). \tag{26.6}$$

The total intensity received by the detector is thus, to a useful degree of approximation, proportional to

$$q(\xi, \phi) = \int\!\!\!\int\!\!\!\int_{-\infty}^{\infty} f(x', y', z)\Phi([(\xi' - \xi)^2/\Theta_1^2 + z^2/\Theta_2^2]/(\eta - \hat{r})^2)\, d\xi'\, d\eta\, dz/(\eta - \hat{r})^2, \tag{26.7}$$

where x' and y' are defined by (25.5), and the infinite limits are included merely for analytical convenience – recall that the image has been required to satisfy (25.1), so that there is nothing to worry about as $\eta \to \hat{r}$.

As with transmission CT, the resolution of an imaging system whose theoretical basis is (26.7) is limited by the variations in the z-direction of the quantity which it is desired to image. It is again necessary to assume that

$$f(x, y, z) \simeq f(x, y, 0) = f(x, y) \tag{26.8}$$

within the beams of the detector array. Note that the integrand in (26.7) is significantly different when ϕ is replaced by $(\phi + \pi)$, unlike the integrand in (9.4). Measurements must therefore be made throughout a range of ϕ of extent 2π, as compared with conventional CT (see §25) for which a range of extent π is sufficient.

The reader will have remarked that the above discussion describes classical radiometry. Consequently, the *radiometric imaging problem* – or equivalently, the *disembodied emission CT problem* – is here posed as: given $q(\xi; \phi)$, for $-\hat{r} < \xi < \hat{r}$ and $0 \le \phi < 2\pi$, reconstruct $f(x, y)$; on the understanding that $q(\xi; \phi)$ is defined by (26.7) as qualified by (26.8).

Substituting (25.5) and (26.8) into (26.7), and introducing the new variables of integration $t = (\xi' - \xi)/(\eta - \hat{r})$ and $\zeta = z/(\eta - \hat{r})$, gives

$$
q(\xi; \phi) = \int\int\int_{-\infty}^{\infty} f([\xi + (\eta - \hat{r})t]\cos(\phi) - \eta \sin(\phi), [\xi + (\eta - \hat{r})t]\sin(\phi)
$$

$$
+ \eta \cos(\phi))
$$

$$
\times \Phi(t^2/\Theta_1^2 + \zeta^2/\Theta_2^2) \, dt \, d\eta \, d\zeta. \tag{26.9}
$$

The functional dependence of Φ upon its argument is always known in practice, either from theory or from radiation pattern measurements on the detector array. The ζ-integration can therefore be performed immediately because, in the integrand of (26.9), ζ only appears in the argument of Φ:

$$
\int_{-\infty}^{\infty} \Phi(t^2/\Theta_1^2 + \zeta^2/\Theta_2^2) \, d\zeta = Y(t), \tag{26.10}
$$

where the functional dependence of Y upon its argument can, of course, be considered as given.

Repeated appeal is made from now on to the notation introduced in §6 and the results established there. Substituting (26.10) into (26.9) and expressing $f(\cdot, \cdot)$ in terms of its spectrum $F(\cdot, \cdot)$ leads to

$$
q(\xi; \phi) = \int\int\int_{-\infty}^{\infty} \exp(-i \, 2\pi\{([(\xi + (\eta - \hat{r})t]\cos(\phi) - \eta \sin(\phi))u
$$

$$
+ ([\xi + (\eta - \hat{r})t]\sin(\phi) + \eta \cos(\phi))v\})F(u, v)Y(t) \, dt \, d\eta \, du \, dv, \tag{26.11}
$$

the one-dimensional Fourier transform of which is

$$
\mathfrak{Q}(\rho; \phi) = \mathsf{F}_{(1\rho)}\{q_{(\xi)}(\xi; \phi)\}
$$

$$
= \int\int\int\int_{-\infty}^{\infty} \exp(i \, 2\pi\{\rho\xi - ([\xi + (\eta - \hat{r})t]\cos(\phi) - \eta \sin(\phi))u
$$

$$
+ ([\xi + (\eta - \hat{r})t]\sin(\phi) + \eta \cos(\phi))v\})F(u, v)Y(t) \, dt \, d\eta \, dv \, d\xi. \tag{26.12}
$$

It is now convenient to introduce the Cartesian coordinates u' and v' rotated by the angle ϕ in Fourier space with respect to the coordinates u and v:

$$u' = u\cos(\phi) + v\sin(\phi), \qquad v' = -u\sin(\phi) + v\cos(\phi). \qquad (26.13)$$

Substituting (26.13) into (26.12) and integrating with respect to ξ gives

$$\mathfrak{Q}(\rho;\phi) = \int\!\!\!\int\!\!\!\int\!\!\!\int_{-\infty}^{\infty} \exp(-i\,2\pi[(\eta-\hat{r})tu' + \eta v'])\,\delta(u'-\rho)$$

$$\times F(u'\cos(\phi) - v'\sin(\phi), u'\sin(\phi) + v'\cos(\phi))Y(t)\,dt\,d\eta\,du'\,dv'$$

$$= \int\!\!\!\int_{-\infty}^{\infty} F(\rho\cos(\phi) - v'\sin(\phi), \rho\sin(\phi) + v'\cos(\phi))$$

$$\times \exp(i\,2\pi\rho\hat{r}t)\,\delta(v' + \rho t)Y(t)\,dt\,dv'$$

$$= \int_{-\infty}^{\infty} F([\cos(\phi) + t\sin(\phi)]\rho, [\sin(\phi)$$

$$- t\cos(\phi)]\rho)\exp(i\,2\pi\rho\hat{r}t)Y(t)\,dt. \qquad (26.14)$$

When $F(\cdot,\cdot)$ is expressed in terms of polar coordinates it is seen that

$$F([\cos(\phi) + t\sin(\phi)]\rho, [\sin(\phi) - t\cos(\phi)]\rho) = F((1+t^2)^{\frac{1}{2}}\rho; \phi - \psi(t)), \qquad (26.15)$$

where

$$\tan(\psi(t)) = t. \qquad (26.16)$$

On writing

$$\mathfrak{Q}(\rho;\phi) = \sum_{l=-\infty}^{\infty} \mathfrak{Q}_l(\rho)\exp(i\,l\rho), \qquad (26.17)$$

and invoking the first equation in (9.7), the lth Fourier coefficient (as defined in §10) of (26.14) becomes

$$\mathfrak{Q}_l(\rho) = \int_{-\infty}^{\infty} F_l((1+t^2)^{\frac{1}{2}}\rho)\exp(i[2\pi\rho\hat{r}t - l\psi(t)])Y(t)\,dt. \qquad (26.18)$$

Since $q(\xi;\phi)$ is given, $\mathfrak{Q}(\rho;\phi)$ can be immediately computed from it – refer to the first line of (26.12). Evaluating the Fourier coefficients of $\mathfrak{Q}(\rho;\phi)$ gives the $\mathfrak{Q}_l(\rho)$. So, $F_l(\cdot)$ is the only unknown in (26.18), which is seen to be a Fredholm integral equation of the first kind. Whilst its solution is much less convenient, from a computational point of view, than inverting a Fourier integral, for instance, it can nevertheless be obtained straightforwardly, in the manner outlined in the following paragraph.

Suppose $g(x)$ is given on a segment, $x_1 < x < x_2$, of the real line, and it is required to find $f(x)$ on the same segment, when the connection between these two functions is

$$g(x) = \int_{x_1}^{x_2} \Re(x, \eta) f(\eta) \, d\eta, \tag{26.19}$$

where $\Re(x, \eta)$ is the given *kernel* of the integral equation. Choose two sets $\{\mathfrak{A}_m(x); \ m = 1, 2, \ldots, \mathfrak{M}\}$ and $\{\mathfrak{B}_n(x); \ n = 1, 2, \ldots, \mathfrak{M}\}$ of what are here called *basis functions* (the optimum forms for these two sets of functions depend critically upon the envisaged application – this question can only be appreciated after detailed study, as is made clear in the references quoted in the introductory comments to this chapter). Next define

$$a_m = \int_{x_1}^{x_2} g(x) \mathfrak{A}_m(x) \, dx = \int_{x_1}^{x_2} \Re_m(\eta) f(\eta) \, d\eta, \tag{26.20}$$

$$\Re_m(\eta) = \int_{x_1}^{x_2} \Re(x, \eta) \mathfrak{A}_m(x) \, dx. \tag{26.21}$$

Then define

$$\Re_{mn} = \int_{x_1}^{x_2} \Re_m(\eta) \mathfrak{B}_n(\eta) \, d\eta, \tag{26.22}$$

so that (26.19) reduces to

$$a_m = \sum_{n=1}^{\mathfrak{M}} \Re_{mn} b_n \quad \text{for } n \in \{1, 2, \ldots, \mathfrak{M}\}, \tag{26.23}$$

with $f(x)$ having been approximated by

$$f(x) \simeq \sum_{n=1}^{\mathfrak{M}} b_n \mathfrak{B}_n(x), \tag{26.24}$$

where the integer $\mathfrak{M}$ is large enough that RHS (26.24) is an adequate (by whatever accuracy criterion is being adopted) representation of $f(x)$. The only unknowns in (26.23) are the b_n. Since (26.23) is a system of $\mathfrak{M}$, linear, inhomogeneous, algebraic equations in $\mathfrak{M}$ unknowns, the b_n can be recovered from it by matrix inversion – i.e. writing (26.23) in the shorthand notation

$$\mathbf{A} = \mathbf{K}\mathbf{B}, \tag{26.25}$$

where the components of the vectors $\mathbf{A}$ and $\mathbf{B}$ are the a_m and the b_n respectively, and $\mathbf{K}$ is a square matrix whose elements are the $\Re_{mn}$, the solution is

$$\mathbf{B} = \mathbf{K}^{-1} \mathbf{A}. \tag{26.26}$$

Having found sufficient of the $F_l(\rho)$ to resolve the image as faithfully as is required, by inverting the same number of the integral equations (26.18), the spectrum is reconstructed with the aid of the first equation in

(9.7) so that $f(x, y)$ can itself be reconstructed by Fourier transformation. So, the radiometric imaging problem is solved.

As already intimated, the radiometric imaging problem is 'disembodied' because it takes no account of any material medium in which the sources may be embedded. Such a medium usually attenuates the radiation. Striking examples of this are provided by the diagnostic science of *nuclear medicine* – typically, a patient is fed or injected with a dose of medication labelled with a radio-isotope in order to inspect, say, a defective organ or a malignant growth. The above formulation of the problem is then incomplete. Suppose that $\mu = \mu(x, y)$ is the *attenuation coefficient* per unit length of the medium for the radiation emitted by the sources. Then, re-examining the arguments leading to (26.7) and (26.8) shows that the total intensity received by the aforementioned detector is proportional to

$$q(\xi; \phi) = \int\limits_{-\infty}^{\infty}\int\int f(x', y')\Phi([(\xi' - \xi)^2/\Theta_1^2 + z^2/\Theta_2^2]/(\eta - \hat{r})^2)$$

$$\times \exp\left(-\int\limits_{\eta}^{\hat{r}} \mu(x'', y'') \, d\eta'\right) d\xi' \, d\eta \, dz/(\eta - \hat{r})^2, \qquad (26.27)$$

where x'' and y'' are the same as the x' and y' defined by (25.5), except that ξ' is replaced by ξ'' and η is replaced by η'.

The *embodied emission CT problem* – or *single-photon emission CT problem* (SPECT), as it is now called in medical contexts – is posed as: given $q(\xi; \phi)$, for $-\hat{r} < \xi < \hat{r}$ and $0 \le \phi < 2\pi$, reconstruct $f(x, y)$; on the understanding that $q(\xi; \phi)$ is defined by (26.27).

There is no known direct solution to the SPECT problem. An iterative solution is outlined in §32. There is a very approximate approach sometimes adopted in practice. It is predicated on being able to do two things: (a) make a meaningful estimate $\bar{\mu}$ of the average attenuation coefficient throughout the cross-section, and (b) physically measure (e.g. with callipers) the actual perimeter of the cross-section. One obtains from (b) the width $\mathrm{w}(\xi; \phi)$ of the cross-section along any ray parallel to the η-axis and distant ξ from it. At the same time one determines the η-coordinates, denoted here by $\eta_-(\xi; \phi)$ and $\eta_+(\xi; \phi)$, of the points where any such ray enters and leaves the cross-section. Note first that

$$\mathrm{w}(\xi; \phi) = \eta_+(\xi; \phi) - \eta_-(\xi; \phi), \qquad (26.28)$$

and second that the exponential factor in (26.27) can be rewritten:

$$\exp\left(-\int\limits_{\eta}^{\hat{r}} \mu(x'', y'') \, d\eta'\right) = \exp\left(-\int\limits_{\eta}^{\eta_+(\xi;\phi)} \mu(x'', y'') \, d\eta'\right). \qquad (26.29)$$

It is now convenient to define

$$q(\xi; \phi, \hat{r}) = q(\xi; \phi), \qquad (26.30)$$

where the latter is defined by (26.27), and

$$\chi(\xi', \xi, \eta, \hat{r}) = (\eta - \hat{r})^{-2} f(x', y') \Phi([(\xi' - \xi)^2/\Theta_1^2 + z^2/\Theta_2^2]/(\eta - \hat{r})^2).$$
$$(26.31)$$

When the detector array is placed with its face in the plane defined by $\eta = -\hat{r}$, and facing the cross-section, of course, the total intensity received by the detector whose beam axis is distant ξ from the η-axis is proportional to

$$q(\xi; \phi, -\hat{r}) = \iiint_{-\infty}^{\infty} \chi(\xi', \xi, \eta, -\hat{r}) \left(\exp\left(- \int_{\eta_-(\xi;\phi)}^{\eta} \mu(x'', y'') \, d\eta' \right) \right) d\xi' \, d\eta \, dz.$$
$$(26.32)$$

On invoking the aforementioned estimate for the average attenuation coefficient, the following approximation is introduced:

$$\exp\left(- \int_{\eta}^{\eta_+(\xi;\phi)} \mu(x'', y'') \, d\eta' \right) + \exp\left(- \int_{\eta_-(\xi;\phi)}^{\eta} \mu(x'', y'') \, d\eta' \right) \approx 1 - \bar{\mu} \mathfrak{w}(\xi; \phi).$$
$$(26.33)$$

It is finally assumed that the average of the intensities measured from opposite directions can be usefully interpreted in the following approximate fashion:

$$[q(\xi; \phi, \hat{r}) + q(\xi; \phi, -\hat{r})]/2[1 - \bar{\mu} \mathfrak{w}(\xi; \phi)]$$

$$\approx \iiint_{-\infty}^{\infty} [\chi(\xi', \xi, \eta, \hat{r}) + \chi(\xi', \xi, \eta, -\hat{r})] \, d\xi' \, d\eta \, dz. \qquad (26.34)$$

Inspection of (26.7) and (26.27) through (26.33) shows that

$$\text{RHS } (26.34) = [q(\xi; \phi, \hat{r}) + q(\xi; \phi, -\hat{r})]/2, \qquad (26.35)$$

where

$$q(\xi; \phi, \hat{r}) = q(\xi; \phi), \qquad (26.36)$$

so that the average of $q(\xi; \phi, \hat{r})$ and $q(\xi; \phi, -\hat{r})$ can be manipulated in virtually the same manner as previously described for $q(\xi; \phi)$ to yield a reconstruction of $f(x, y)$. The approximations leading up to (26.34) are very crude, but they nevertheless can provide useful results in practice. It is appropriate, therefore, to call

$$p_{\text{SPECT}}(\xi; \phi) = \text{RHS } (26.34) \qquad (26.37)$$

the SPECT-projection at angle ϕ.

27   The scattered CT problem – the echo-location problem

Suppose a parallel beam of radiation probes a body or a region of space and a separately positioned detector senses the scattered energy. If one is interested in a single cross-section (defined by $z = 0$) through the body or region, and if the radiation is characterized by straight rays, then there is no need to be concerned with other than the x, y-plane, within which the scattering coefficient (per unit volume) is here denoted by $f(x, y)$. Suppose also that it is intended to make sufficient measurements to permit a faithful image of the distribution of the scattering coefficient throughout the cross-section to be reconstructed. This situation relates to several important practical applications:

(i) Radar (bistatic as well as monostatic): when the region is homogeneous and the scatterers are isolated.

(ii) Sonar (bistatic as well as monostatic) and medical echography: when the inhomogeneities (e.g. variations in the acoustic/ultrasonic refractive index) of the region are slight and the scatterers are isolated. It must be noted, however, that wavefront distortion (equivalent to the rays being 'curved') is appreciable in perhaps most sonar-medical-ultrasonics applications (the ramifications of this for CT are discussed further in §32).

(iii) Compton-scatter imaging: when X-rays probe a body and the energy deflected by Compton-scattering is detected.

The scatterers are taken to be confined within a circle of radius $\hat{r}$ – i.e. (25.1) applies. The beam axis of a collimated source of radiation is directed parallel to the y-axis and distant X from it. The source is positioned outside the circle of radius $\hat{r}$. This source generates what is here called the *probing beam*. A radiation detector, also outside the circle, has its beam axis directed parallel to the negative ξ-axis and distant η from it (in this section the angle ϕ is taken as fixed, rather than variable as in §§25 and 26). The detector senses scattering from the point – called the *scattering point* whose x, y-coordinates are written (X, Y) – whose ξ, η-coordinates are $(X \sec(\phi) + \eta \tan(\phi), \eta)$. Note that

$$Y = X \tan(\phi) + \eta \sec(\phi). \tag{27.1}$$

Since the scattering is at an angle $(\pi/2 - \phi)$ away from the direction of the probing beam, the image should strictly be written as $f(x, y; \phi)$ rather than merely $f(x, y)$. The latter form is nevertheless retained here for both simplicity and continuity of notation, but the implicit dependence upon ϕ should not be forgotten. The point is that most scattering processes exhibit appreciable variations of scattering amplitude with scattering angle.

For an *ideal scatter-imaging* system then, the image is formed immediately. However, this neglects the attenuation through any medium within which the scatterers are embedded (in most instances, of course, the same medium is responsible for both the scattering and the attenuation). Such neglect is usually justified in radar applications. In many sonar and medical-ultrasonic-imaging applications the effects of attenuation can be accounted for to a useful degree of approximation by estimating *a priori* (like for the approximate approach to SPECT described in the latter part of §26) an average attenuation coefficient. Attenuation usually dominates Compton-scatter imaging systems.

Another factor overlooked in the ideal case is the resolution limit set by, first, the finite areas of the probing beam and the detectors, and second, the finite angular widths of both the probing and the detector beams. In Compton-scatter imaging the divergence of the probing beam can reasonably be neglected, but the detector beamwidths cannot be very narrow in practice because the major system limitation is likely to be low photon counts – i.e. a comparatively wide beam is needed to obtain a useful signal-to-noise ratio. The widths of radar and sonar beams can never be neglected in system calculations.

Attenuation is treated explicitly in this section. However, the resolution limitations are overlooked (they are implicit, of course, and can be taken into account in the ways previously discussed in §§25 and 26 – the analysis presented in §28 is also germane in this regard).

The attenuation coefficient (per unit length) of the radiation is denoted, as in §26, by $\mu = \mu(x, y)$. It is convenient to normalize to unity what the radiation intensity would be at the scattering point if μ was identically zero. So, when $\mu \neq 0$, this intensity becomes

$$\Im_s = \exp\left(- \int_{-\infty}^{Y} \mu(x, y)\, dy\right), \tag{27.2}$$

where the infinite lower limit merely implies that the probing beam traverses the whole cross-section, and

$$Y = x \tan(\phi) + \eta \sec(\phi). \tag{27.3}$$

The energy emanating from the scattering point spreads out as from a 'point source', so that it weakens with the inverse square of the distance as well as by attenuation. As in §26, therefore, the detected intensity depends upon the position of the detector, along the ray parallel to the negative ξ-axis and distant η from it. In order to be specific, a detector array is postulated with its face parallel to the η-axis and distant $\hat{r}$ from it (the beam axis of each detector is parallel to the negative ξ-axis). Consequently, the detected intensity sensed by the detector viewing the scattering point is found by multiplying $\Im_s$, as given by (27.2), by the scattering coefficient $f(x, y)$, the 'spreading' factor $1/(\hat{r} - \Xi)^2$ where

$$\Xi = X \sec(\phi) + \eta \tan(\phi), \tag{27.4}$$

and the attenuation

$$\exp\left(-\int_{\Xi}^{\infty}\mu(x, y)\, d\xi\right),$$

where the infinite upper limit merely implies that the scattered energy is detected outside the cross-section.

The detected intensity, sensed by the detector viewing the scattering point, can thus be written as

$$p_+(X, \eta; \phi) = (f(X, Y)/(\hat{r} - \Xi)^2)\exp\left(-\int_{-\infty}^{Y}\mu(x, y)\, dy + \int_{\Xi}^{\infty}\mu(x, y)\, d\xi\}\right),$$

$$(27.5)$$

where the reason for appending the subscript + is made clear below.

By varying the x-coordinate of the probing source and, for each position of the source, recording the outputs of all detectors in the array, values of $p_+(X, \eta; \phi)$ are obtained for all relevant values of X and η, for the one fixed value of ϕ. It transpires that this does not represent sufficient data to enable the image to be reconstructed. It is also necessary to place a probing source and a detector array opposite to the previous ones. The new probing source is positioned at some $y > \hat{r}$ with its beam axis directed parallel to the negative y-axis and distant X from it. Consider that detector in the new array which is positioned at $\eta = -\hat{r}$ with its beam axis parallel to the positive ξ-axis and distant η from it. The point both irradiated by the probing beam and viewed by the detector is the same as the previous scattering point. Furthermore, the scattering angle is the same, so that the scattering coefficient remains $f(X, Y; \phi) = f(X, Y)$. So, introducing a $-$ subscript to distinguish the new detected intensity from the previous one, it is seen that

$$p_-(X, \eta; \phi) = (f(X, Y)/(\hat{r} + \Xi)^2)$$

$$\times \exp\left(-\left\{\int_{\infty}^{Y}\mu(x, y)\, d(-y) + \int_{\Xi}^{-\infty}\mu(x, y)\, d(-\xi)\right\}\right), \quad (27.6)$$

where $d(-y)$ and $d(-\xi)$ emphasize that the probing beam and the scattered ray (this is the ray that, of all those scattered from (X, Y), is detected) travel in the negative y- and ξ-directions respectively.

The *scattered CT problem* – or the *echo-location problem* – is here posed as: given $p_\pm(X, \eta; \phi)$, for $-\hat{r} < X < \hat{r}$ and $-\hat{r} < \eta < \hat{r}$, for some fixed value of ϕ, reconstruct $f(x, y)$; on the understanding that $p_+(X, \eta; \phi)$ and $p_-(X, \eta; \phi)$ are defined by (27.5) and (27.6) respectively.

Since the position of the scattering point is known, the given data permit the calculation of

$$p(X, \eta; \phi) = p_+(X, \eta; \phi)p_-(X, \eta; \phi)(\hat{r}^2 - \Xi^2)^2. \quad (27.7)$$

Referring to (27.5) and (27.6), and recognizing that

$$\int_\infty^Y \mu(x, y)\, d(-y) = \int_Y^\infty \mu(x, y)\, dy \quad \text{and} \quad \int_\Xi^{-\infty} \mu(x, y)\, d(-\xi) = \int_{-\infty}^\Xi \mu(x, y)\, d\xi,$$

(27.8)

confirms that

$$p(X, \eta; \phi) = l(X; 0)l(-\eta; -\pi/2 + \phi)(f(X, Y))^2 \qquad (27.9)$$

where

$$l(\xi; \phi) = \exp\left(-\int_{-\infty}^\infty \mu(x, y)\, d\eta\right). \qquad (27.10)$$

The integral on RHS (27.10) is along the straight line parallel to the η-axis and distant ξ from it.

For an ideal scatter-imaging system (as defined earlier in this section), $l(\xi; \phi) \equiv 0$, so that

$$f(X, Y) = (p(X, \eta; \phi))^{\frac{1}{2}}, \qquad (27.11)$$

as follows from (27.9). As noted earlier, the scatter-imaging problem is then solved immediately – by analogy with §26, this should be called the *disembodied scattered CT problem* because the scatterers must hang in free space without the support of a material medium. This, of course, relates closely to the conventional radar situation.

Reference to §25 indicates that (27.10) can be rewritten as

$$l(\xi; \phi) = \exp(-p_\mu(\xi; \phi)), \qquad (27.12)$$

where $p_\mu(\xi; \phi)$ is a conventional projection through the material medium causing the attenuation. So, besides measuring the scattering from the probing beam with the aforementioned detector array, a further detector should be placed at some point $(X, y > \hat{r})$ with its beam axis directed along parallel to the negative y-axis and distant X from it. By moving this *ancillary detector* along with the probing beam, $l(X; 0)$ is obtained for $-\hat{r} < X < \hat{r}$. If conventional CT measurements are made along all relevant rays parallel to the ξ-axis, using a third probing beam and a second ancillary detector, then $l(-\eta; -\pi/2 + \phi)$ is obtained. It is therefore worthwhile posing the *embodied scattered CT problem* as: given $p_\pm(X, \eta; \phi)$, $l(X; 0)$ and $l(-\eta; -\pi/2 + \phi)$, for $-\hat{r} < X < \hat{r}$ and $-\hat{r} < \eta < \hat{r}$ for some fixed value of ϕ, reconstruct $f(x, y)$; on the understanding that $p_+(X, \eta; \phi)$, $p_-(X, \eta; \phi)$, and $l(\xi; \phi)$ are defined by (27.5), (27.6) and (27.10) respectively. Inspection of (27.7) and (27.9) reveals that $f(x, y)$ can be reconstructed immediately from the given data, because the value of Ξ can be calculated for each scattering point from the known geometry of the scatter-imaging system.

28   The synchronized CT problem

This section is concerned with imaging systems which inherently incorporate timing, or *synchronization,* information that permits 'immediate
localization' of sources or scatterers. The situations envisaged here are
similar to those examined in §27, in which the chief emphasis is on
devising means for compensating for the attenuation experienced by the
radiation on its passage through the cross-section being imaged. In this
section, however, the interest is in systems whose main limitation is poor
synchronization, due either to the data being incomplete or to an inherently low signal-to-noise ratio. Two types of system are discussed.

The first type of system is monostatic echo-location (e.g. radar or sonar)
in which the same antenna (or transducer) is used for both transmission
and reception. The scatterers are assumed to lie in the x, y-plane, and to
be embedded in a homogeneous medium (i.e. the standard radar situation, but not often applicable to sonar or medical echography). The
radiation can therefore be usefully approximated as bundles of straight
rays. The medium is taken to be effectively lossless. Even though there is
thus no attenuation of the rays, their *spreading loss* (i.e. decrease of
radiation intensity with inverse square of distance) must be taken into
account. It is now necessary to introduce the imprecise but important
concept of a *resolution cell,* which is the effective size of the smallest
region of the cross-section that can be isolated by the imaging system.
This size is fixed by the antenna beamwidth and the system range
resolution. The resolution is here supposed poor enough that there are
almost always several individual scatterers within a single resolution cell.
So, the amplitude of the detected radiation is sensitive to 'interference'
between the individual scattered signals. The distribution of scatterers
(more precisely, the distribution of *scattering amplitude*) throughout the
cross-section can therefore not only *not* be positive, it must in general be
complex. The distribution is denoted by $\mathfrak{f}(\mathbf{x})$, rather than $f(\mathbf{x})$, to emphasize this. The size of the image is conveniently normalized by requir-
ing that

$$\mathfrak{f}(x, y) = \mathfrak{f}(r; \theta) = 0 \qquad \text{for } r > \hat{r}, \qquad (28.1)$$

which is seen to be in keeping with (25.1).

To reconstruct $\mathfrak{f}(x, y)$ faithfully, a range of measurements is needed, of
course. A particular measurement is made with the centre of the antenna's radiating aperture (here called the *phase centre*) placed at the
point whose r, θ-coordinates are $(\hat{r}; \alpha)$, where α can have any value in the
range 0 to 2π. The antenna's beam axis (here referred to as the τ-axis)
can be swept so as to cover the whole cross-section. The angle, which
ranges from $-\pi/2$ to $\pi/2$, between the τ-axis and the straight line from
$(\hat{r}; \alpha)$ to the origin of the r, θ-coordinates is denoted by β.

Practical radar beamwidths are narrow enough that the form of the antenna beam (again denoted by Φ) can be simplified as in (26.6). Denoting by σ the Cartesian coordinate in the x, y-plane perpendicular to τ, with the origin of the σ, τ-coordinates at the antenna's phase centre, the beam can thus be reasonably represented by

$$\Phi = \Phi(\sigma, \tau) = \Phi([\sigma^2/\Theta_1^2 + z^2/\Theta_2^2]/\tau^2), \tag{28.2}$$

where z is the Cartesian coordinate perpendicular to the x, y-plane, and Θ_1 and Θ_2 are the effective beamwidths in planes parallel and perpendicular, respectively, to the x, y-plane.

Because signals from individual scatterers within a single resolution cell interfere, it is necessary to analyse actual signals (i.e. quantities that can possess phases, such as electric fields or currents) rather than powers or intensities. So, the spreading loss is proportional to the first, rather than the second, inverse power of the distance. Consider any point (σ', τ') – using the σ, τ-coordinates – within the resolution cell centred on the point $(0, \tau)$. The inherent resolution of almost any practical echo-location system is such that the amplitude of the transmitted signal at (σ', τ') can be assumed proportional to Φ/τ, which accords with the accuracy implied by the representation (28.2) for the antenna beam. The signal scattered back to the antenna from a point scatterer, of unit scattering amplitude and situated at (σ', τ'), is then proportional to $(\Phi/\tau)^2$. The *range resolution* (i.e. the resolution of the system in the τ-direction) is conveniently characterized by a one-dimensional psf $h(\tau)$. Consequently, the signal received by the antenna from all the scatterers within the resolution cell centred at $(0, \tau)$ is proportional to

$$s(\tau; \beta; \alpha) = \int\!\!\!\int\!\!\!\int_{-\infty}^{\infty} f(x, y)(\Phi(\sigma', \tau'))^2 h(\tau - \tau') \, d\sigma' \, d\tau' \, dz/\tau^2, \tag{28.3}$$

where the x, y and σ', τ'-coordinates are related by

$$\begin{aligned} x &= \sigma' \sin(\beta - \alpha) - \tau' \cos(\beta - \alpha) + \hat{r} \cos(\alpha), \\ y &= \sigma' \cos(\beta - \alpha) + \tau' \sin(\beta - \alpha) + \hat{r} \sin(\alpha). \end{aligned} \tag{28.4}$$

As already pointed out, the kind of echo-location system being examined here comes closest in practice to radar. Now, radar *targets* (the technical name for radar scatterers) are usually moving, which means that the aforementioned range of measurements must be completed before the form of $f(x, y)$ can change. For a single value of α, the intervals $0 < \tau < 2\hat{r}$ and $-\pi/2 < \beta < \pi/2$ can together be covered rapidly enough by employing electronic beam-steering (e.g. phased-array radar). It is quite impractical, however, to think of disposing electronically-scanned antennas all around the circle of radius $\hat{r}$; but there could be a few antennas, say N of them, positioned at particular points on this circle. Accordingly, the *first syn-chronized CT problem* – or the *low-resolution monostatic echo-location problem* – is here posed as: given $s(\tau; \beta; \alpha)$, for $0 < \tau < 2\hat{r}$ and $-\pi/2 < \beta < \pi/2$, and for a discrete set $\{\alpha_n; n = 1, 2, \ldots, N\}$ of values of α, reconstruct $f(x, y)$; on the understanding that $s(\tau; \beta; \alpha)$ is defined by (28.3).

Because the targets lie only in the cross-section being imaged, the z-integration in (28.3) can be carried out immediately, provided the beam shape is given *a priori* (which it always is in practice because the antenna beam is measured before any system is put into operation). By analogy with (26.10) it is convenient to define

$$Y(t) = \int_{-\infty}^{\infty} (\Phi(t^2/\Theta_1^2 + \zeta^2/\Theta_2^2))^2 \, d\zeta. \tag{28.5}$$

Although the two definitions of $Y(t)$ differ, it is pointless introducing new notation because no confusion need result. Similar manipulations to those by which (26.7) is transformed into (26.9) permits (28.3) to be re-expressed, with the aid of (28.4) and (28.5), as

$$s(\tau; \beta; \alpha) = \int_{-\infty}^{\infty} \int Y(t)$$

$$\times f(\tau t \sin(\beta - \alpha) - \tau' \cos(\beta - \alpha) + \hat{r} \cos(\alpha), \tau t \cos(\beta - \alpha)$$
$$+ \tau' \sin(\beta - \alpha) + \hat{r} \sin(\alpha)) h(\tau - \tau') \, dt \, d\tau'. \tag{28.6}$$

In the ideal case, when the resolution is perfect, implying that

$$Y(t) = h(t) = \delta(t), \tag{28.7}$$

each resolution cell becomes infinitesimal, so that (28.6) reduces to

$$s(\tau; \beta; \alpha) = f(\hat{r} \cos(\alpha) - \tau \cos(\beta - \alpha), \hat{r} \sin(\alpha) + \tau \sin(\beta - \alpha)). \tag{28.8}$$

Therefore, the complete image is immediately recovered by measuring $s(\tau; \beta; \alpha)$ throughout the intervals $0 < \tau < 2\hat{r}$ and $-\pi/2 < \beta < \pi/2$, for any single value of α.

When (28.7) is not a reasonable representation of reality, as it rarely is in practice, one has to resort to numerical estimation procedures. It is sometimes useful to model $f(x, y)$ as a discrete assemblage of point targets, e.g.

$$f(x, y) = \sum_{j=1}^{J} A_j \delta(x - x_j) \delta(y - y_j), \tag{28.9}$$

where the positive integer J and the scattering amplitudes A_j and positions (x_j, y_j) of the individual targets are to be estimated from the given data. Knowing the values of Θ_1 and the effective width of a resolution cell, each $\mathfrak{s}(\tau; \beta; \alpha_n)$ can be regarded as K, say, independent pieces of data. So, there are NK independent pieces in all. On substituting (28.9) into (28.6), for all N values of α_n, all NK pieces of data can be combined into a single column vector, $\mathbf{D}$ say. While $\mathbf{D}$ is linearly related to the A_j, it is non-linearly related to the (x_j, y_j), which means that one of the many existing non-linear programming 'search' techniques must be invoked if it is to be possible to estimate the unknowns on RHS (28.9)

The second type of system examined in this section is *positron-emission tomography* (PET) which images finite distributions of what are here called *pe-sources*. These sources generate *positron–electron pairs* (pep) that spontaneously 'annihilate'. Each annihilation *event* produces two photons which are emitted in diametrically opposed directions (to conserve both energy and momentum). Coincidence circuitry not only identifies the lines through the cross-section along which the photons travel, it is also capable of estimating where on these lines the events take place. Since pe-sources are indisputably spatially incoherent, it is convenient to return to the symbol f employed in §§25 through 27, so that the symbol $\mathfrak{f}$ used previously in this section can be discarded. The distribution of pe-sources over the cross-section is therefore denoted by $f(x, y)$, whose size is again conveniently normalized by invoking (25.1). It is perfectly feasible in practice to surround the cross-section with photon detectors, all pairs of which are interconnected with coincidence circuits. So, it makes sense to postulate a 'complete' set of measurements. It must be emphasized, however, that all these measurements are always significantly noisy in practice.

Consider a pep existing at the point whose σ, τ-coordinates are $(0, \tau)$. Suppose that when this pep annihiliates it emits photons which travel along the positive and negative τ-axes and are detected at the points whose coordinates, in the r, θ-system, are $(\hat{r}; \alpha)$ and $(\hat{r}; \alpha + \pi - 2\beta)$. If pe-sources are embedded in a material medium possessing a photon attenuation coefficient (per unit length of travel) $\mu(x, y)$, the probability that either photon is individually detected is proportional to

$$\exp\left(\pm \int_{\pm\infty}^{\tau} \mu(x, y) \, d\tau' \right),$$

where the $+$ sign refers to one photon and the $-$ sign to the other, and where the infinite lower limit on the integral merely emphasizes that the photons have to escape from the material medium before they can be detected. The x, y-coordinates are related to the coordinate τ' by (28.4), but with $\sigma' = 0$, i.e.

$$x = \hat{r} \cos(\alpha) - \tau' \cos(\beta - \alpha) \quad \text{and} \quad y = \hat{r} \sin(\alpha) + \tau' \sin(\beta - \alpha)$$
$$(28.10)$$

The probability that both photons are detected simultaneously (here called *coincidence-detection*) is proportional to the product of the individual probabilities, i.e. it is proportional to

$$l(;\beta;\alpha) = \exp\left(-\int_{-\infty}^{\infty} \mu(x, y)\, d\tau'\right) = \exp(-p_\mu(;\beta;\alpha)), \quad (28.11)$$

where $p_\mu(;\beta;\alpha)$ is a conventional projection (see §9) through the material medium causing the attenuation, along the ray coincident with the τ-axis; and where $l(\cdot)$ is analogous to the quantity identified by the same symbol in §27. So, besides having the capability for detecting the photons emitted by pe-sources in all directions in the x, y-plane, a PET system should also be instrumented so that preliminary measurements of $p_\mu(;\beta;\alpha)$ can be made for $-\pi/2 < \beta < \pi/2$ and $0 < \alpha < 2\pi$. The radiation employed for the preliminary measurements should consist of photons of the same energy as those emitted by the pe-sources.

As pointed out previously, the aforementioned coincidence circuitry estimates where on the straight line the annihilation event takes place. However, this estimation tends to be rather crude in practice, and it is characterized by a broad impulse response. Since photons of the energy employed in PET systems always travel at virtually the speed of light, the impulse response can be represented by a one-dimensional psf, which is here denoted by $h(\cdot)$. So, the total probability of coincidence-detection of events, originating within the volume element centred at $(0, \tau)$ and emitting photons along the positive and negative τ-axes, is proportional to the density $f(x, y)$ of the pe-sources in the neighbourhood of $(0, \tau)$ and to the attenuation factor $l(;\beta;\alpha)$ defined by (28.11), all blurred by the psf. Since detection probabilities are directly proportional to the photon fluxes incident upon detectors (provided the latter do not approach saturation, which they are very unlikely to in a practical PET system), the measured output of the coincidence circuitry, connected to the detectors sensing photons travelling along the positive and negative τ-axes, is proportional to

$$q(\tau;\beta;\alpha) = l(;\beta;\alpha) \int_{-\infty}^{\infty} h(\tau - \tau')$$

$$\times f(\hat{r}\cos(\alpha) - \tau'\cos(\beta - \alpha), \hat{r}\sin(\alpha) + \tau'\sin(\beta - \alpha))\, d\tau'. \quad (28.12)$$

The physical reason for the factor $l(;\beta;\alpha)$ being outside the integral is that there is no preferred direction for the emission of photons by pep.

The *second synchronized CT problem* – or the *PET problem* – is here posed as: given $l(;\beta;\alpha)$ and $q(\tau;\beta;\alpha)$, for $0 < \tau < 2\hat{r}$, $-\pi/2 < \beta < \pi/2$ and $0 < \alpha < 2\pi$, reconstruct $f(x, y)$; on the understanding that $l(;\beta;\alpha)$ and $q(\tau;\beta;\alpha)$ are defined by (28.11) and (28.12) respectively; it is further understood that the given $q(\tau; \beta; \alpha)$ has an inherently low signal-to-noise ratio.

The first step in the solution to the PET problem is to form the quantities

$$\tilde{q}(\tau; \beta; \alpha) = (q(\tau; \beta; \alpha))/l(; \beta; \alpha), \tag{28.13}$$

which represent what the coincidence circuits would measure if there was no attenuation due to the medium in which the pe-sources are embedded. The point is that the aforesaid preliminary measurements allow the effect of the attenuation to be removed.

The quantity $\tilde{q}(\tau; \beta; \alpha)$, for particular fixed values of the parameters τ, β, and α, is characterized by a single ray through the cross-section. It is now convenient to take the $\tilde{q}(\tau; \beta; \alpha)$, for all available values of the parameters, and re-order them, collecting those which, for a particular value of the angle ϕ introduced in the second paragraph of §9, are characterized by rays parallel to the η-axis. Repeating this for the other values of ϕ, the $\tilde{q}(\tau; \beta; \alpha)$ are thus rearranged into sets which closely correspond to conventional projections (see §§9 and 25). This kind of rearrangement (often called *rebinning*) of data into sets characterized by parallel rays is adopted in some commercial conventional-CT machines that actually measure 'fan-beam projections' (see §25) – the point is that, while fan-beams can be more convenient measurement-wise, parallel-beams have certain computational advantages (see §33).

Invoking the ξ, η-coordinates introduced in §9, and noting that the position of the origin of the τ-axis is unimportant, in principle, in so far as the integral in (28.12) is concerned because the photons must escape from the cross-section before they are detected, it is seen that the re-ordering of the $\tilde{q}(\tau; \beta; \alpha)$ can be expressed by

$$\tilde{p}(\tau, \xi; \phi) = \tilde{q}(\tau; \beta; \alpha)$$

$$= \int_{-\infty}^{\infty} h(\tau - \eta) f(\xi \cos(\phi) - \eta \sin(\phi), \xi \sin(\phi) + \eta \cos(\phi)) \, d\eta,$$

$$\tag{28.14}$$

where the symbol $\tilde{p}$ is introduced by analogy with the symbol p employed in §9 for the conventional projection – the superscript tilde is appended because a further projection-like quantity has to be introduced, in (28.16) below. The η-axis is chosen parallel to the τ-axis, so that

$$\phi = \alpha - \beta. \tag{28.15}$$

On defining

$$p(\omega, \xi; \phi) = \mathsf{F}_{(1\omega)}\{\tilde{p}_{(\tau)}(\tau, \xi; \phi)\}, \tag{28.16}$$

and appealing to the convolution theorem (7.3), the Fourier transform of (28.14) with respect to τ becomes

$$p(\omega, \xi; \phi) = H(\omega) \int_{-\infty}^{\infty} \exp(i\, 2\pi\omega\eta)$$

$$\times f(\xi \cos(\phi) - \eta \sin(\phi), \xi \sin(\phi) + \eta \cos(\phi))\, d\eta, \qquad (28.17)$$

where $H(\omega) = \mathbf{F}\{h(\tau)\}$.

There are two interesting limiting cases. The first, for which $h(\tau) = \delta(\tau)$, characterizes *ideal PET* because RHS (28.14) then reduces to $f(X, Y)$, where $X = (\xi \cos(\phi) - \tau \sin(\phi))$ and $Y = (\xi \sin(\phi) + \tau \cos(\phi))$, so that the PET measurements give the image directly provided sufficiently wide ranges of τ, β, and α are covered during the measurements. The second limiting case occurs when the range resolution is negligible, i.e. $h(\tau) = 1$ which implies that $H(\omega) = \delta(\omega)$. Inspection of (28.17) then shows that $p(\omega, \xi; \phi)$ only exists for $\omega = 0$, in which case $p(\omega, \xi; \phi)$ is equivalent to a conventional projection $p(\xi; \phi)$, as defined in §9. This is the rationale behind the choice of the notation $p(\omega, \xi; \phi)$, which is appropriately called a *PET-projection*.

By analogy with (9.6) it is instructive to define

$$F(\omega, \rho; \phi) = \mathbf{F}_{(1\rho)}\{p_{(\xi)}(\omega, \xi; \phi)\}, \qquad (28.18)$$

which, when applied to (28.17), gives

$$F(\omega, \rho; \phi) = H(\omega) \iint_{-\infty}^{\infty} \exp(i\, 2\pi[\omega\eta + \rho\xi])$$

$$\times f(\xi \cos(\phi) - \eta \sin(\phi), \xi \sin(\phi) + \eta \cos(\phi))\, d\xi\, d\eta$$

$$= H(\omega) \iiiint_{-\infty}^{\infty} F(u, v)$$

$$\times \exp(i\, 2\pi\{[\rho - u \cos(\phi) - v \sin(\phi)]\xi$$

$$+ [\omega + u \sin(\phi) - v \cos(\phi)]\eta\})\, du\, dv\, d\xi\, d\eta$$

$$= H(\omega) \iiiint_{-\infty}^{\infty} F(u' \cos(\phi) - v' \sin(\phi), u' \sin(\phi) + v' \cos(\phi))$$

$$\times \exp(i\, 2\pi\{[\rho - u']\xi + [\omega - v']\eta\})\, du'\, dv'\, d\xi\, d\eta, \qquad (28.19)$$

where (6.11) and (6.12) have been invoked and the u', v'-coordinates are defined in terms of the u, v-coordinates by

$$u' = u \cos(\phi) + v \sin(\phi) \quad \text{and} \quad v' = -u \sin(\phi) + v \cos(\phi). \qquad (28.20)$$

With the aid of (6.3) the ξ and η integrations can be effected immediately in (28.19), which then reduces to

$$F(\omega, \rho; \phi) = H(\omega) \int\!\!\int_{-\infty}^{\infty} F(u' \cos(\phi) - v' \sin(\phi), u' \sin(\phi) + v' \cos(\phi))$$

$$\times \delta(u' - \rho)\, \delta(v' - \omega)\, du'\, dv'$$

$$= H(\omega) F(\rho \cos(\phi) - \omega \sin(\phi), \rho \sin(\phi) + \omega \cos(\phi)),$$

$$= H(\omega) F((\rho^2 + \omega^2)^{\frac{1}{2}}; \phi + \psi), \tag{28.21}$$

where $F(\cdot)$ has been re-expressed as a function of polar variables, (6.5) through (6.8) have been used and

$$\psi = \arc\tan(\omega/\rho). \tag{28.22}$$

It is emphasized that RHS (28.21) is the spatial frequency content of a single PET-projection, whereas $F(\rho; \phi)$ is the corresponding content of a single conventional projection. So, while a conventional projection merely relates to a single line through the origin of Fourier space, a PET-projection is characterized by two sectors whose widths are the effective width of $H(\omega)$. The information to be derived from successive PET-projections is thus seen to overlap in Fourier space, thereby introducing redundancy that improves the overall signal-to-noise ratio – i.e. at each point $(\rho; \phi)$ one is provided with several estimates of $F(\rho; \phi)$ which can be averaged to obtain a better final estimate. There is another intriguing possibility. Since each $F(\omega, \rho; \phi)$ spans an area of Fourier space, rather than merely a line as in the conventional case, point samples of $F(\rho; \phi)$ can be constructed on a rectangular grid. This allows the image to be formed by direct Fourier reconstruction without encountering the interpolatory difficulties experienced with conventional CT (refer to §§25 and 33).

29 Discrete projections

In practice it is, of course, only possible to measure a finite number of projections. As in §14, therefore, it is important to distinguish between the true image $f(\mathbf{x})$ and the best possible version of it, denoted here by $\hat{f}(\mathbf{x})$, which can be reconstructed from the given data. It is also necessary to distinguish between $F(\mathbf{u})$ and the spectrum, denoted by $\hat{F}(\mathbf{u})$, of $\hat{f}(\mathbf{x})$.

The *discrete projection problem* is here posed as: given $p(\xi; \phi_n)$, for $-\hat{r} < \xi < \hat{r}$ and $0 \le \phi_n < \pi$, with $n \in \{1, 2, \ldots, N\}$ and all ϕ_n different, reconstruct $\hat{f}(\mathbf{x})$; on the understanding that $p(\xi; \phi)$ is defined by (9.4).

It is standard image reconstruction practice to form $\hat{f}(\mathbf{x})$ by subjecting the N given projections to the modified-back-projection procedure (see

§33). It is therefore important to answer the general question: how degraded can $\hat{f}(\mathbf{x})$ be expected to be in comparison with $f(\mathbf{x})$? The answer is conveniently couched in terms of the spatial frequency content of $\hat{f}(\mathbf{x})$. This content is characterized by the circle, of radius $\hat{\rho}$ say, in Fourier space outside of which $\hat{F}(\rho; \phi)$ is negligible, i.e.

$$\hat{F}(\rho; \phi) \simeq 0 \qquad \text{for } \rho > \hat{\rho}. \tag{29.1}$$

The projection theorem (9.5) shows that $p(\xi; \phi_n)$ characterizes $F(\rho; \phi)$ along the straight line through the origin of Fourier space inclined at the angle ϕ_n to the u-axis. So, the given projections characterize Fourier space on (irregularly spaced, in general) 'spokes', each of which corresponds to one of the ϕ_n (and to $(\phi_n + \pi)$, of course). The sampling of Fourier space is conditioned by the furthest separation of spokes on the circle $\rho = \hat{\rho}$. When the projections are equally spaced, e.g.

$$\phi_n = n\pi/N, \tag{29.2}$$

this separation is $\hat{\rho}\pi/N$. Strict reliance on the sampling theorem (§10) then demands that $2\hat{r}$ be no more than $N/\hat{\rho}\pi$. Most of Fourier space is sampled more closely than $\hat{\rho}\pi/N$, of course. In fact, the sample spacing is proportional to ρ. It has therefore become accepted that the effective overall sample spacing is roughly $\hat{\rho}\pi/2N$ – refer back to (ii) in §25. This implies that

$$\hat{r} \le N/\hat{\rho}\pi. \tag{29.3}$$

From a practical viewpoint, the discrete projection problem is probably the most important of all theoretical CT problems. Since 'resolution' is central to image quality, it is worthwhile confirming (29.3) by an alternative line of reasoning.

It is convenient to express $p(\xi; \phi)$ as a Fourier series, as in (9.7). Because of (9.15), the N given projections are sufficient to allow $2N$ of the $p_l(\xi)$ to be estimated throughout $-\hat{r} < \xi < \hat{r}$, where the size of the image is set by (25.1). On (arbitrarily) choosing the integer l to run from $(-N+1)$ to N, the given projections are assumed to be interpolatable throughout $0 \le \phi < 2\pi$ according to the formula

$$p(\xi; \phi) \simeq \sum_{l=-N+1}^{N} p_l(\xi) \exp(\mathrm{i}\, l\phi), \tag{29.4}$$

where, for each $n \in \{1, 2, \ldots, N\}$ and all ξ throughout $-\hat{r} < \xi < \hat{r}$,

$$p(\xi; \phi_n) = \sum_{l=-N+1}^{N} p_l(\xi) \exp(\mathrm{i}\, l\phi_n). \tag{29.5}$$

The $p_l(\xi)$ are found by treating (29.5) as a system of $2N$ linear, in-homogeneous, algebraic equations (for each value of ξ) which are solved by (for example) matrix inversion. If the ϕ_n are equally spaced, so that (29.2) applies, the $p_l(\xi)$ can be straightforwardly evaluated with the aid of the D.F.T. (and its efficient algorithmic realization, the F.F.T. – refer to §12).

On approximating the reconstructable image by a Fourier series similar to that introduced in (29.4), i.e.

$$\hat{f}(r; \theta) \simeq \sum_{l=-N+1}^{N} \hat{f}_l(r) \exp(i\, l\theta), \tag{29.6}$$

it follows from (9.7), (9.8), (9.10) and (9.13), with F replaced by $\hat{F}$, and from (29.1) that

$$\hat{f}_l(r) = 2\pi(-i)^l \int_0^{\hat{\rho}} \mathbf{F}_{(1\rho)}\{p_l(\xi)\} J_l(2\pi\rho r)\rho \, d\rho, \tag{29.7}$$

where the notation introduced in (6.17) has been adapted here (in an obvious manner) for functions of a single variable.

The Bessel function $J_\nu(x)$ is approximately proportional to a sinusoid of the real, positive variable x divided by $x^{\frac{1}{2}}$, when $x > |\nu|$. For $x < |\nu|$, with ν real, $J_\nu(x)$ tends monotonically and rapidly to zero as x decreases. Consequently, RHS (29.7) can be expected to be small in general for any values of $|l|$ which happen to exceed the largest permitted value of $2\pi\rho r$. Inspection of (25.1), (29.4), and (29.7) indicates that the maximum values of r, $|l|$ and ρ are $\hat{r}$, N, and $\hat{\rho}$ respectively. It follows that RHS (29.7) can in general be expected to have significant value for all $l \in \{-N+1, -N+2, \ldots, N\}$ only if

$$N - \tilde{n} < 2\pi\hat{\rho}\hat{r}, \tag{29.8}$$

where $\tilde{n}$ is an integer large enough to ensure that $J_N(2\pi\hat{\rho}\hat{r})$ truly is small, according to whatever criterion is appropriate in any particular application. Irrespective of the latter, $\tilde{n}$ is much less than N (when the latter is large), so that the limit on $2\hat{r}$ is roughly $N/\hat{\rho}\pi$. Since $J_N(2\pi\hat{\rho}r)$ tends rapidly towards zero as r decreases, when $r/\hat{r}$ is less than roughly $(1 - 2/N)$, it is reasonable to double the limit on $\hat{r}$. So, (29.3) is once again obtained.

Whenever the projection data are imperfect (for whatever reasons – refer also to §§30 through 32), the available data can be represented in Fourier space by

$$\hat{F}(\mathbf{u}) = F(\mathbf{u})\Im(\mathbf{u}), \tag{29.9}$$

where $\Im(\mathbf{u})$ is here called the *(spectral) imperfection factor*. It is appropriate to call its Fourier transform the *imperfect delta function* $\tilde{\delta} = \tilde{\delta}(\mathbf{x})$, i.e.

$$\tilde{\delta} = \mathbf{F}\{\Im\}. \tag{29.10}$$

It then follows from the convolution theorem (7.3) that

$$\hat{f}(\mathbf{x}) = f(\mathbf{x}) \odot \bar{\delta}(\mathbf{x}). \tag{29.11}$$

It is convenient to normalize $\Im(\mathbf{u})$ such that

$$\Im(0) = 1, \tag{29.12}$$

so that $\bar{\delta}(\mathbf{x}) \to \delta(\mathbf{x})$ when the data are perfect, i.e. when $\Im(\mathbf{u}) \to 1$.

In practice, the most obvious manifestation of incompleteness of the given projection data is the appearance of multiple 'streaks' across the whole extent of $\hat{f}(\mathbf{x})$. When (29.3) holds, this 'streaking' is usually unobjectionable. It is worth emphasizing that the streaks are due to the 'holes' in $\Im(\mathbf{u})$ – i.e. the fact that $\Im(\mathbf{u})$ is effectively zero throughout an appreciable part of the region of Fourier space for which $\rho < \hat{\rho}$.

By evaluating RHS (29.4) for values of ϕ different from any of the given ϕ_n, projections other than the given ones are estimated. These are appropriately called *interpolated projections*. Making use of the latter modifies $\Im(\mathbf{u})$, and consequently $\bar{\delta}(\mathbf{x})$, but there is no unequivocal evidence for any significant improvement in resolution, although the visual appearance of $\hat{f}(\mathbf{x})$ is sometimes enhanced thereby.

Even though much effort has been expended on devising interpolation schemes, most of the results have been disappointing. It is usually well nigh impossible to 'fill in' the 'holes' in $\Im(\mathbf{u})$. Interpolation only seems to be successful when 'strong' *a priori* information is available and can be readily incorporated into the reconstruction procedure. A useful example occurs when it is known that part of the image is of uniform intensity. Suppose, for instance, that $f(\mathbf{x})$ is given by (21.1), under the constraint (21.4), on the understanding that

$$h(\mathbf{x}) = \bar{h} = \text{a constant} \tag{29.13}$$

throughout Ω_h. It is convenient here to let the shapes of the frames Ω_g and Ω_h be arbitrary. The information implicit in (29.13) is useful even if the size of Ω_h is not given. It often transpires that the boundary of Ω_h can be visually recognized, with useful precision, even when the data are so incomplete that $\hat{f}(\mathbf{x})$ is contaminated by severe 'streaking'. An estimate $\hat{h}$ of $\bar{h}$ can then be obtained by integrating $\hat{f}(\mathbf{x})$ throughout the interior of (the estimate of) Ω_h and dividing by its area. There is no rigorous explanation of this, of course, but the following argument suggests why it should not perhaps be unexpected.

The difference between $\hat{h}$ and $\bar{h}$ tends to decrease as Ω_h gets larger. So, it is useful to think of Ω_h occupying an empty pocket in Ω_g – remember that Ω_g and Ω_h are physically separated because they satisfy (21.4). The phase of $(G(\mathbf{u})H^*(\mathbf{u}))$ must vary appreciably with $\mathbf{u}$, therefore, and the more so the larger is w_h, which is the narrowest width of the convex envelope of Ω_h. On referring to the definition (7.17) it is seen that similar phase variations can be expected for $(G(\mathbf{u})\delta_h^*(\mathbf{u}))$. It follows that the quantity

$$\varepsilon = \int\!\!\int\limits_{-\infty}^{\infty} G(\mathbf{u})\,\delta_h^*(\mathbf{u})\,d\Sigma(\mathbf{u}) \qquad (29.14)$$

is likely to be small in some meaningful sense (because the oscillations of the integrand must lead to a fair amount of cancellation). Now note from (6.13), (7.3), (7.17), (29.9) and (29.11) that

$$\int\!\!\int\limits_{\Omega_h} \hat{f}(\mathbf{x})\,d\sigma(\mathbf{x}) = \int\!\!\int\limits_{\Omega_h} \int\!\!\int\limits_{-\infty}^{\infty} F(\mathbf{u})\Im(\mathbf{u})\exp(-\mathrm{i}\,2\pi\mathbf{u}\cdot\mathbf{x})\,d\Sigma(\mathbf{u})\,d\sigma(\mathbf{x})$$

$$= \int\!\!\int\limits_{-\infty}^{\infty} F(\mathbf{u})\Im(\mathbf{u})\,\delta_h^*(\mathbf{u})\,d\Sigma(\mathbf{u}). \qquad (29.15)$$

Substituting the Fourier transform of (21.1) into (29.15) gives

$$\int\!\!\int\limits_{\Omega_h} \hat{f}(\mathbf{x})\,d\sigma(\mathbf{x}) = \varepsilon' + \int\!\!\int\limits_{-\infty}^{\infty} H(\mathbf{u})\,\delta_h^*(\mathbf{u})\Im(\mathbf{u})\,d\Sigma(\mathbf{u}), \qquad (29.16)$$

where

$$\varepsilon' = \int\!\!\int\limits_{-\infty}^{\infty} G(\mathbf{u})\,\delta_h^*(\mathbf{u})\Im(\mathbf{u})\,d\Sigma(\mathbf{u}). \qquad (29.17)$$

The 'holes' in $\Im(\mathbf{u})$ must in general prevent there being as much cancellation in the integrand of RHS (29.17) as there is in the integrand of RHS (29.14). Nevertheless, $|\varepsilon'|$ can often be expected to be significantly smaller than the magnitude of the second term on RHS (29.16). Inspection of (7.17) and (29.13) reveals that

$$H(\mathbf{u}) = \bar{h}\delta_h(\mathbf{u}). \qquad (29.18)$$

so that (29.16) reduces (approximately) to

$$\int\!\!\int\limits_{\Omega_h} \hat{f}(\mathbf{x})\,d\sigma(\mathbf{x}) \simeq \bar{h} \int\!\!\int\limits_{-\infty}^{\infty} |\delta_h(\mathbf{u})|^2\,\Im(\mathbf{u})\,d\Sigma(\mathbf{u}). \qquad (29.19)$$

The energy conservation theorem (6.22) and the definition (7.17) show that

$$\int\!\!\int\limits_{-\infty}^{\infty} |\delta_h(\mathbf{u})|^2\,d\Sigma(\mathbf{u}) = \mathfrak{A}_h = \text{area of } \Omega_h. \qquad (29.20)$$

Unless the convex envelope of Ω_h is circular, the main lobe of $\delta_h(\mathbf{u})$ is asymmetric. Its maximum effective width is $1/w_h$, because (as has already been stated) w_h is the narrowest width of the convex envelope of Ω_h. This means that $|\delta_h(\mathbf{u})|^2$ is small for $|\mathbf{u}| > 1/\omega_h$. When N equispaced projections are given, the sample spacing on the circle of radius $1/\omega_h$ centred on the origin of Fourier space is roughly π/Nw_h. To satisfy the sampling theorem, the spacing should not exceed $1/w_h$ – refer to §10. It is thus seen that n need only be 4. It follows that the integral on RHS (29.19) can effectively reduce to that on LHS (29.20) even when the 'holes' in $\Im(\mathbf{u})$ are comparatively huge.

The previously introduced quantity $\hat{\bar{h}}$ is defined by

$$\hat{\bar{h}} = (1/\hat{\mathfrak{A}}_h) \iint_{\hat{\Omega}_h} \hat{f}(\mathbf{x}) \, d\sigma(\mathbf{x}), \tag{29.21}$$

where $\hat{\mathfrak{A}}_h$ is the area of the visual estimate $\hat{\Omega}_h$ of Ω_h. If $N > 3$, one can conclude that $\hat{\bar{h}}$ is close to $\bar{h}$, provided $\hat{\Omega}_h$ is a reasonable estimate of Ω_h.

30 Incomplete projections

It is sometimes either impossible, inconvenient, or undesirable to gather projection data over the whole cross-section of a body. There are all sorts of reasons for this. For example, the species of particle beam (or wave motion, or whatever) being employed to measure the projections may not be able to penetrate part of the body. Another instance arises when ionizing radiations (e.g. X-rays) are used for repeated probing of biological tissues (e.g. people). It can then be advantageous to reduce the radiation dose by confining the beam to the part of the cross-section which is of chief concern, provided this does not prevent an acceptable image being reconstructed.

The incomplete projection $p_{\text{inc}}(\xi; \phi)$ is defined by

$$
\begin{aligned}
p_{\text{inc}}(\xi; \phi) &= p(\xi; \phi) && \text{for } \xi \in \mathbf{L}^+(\phi), \\
&= 0 && \text{for } \xi \in \mathbf{L}^-(\phi),
\end{aligned}
\tag{30.1}
$$

where $\mathbf{L}^+(\phi)$ identifies the values of ξ for which projection data are gathered. It is to be understood that (25.1) defines the size of the cross-section, and that $\mathbf{L}^+(\phi)$ and $\mathbf{L}^-(\phi)$ are non-intersecting segments of the ξ-axis. The union of $\mathbf{L}^+(\phi)$ and $\mathbf{L}^-(\phi)$ spans the interval $-\hat{r} < \xi < \hat{r}$.

In many applications, the projections are sampled (rather than recorded continuously) along the ξ-axis. So, the notations $\mathbf{L}^+(\phi)$ and $\mathbf{L}^-(\phi)$ could be invoked to make this explicit. As remarked in (iii) of §25, however, this is not usually a major practical limitation, and it is not mentioned any further here.

Two classes of incomplete projections are discussed in detail in this section. The first is the class of *hollow projections*, which are here characterized by

$$\mathbf{L}^+(\phi) = \mathbf{L}^+ \sim r^+ < |\xi| < \hat{r}, \tag{30.2}$$

where r^+ is a real, positive constant and $\sim$ here means 'corresponds to'. There are no measured data in the middle of the projections, which is, of course, why they are called hollow. Provided the $p_{inc}(\xi; \phi)$ are given for enough values of ϕ to satisfy the sampling constraints applying in any particular application, $f(r, \theta)$ can in principle be reconstructed perfectly for $r^+ < r < \hat{r}$, as is explained below.

The resolution of the image is necessarily finite, and can be quantitatively represented by some length Δ. It follows from (9.4) and (25.1) that $p(\hat{r} - \Delta/2; \phi)$ and $p(-\hat{r} + \Delta/2; \phi)$, respectively, are directly proportional to $f(\hat{r} - \Delta/2; \phi)$ and $f(\hat{r} - \Delta/2; \phi + \pi)$. So, the 'outermost' point samples of the image can be estimated immediately. After appropriately subtracting these point samples from the given projections, the resulting projections are again hollow. They are, however, characterized by

$$\mathbf{L}^+ \sim r^+ < |\xi| < \hat{r} - \Delta. \tag{30.3}$$

Consequently, $f(\hat{r} - 3\Delta/2; \theta)$ can be immediately estimated in the same way throughout $0 \le \theta < 2\pi$. This process can (in principle) be continued until all point samples of $f(r; \theta)$ are recovered within the domain defined by $r^+ < r < \hat{r}$ and $0 \le \theta < 2\pi$.

The procedure outlined in the previous paragraph can actually work with perfect, computer-generated data. However, the addition of tiny amounts of 'noise' makes it go completely haywire. It is one of those 'unwrapping' algorithms which are useless in practice (it is far more error-sensitive, for instance, than the somewhat related KT-processing technique described in §37). What is effective, is to 'complete' the given projections by smoothly 'joining' the 'breaks' at $r = \pm r^+$. Satisfactory results are obtained by invoking the simple edge-extension principle introduced in §15 (it is not feasible to employ overlapped-edge-extension, which requires data on a rectangular grid of points, because projections are defined on a radial grid – a similar difficulty is discussed at length in §33). All that need be done is ensure that $p(\xi; \phi)$ and $\partial p(\xi; \phi)/\partial \xi$ are continuous at $\xi = \pm r^+$ and to force all the 'completed' projections to satisfy the primary consistency condition (9.19).

Complementary to the class of hollow projections is the class of *truncated projections*, which are so called because they are 'cut off' at their extremities. They are here characterized by

$$\mathbf{L}^+(\phi) = \mathbf{L}^+ \sim |\xi| < r^+. \tag{30.4}$$

Theoretically, it is impossible to reconstruct the image from truncated projections, even inside the circle of radius r^+. Satisfactory results are nevertheless obtained in practice by 'completing' the projections in much the same way as already described for hollow projections. Paradoxically, perhaps, the results tend to be even better than those obtained with hollow projections, provided the convex envelope of the cross-section is given (as it often is in practical applications) – this permits the 'completion' of each given projection to be consistent with the actual geometry of the body under examination. The primary consistency condition (9.19) should ideally be satisfied, with $p(\xi; \phi)$ and $\partial p(\xi; \phi)/\partial \xi$ being continuous at $\xi = \pm r^+$. Furthermore, each $p(\xi; \phi)$ should be set to zero for the values of ξ which define the convex envelope.

It must, of course, be understood that the reconstructed images are only faithful for $r > r^+$ in the case of hollow projections, and for $r < r^+$ in the case of truncated projections.

31 Distorted projections

Projections are never ideal in practice. Although various imperfections are mentioned in passing in previous sections of this chapter, they are peripheral to the main argument. The distortions considered in this section are its dominant theme.

Diffraction is necessarily associated with the propagation of macroscopic wave motion (e.g. radio waves, ultrasound, microwaves, acoustics, seismic waves) unless the wavelength is extremely small compared with the scale of all significant inhomogeneities in the propagation medium. Notwithstanding this, the transmission of wavefronts through inhomogeneous media can often be accurately described in terms of rays, even when the linear dimensions of a typical inhomogeneity are little larger than the wavelength. The reason for this is that, while diffraction must be invoked to account for scattering and for the major part of reflection and reverberation, the onward passage of a wavefront requires only refraction for most of its explanation. Diffraction can be almost completely forgotten when considering the X-rays and gamma rays that figure in the kinds of application envisaged in previous sections of this chapter. [It is worth noting that the intensity of a diffracted beam of X-rays is miniscule compared with that of the incident beam, which makes the achievements of X-ray crystallographers all the more remarkable!]

As far as 'image reconstruction from projections' goes, the key property of X-rays and gamma rays is that they can virtually always be described in terms of straight rays. Many of the media that support macroscopic wave motion exhibit significant refractive index variations, implying that the rays tend to be curved. The equivalent of a projection can still be measured of course. One can, for instance, estimate the time taken by a pulse to traverse a cross-section (thereby providing what is sometimes called a *delay projection*) or one can record the attenuation suffered by the pulse (giving a more conventional kind of projection). The ray curvature, however, causes the projection to be distorted, in the sense that it is no longer describable by an integral along a straight line. In fact, it is appropriate to define the *distorted projection* $p_{dist}(\xi; \phi)$ at the angle ϕ by

$$p_{dist}(\xi; \phi) = \int_{s(\xi;\phi)} f(x, y) \, ds, \tag{31.1}$$

where $s(\xi; \phi)$ identifies a ray path from a particular source to a particular sink, both distant ξ from the η-axis, and ds denotes the line element along this (curved) path. The image $f(x, y)$ can represent the distribution over the cross-section of either the refractive index or the attenuation coefficient of the medium supporting the wave motion. RHS (31.1) reduces to RHS (9.4) when the ray path is a straight line. Some implications of (31.1) are examined in §32.

It can happen that the true image alters significantly during the time taken to measure the full set of given projections. This always occurs when a conventional X-ray CT machine forms an image of a cross-section through the thorax. It is sometimes practicable to inhibit variations due to respiration, by using drugs for instance, or when a (healthy enough) patient's breath is held, but there is no way to avoid the density changes caused by the beating of the patient's heart. In all situations involving appreciable movement by parts of a body during the time taken to gather the projection data, the distribution of 'density' throughout the cross-section varies with the projection angle. The distorted projection at angle ϕ is then given by

$$p_{dist}(\xi; \phi) = \int_{-\infty}^{\infty} f(x, y; \phi) \, d\eta, \tag{31.2}$$

instead of by (9.4).

It can be helpful to rewrite (31.2) in the form

$$p_{dist}(\xi; \phi) = p(\xi; \phi) + p_{err}(\xi; \phi), \tag{31.3}$$

where $p(\xi; \phi)$ is given by (9.4), with $f(x, y)$ being the 'average image' (i.e. what the image would be if the movement could be prevented), and where the *error projection* $p_{err}(\xi; \phi)$ is defined by

$$p_{err}(\xi; \phi) = \int_{-\infty}^{\infty} (f(x, y; \phi) - f(x, y)) \, d\eta. \tag{31.4}$$

Density variations can occur unbeknownst to the experimenter. Alternatively, the experimenter may be aware of their occurrence, but know of no algorithm for correcting the distortions due to them. In general, the only useful way to proceed is to treat the given projections as ideal when forming the reconstructed image $\hat{f}(x, y)$. By analogy with the projection theorem (9.5), or (9.6), this implies that the available estimate of the spectrum is given by

$$\hat{F}(\xi; \phi) = \mathbf{F}_{(1\rho)}\{p_{\text{dist}(\xi)}(\xi; \phi)\}. \tag{31.5}$$

There are two archetypal *density variation artefacts*. The first comes about when the density of a stationary part of the cross-section changes. If the boundary of this part is apparent in $f(x, y)$ it is liable to be so in $\hat{f}(x, y)$. The human visual system is often successful in separating this boundary from the mass of streaks due to the presence of $p_{\text{err}}(\xi; \phi)$ on RHS (31.3). There is, of course, no way of 'proving' this mathematically, but it is true nevertheless! This must somehow be closely related to the reasoning given in §29 as to why $\hat{h}$ tends to be a good estimate of $\bar{h}$, but no precise argument can be offered.

The second archetypal density variation artefact arises when there is movement of part of a cross-section, but the density distribution of the part itself does not alter. Because the image enters linearly into the projection integrals – e.g. RHSs of (9.4), (31.2) and (31.4) – the artefact can be fully characterized from consideration of a single point (or pixel) within the cross-section. Since it is irrelevant where the point is positioned, no generality is lost by examining the particular point image

$$g(x, y) = \delta(x)\,\delta(y) = \delta(\xi)\,\delta(\eta), \tag{31.6}$$

which takes the form

$$g(x, y; \phi) = \delta(\xi - \alpha(\phi))\,\delta(\eta - b(\phi)) \tag{31.7}$$

when the point moves, where $a(\phi)$ and $b(\phi)$ are arbitrary real functions of ϕ. Substituting (31.7) into (31.2) gives

$$p_{\text{dist}}(\xi; \phi) = \delta(\xi - a(\phi)), \tag{31.8}$$

so that motion parallel to the rays does not introduce any distortion. Reference to (31.5) then shows that the available estimate $\hat{G}(\rho; \phi)$ of the spectrum of $g(x, y)$ is

$$\hat{G}(\rho; \phi) = \exp(i\,2\pi a(\phi)\rho). \tag{31.9}$$

The two-dimensional Fourier transform $\hat{g}(r; \theta)$ of $\hat{G}(\rho; \phi)$ can be thought of as a psf characterizing the blurring of the point due to its motion. Nothing very useful of a general nature can be deduced from (31.9). Special cases can be evaluated numerically, of course, and they suggest very strongly that the blurring is spread over an area close to that covered by the motion of the point. This accords with practical experience, in that motion artefacts tend to be confined to where movement is localized, provided the motion is nondescript or irregular, or (quasi)random. The normal cyclic motion of the beating heart in its cavity is a good instance. Even though the heart beats steadily (hopefully!) it wobbles (even almost rattles in some cases) as it pumps the blood through it, and the relative motions of the several great vessels are quite complicated. As a consequence, conventional X-ray CT images of cross-sections through the thorax intersecting the heart show a more or less uniform 'blur' over the cardiac region, but there is rarely any artefact of the motion outside this region. Certain simple, regular motions can induce severe artefacts, especially if there are any abrupt startings and stoppings. [It may be that 'smoothness' of the motion is more important than being 'nondescript' as regards the avoidance of 'motion artefacts' – these points are illustrated in Example V.] Another interesting facet of images reconstructed from projections through objects containing moving parts is that it transpires that faint versions of the true image can be apparent through the blurring when the movement is restricted to only a fraction of the time taken to gather the projection data, which may well explain much of the spectacular clinical success of X-ray CT – it seems very unlikely that patients can be kept as still as a simple-minded consideration of the motion artefact problem might require. The upshot of all this is that there tends to be faithful reconstruction of all parts of images which are not in the immediate neighbourhoods of parts that are moving.

The gist of §§29 and 30 and of the above discussion of density variation artefacts has led to the development of a simple but powerful algorithm for reconstructing images of the beating heart. The patient's electrocardiagram (ECG, or EKG) must be recorded while the projections are being measured, so that it is known which projections correspond to the various epochs of the heart cycle. The algorithm involves removing (notionally!) the heart from its cavity and reconstructing its image for each epoch separately. It is thus reminiscent of Aztec ceremonies, but the computer is kinder than those ancient Central Americans because it can return the heart unharmed to its proper place!

The problem of imaging the beating heart is compounded by the effective density of blood being little different from that of cardiac muscle. This means that X-ray contrast material must be added to the blood. Furthermore, for a variety of technical reasons, the concentration of contrast material in the blood can be expected to vary significantly throughout the time needed to gather the projection data. It is appropriate to divide each period of the ECG into M short, separated intervals defining the epochs of the heart cycle, where M depends upon the total number $\hat{N}$ of projections that can be measured, which is mainly governed by how long the patient's respiration can be suspended. If N_m is the number of discrete projections measured for the mth epoch, then

$$\hat{N} = \sum_{m=1}^{M} N_m. \tag{31.10}$$

There is never any difficulty in arranging for $\hat{N}$ to be large enough to permit the reconstructed image – which is of radius $\hat{r}$, because (25.1) is taken to define the size of the whole true image – to be faithful outside the part of the cross-section containing the heart. It is convenient to place the origin of coordinates at the centre of the circle, or radius r^+, which encloses all of the blurring due to heart motion and to variations in the concentration of contrast material. This image is appropriately denoted by $f(r; \theta)$ for $r > r^+$. On denoting the ideal image of the whole cross-section during the mth epoch by $f_m(r; \theta)$, it follows that

$$f_m(r; \theta) = f(r; \theta) \qquad \text{for } r > r^+. \tag{31.11}$$

When the nth projection for the mth epoch is measured, the density is distorted to $f_{m,n}(r; \theta)$, say; but note that

$$f_{m,n}(r; \theta) = f(r; \theta) \qquad \text{for } r > r^+, \tag{31.12}$$

which implies that

$$p_{\text{dist}}(\xi; \phi_{m,n}) = \int_{-\infty}^{\infty} f_{m,n}(r; \theta)\, d\eta, \tag{31.13}$$

where $\phi_{m,n}$ is the angle at which the nth projection for the mth epoch is measured. Since, for any fixed m, the $f_{m,n}(r; \theta)$ can be expected to vary appreciably with n, the recoverable image $\hat{f}_m(r; \theta)$ for the mth epoch can be expected to differ significantly from $f_m(r; \theta)$ for $r < r^+$.

The *Aztec CT problem* is here posed as: given $p_{\text{dist}}(\xi; \phi_{m,n})$, for $-\hat{r} < \xi < \hat{r}$ and $0 \leq \phi < \pi$, with $1 \leq m \leq M$ and $1 \leq n \leq N_m$, reconstruct all M of the $\hat{f}_m(r; \theta)$; on the understanding that all of these quantities are defined as in the previous paragraph.

A 'simple-minded' approach is to submit the $p_{\text{dist}}(\xi; \phi_{m,n})$, for each m and for $n \in \{1, 2, \ldots, N_m\}$, to the conventional modified-back-projection procedure (see §33). One drawback of this is that N_m is not usually large enough to provide an acceptable resolution for the reconstructed image. The inequality (29.3) requires that the spectrum should be effectively cut off at

$$\hat{\rho} = N_m / \pi \hat{r}. \tag{31.14}$$

The 'Aztec solution' is based on (31.11) and (31.12) which emphasize that the modified-back-projection procedure, when applied to all $\hat{N}$ of the given projections, recovers $f(r; \theta)$ faithfully for $r > r^+$. Having carried out this procedure, projections $p_{\text{comp}}(\xi; \phi_{m,n})$ are calculated for the *hollow image*, which is $f(r; \theta)$ for $r > r^+$ and zero for $r < r^+$. This is the 'Aztec operation', because it corresponds to 'tearing out the heart'! Each given projection $p_{\text{dist}}(\xi; \phi_{m,n})$ is transformed to the *difference projection*

$$p_{\text{diff}}(\xi; \phi_{m,n}) = p_{\text{dist}}(\xi; \phi_{m,n}) - p_{\text{comp}}(\xi; \phi_{m,n}), \tag{31.15}$$

which is seen to be what would be measured for the heart alone if it could be perfectly preserved in full working order after being 'torn out of its cavity'. It is now appropriate to submit the $p_{\text{diff}}(\xi; \phi_{m,n})$, for each m and for $n \in \{1, 2, \ldots, , N_m\}$, to the modified-back-projection procedure. The cut-off in Fourier space is given by

$$\hat{\rho} = N_m/\pi r^+, \tag{31.16}$$

which is larger than that given by (31.14) because r^+ is less than $\hat{r}$.

The Aztec solution has another more important advantage. Many of the stationary parts of the thorax, such as the spinal column and the ribs, are of high density, so that they generate strong streaks across images reconstructed from few projections. The contributions of these stationary parts to each given projection is accurately accounted for in the $p_{\text{comp}}(\xi; \phi_{m,n})$, so that $p_{\text{diff}}(\xi; \phi_{m,n})$ is virtually free of their influence. Consequently, the Aztec versions of $f_m(r; \theta)$, besides being about twice as well resolved as their simple-minded counterparts, exhibit fewer and weaker artefacts. It is not surprising, therefore, that details of cardiac anatomy can be recognized more clearly in the Aztec versions of $f_m(r; \theta)$.

32 Iterative image reconstruction

The ways in which the various CT problems examined in previous sections of this chapter are formulated, emphasize that direct solutions can only be constructed if the problems are themselves suitably idealized. While the resulting 'simplified' solutions are useful in several important practical contexts (and, in fact, the ideal transmission CT problem, as posed in §25, is the theoretical basis of recent transformation of diagnostic radiology brought about by X-ray CT), it is still worthwhile looking for ways to improve them.

Significant improvements to the idealized solutions cannot be constructed straightforwardly. In each case it is necessary to adopt an iterative approach. One needs a physical model, here denoted symbolically by M, which can quantitatively predict the measured data, given the form of the true image. It is appropriate here to comment on characteristics of M that apply to particular problems mentioned in previous sections.

Beam-hardening (§25): M has two significant parts. The first is the energy (or temporal frequency) spectrum of the X-ray beam. The second is the variation with body density of the frequency-dependent X-ray attenuation coefficient (remembering that it may in fact be more a matter of Compton-scattering than simple absorption). Given the distribution of density along a particular ray path, the intensity of the emerging X-ray beam can then be calculated accurately.

Attenuation-correction for SPECT (§26): The significant part of M is a knowledge of the density distribution of the medium in which the radio-labelled isotope sources are embedded. This knowledge can only be obtained from prior X-ray transmission CT measurements employing photons in a band of frequencies centred on the frequency of the photons emitted by the radio-labelled sources.

Interpolation (§29): Most of the methods implied in the first three sentences of the paragraph containing (29.13) are themselves essentially iterative (for instance, the maximum entropy technique, which is mentioned in §§15 and 18 and in the introductory comments to Chapter IV, is often invoked in this context). M gives explicit expression to the particular algorithm devised to make each iterate of the image as consistent as possible with the given projections, while incorporating whatever *a priori* information (e.g. image positivity) happens to be specified.

Difference-projections (§31): M has three significant parts. First, it is envisaged that the image amplitude is effectively piecewise constant. Second, it is understood that only few projections are given (so that reconstructed images tend to be severely contaminated by streak artefacts). Third, it is postulated that the boundaries and amplitudes of the regions of constant amplitude can be usefully estimated in the manner intimated in §31.

Ray-bending (§31): M has two significant parts. First, there is the estimation from the given projections of the distribution of refractive index throughout the cross-section. Second, there is the computation of ray paths from such a refractive index distribution.

It is worth emphasizing that M is likely to be *impotent* in part. For instance, when $p_{dist}(\xi; \phi)$ is defined in terms of curved rays, the latter may be bent out of the plane of the cross-section, implying RHS (31.1) is an inadequate description. There is clearly no remedy for this if measurements are made only in the x, y-plane. This difficulty is accommodated in practice by employing detectors that are wider perpendicular to the x, y-plane than they are in the plane. This reduces the resolution markedly, but it is often all that can be sensibly done.

All iterative schemes reduce to the following sequence of steps. An estimate $\hat{f}(\mathbf{x})$ of the image is reconstructed from the data by modified-back-projection (see §33). M is then invoked to estimate the inconsistency of the reconstructed image with the data. M is invoked again to transform the given projections to make them compatible with the ideal model represented by RHS (9.4), with $f(x, y)$ inside the integral replaced by $\hat{f}(\mathbf{x})$. A new version of $f(\mathbf{x})$ is then reconstructed by modified-back-projection of the transformed projections. M is invoked as before to produce a new set of transformed projections. The iterations are continued until the differences between successive versions of $\hat{f}(\mathbf{x})$ are (hopefully) less than some prescribed threshold, which is set by the noise in the data and the uncertainties in M .

Convergence is assured for none of these iterative schemes. It has been found, in fact, that some of them tend to converge at first and then diverge wildly. It is also worth mentioning that most of them are computationally very expensive. Several such schemes are nevertheless useful in practice (see references quoted in the introductory comments to this chapter). When they should be applied, and how many iterations should be attempted, can only be decided on the basis of practical computational experience.

33   Fourier reconstruction, speckle, and back-projection

Although the image frame (as defined in §7) is generally rectangular, it is often convenient to take it to be square. Many expressions become simpler and it is easier to do the algebra without making mistakes. Furthermore, the extra computational effort, which is needed to handle the increased number of point samples of the spectrum that arise when the smaller of the two extents is increased to make the frame square, is rarely objectionable. In this section, therefore, both the x-extent and the y-extent of Ω_f are assumed equal and are given the value L.

The sampling theorem (see §10) shows that $F(u, v)$ is fully specified by its point samples $F_{l,m}$ on the *square grid* of points $(l/L, m/L)$, where the ranges of the integers l and m are large enough that all significant $|F_{l,m}|$ are included. When $L_1 = L_2 = L$, note that (10.6) reduces to

$$F_{l,m} = F(l/L, m/L). \tag{33.1}$$

Given the $F_{l,m}$, the image $f(x, y)$ can be accurately and efficiently reconstructed with the aid of the F.F.T. algorithm (see §12). There is a very real difficulty, however, arising from all CT problems being characterized by spectra that are naturally specified on radial rather than rectangular grids. In this regard, it is instructive to examine the simplest case – i.e. the ideal transmission CT problem posed in §25.

The CT data are represented by the set of given projections. One invokes the projection theorem (9.6) to compute $F(\rho; \phi)$ from the given $p(\xi; \phi)$. So, the spectrum is effectively given on 'spokes' radiating from the origin of Fourier space. Points on these spokes define a *radial grid* – i.e. the set $\{(\rho_\mu; \phi_\nu);$ for all pairs of $\mu \in \{1, 2, \ldots, \hat{\mu}\}$ and $\nu \in \{1, 2, \ldots, \hat{\nu}\}\}$, where $\hat{\mu}$ and $\hat{\nu}$ are large enough that the $F_{l,m}$ can, in principle, be accurately recovered from the $F(\rho_\mu; \phi_\nu)$.

The actual recovery of the $F_{l,m}$ is far from trivial, as can be seen from (10.4) by rewriting $F(u, v)$ as $F(\rho_\mu; \phi_\nu)$, setting $L_1 = L_2 = L$ and replacing

u and v on the RHS by $(\rho_\mu \cos(\phi_\nu))$ and $(\rho_\mu \sin(\phi_\nu))$ respectively. The nature of the sinc functions makes matrix inversion (or any other equivalent procedure) extremely error-sensitive, and far more time-consuming than the aforementioned F.F.T. operation.

A more efficient, but still accurate interpolation procedure can be based on (9.7) through (9.17). The first relation in (9.7) shows that $F(\rho_\mu; \phi)$ can be readily computed provided all significant $F_l(\rho_\mu)$ are available. The latter are obtained from the given projections via the second relation in (9.7) followed by (9.8). All three formulas in (9.7) and (9.8) can be realized computationally with the F.F.T. algorithm, provided the points along the ξ- and ρ-axes and the angles ϕ are equispaced. The ρ_μ cannot be so spaced, unfortunately, because the radial coordinates of points on the square grid are irregularly spaced. Note that

$$\text{radial coordinate of } (l/L, m/L) = (l^2 + m^2)^{\frac{1}{2}}/L. \tag{33.2}$$

Furthermore, very few of the angular coordinates of the square grid points are rational fractions of π, indicating that the computational efficiency of the F.F.T. algorithm cannot be taken advantage of when evaluating the summation in the first relation in (9.7). One can avoid this last difficulty by reconstructing the image via the second relation in (9.7) followed by (9.8), (9.13), and (9.10), all of which can be implemented with equispaced samples along the ξ- and ρ-axes and in the ϕ-direction (assuming the given projections are equispaced in ϕ and regularly sampled along the ξ-axis). The trouble here is that there does not seem to be any algorithm, whose efficiency is comparable to that of the F.F.T., for evaluating RHS (9.13). While this interpolation procedure is preferable to the previously discussed approach based on (10.4), it is still computationally expensive.

One can, of course, resort to crude interpolatory approaches. The simplest is that known as *nearest neighbour interpolation*, in which an estimate $\hat{F}_{l,m}$ of the value of $F_{l,m}$ is equated to the $F(\rho_\mu; \phi_\nu)$ which is nearest to it in the sense that, for l and m specified, $((\rho_\mu \cos(\phi_\nu) - l/L)^2 + (\rho_\mu \sin(\phi_\nu) - m/L)^2)$ is a minimum for $\mu \in \{1, 2, \ldots, \hat{\mu}\}$ and $\nu \in \{1, 2, \ldots, \hat{\nu}\}$. Such interpolation can be effected very efficiently. Only slightly slower is *linear interpolation*, for which $\hat{F}_{l,m}$ is an average of the two nearest $F(\rho_\mu; \phi_\nu)$ inversely weighted according to the respective separations of the two radial grid points from $(l/L, m/L)$. When either of these interpolatory approaches is invoked, it is appropriate to write

$$\hat{F}_{l,m} = F_{l,m} + E_{l,m}. \tag{33.3}$$

The *interpolation error* $E_{l,m}$ is strictly a determinate quantity, but it tends to have a pseudo-random character in practice. When $\hat{F}_{l',m'}$ is substituted for $F_{l',m'}$ in (12.2), with L_1 and L_2 set equal to L in both (12.2) and (12.3), the resulting estimates $\hat{f}_{l,m}$ of the point samples of the image can be written as

$$\hat{f}_{l,m} = f_{l,m} + e_{l,m}, \tag{33.4}$$

where $f_{l,m}$ is the true point sample and $e_{l,m}$ is here called the *transformed interpolation error*.

Each $E_{l,m}$ occupies one pixel in Fourier space, so that it has the character of a two-dimensional delta function. Reference to (6.13) and (6.20), or equivalently (12.2) and (12.4), then shows that there is a contribution to each $e_{l,m}$ from every $E_{l,m}$. The aforementioned pseudo-random character of the latter implies that the $e_{l,m}$ are effectively Rayleigh-distributed – this is another of the many remarks sprinkled throughout this book which, while not meant to be rigorous, are nevertheless of considerable practical significance. It is thus clear that the $e_{l,m}$ must overlay the reconstructed image with 'blotches', which are closely related to the 'speckles' whose properties and formation are discussed in some detail in §34. If the $|E_{l,m}|$ are large enough on the average, the 'speckle-level' can be unduly high. In standard medical CT practice, in fact, crude interpolation in Fourier space is now avoided because the resulting artefacts are usually unacceptable.

If the interpolation could be transferred from Fourier space to image space, a crude procedure could be satisfactory because the maximum distortion would correspond to shifting true pixel values by half a sample spacing. It is therefore worthwhile looking for image-space interpolatory procedures. Such a one is described below. It is the modified-back-projection procedure, which is standard medical X-ray CT practice nowadays. Its practical manifestation is illustrated in Example V. Its theoretical justification is conveniently begun by combining (6.11) and (6.13) to give

$$f(r;\theta) = \int_0^\infty \int_0^\pi F(\rho;\phi) \exp(-i\,2\pi\rho r \cos(\phi-\theta))\rho\, d\phi\, d\rho$$

$$+ \int_0^{-\infty} \int_0^\pi F(-\rho;\phi+\pi) \exp(-i\,2\pi\rho r \cos(\phi-\theta))\rho\, d\phi\, d\rho$$

$$= \int_0^\pi \left(-\int_{-\infty}^0 F(\rho;\phi) \exp(-i\,2\pi\rho r \cos(\phi-\theta))\rho\, d\rho \right.$$

$$\left. + \int_0^\infty F(\rho;\phi) \exp(-i\,2\pi\rho r \cos(\phi-\theta))\rho\, d\rho \right) d\phi, \tag{33.5}$$

because $\cos(\phi+\pi-\theta) = -\cos(\phi-\theta)$ and

$$F(-\rho;\phi+\pi) = F(\rho;\phi), \tag{33.6}$$

by definition. Since the way the ξ, η-coordinates are defined in §9 ensures that $(r\cos(\phi - \theta)) = \xi$, it follows from (33.5) and (9.5) that

$$f(r; \theta) = \int_0^\pi \int_{-\infty}^\infty F(\rho; \phi) \exp(-i\,2\pi\rho\xi)\,|\rho|\,d\rho\,d\phi = \int_0^\pi \tilde{p}(\xi; \phi)\,d\phi, \quad (33.7)$$

where $\tilde{p}(\xi; \phi)$, which is here called the *modified projection* at angle ϕ, is defined by the one-dimensional convolution

$$\tilde{p}(\xi; \phi) = p(\xi; \phi)\,\textcircled{\raisebox{0pt}{$\ominus$}}\,q(\xi; \phi), \quad\quad\quad (33.8)$$

where $q(\xi; \phi)$ is the one-dimensional Fourier transform of $|\rho|$, i.e.

$$q(\xi; \phi) = \int_{-\infty}^\infty \exp(-i\,2\pi\rho\xi)\,|\rho|\,d\rho, \quad\quad (33.9)$$

which is, of course, exceedingly ill-behaved on its own account. When convolved with $p(\xi; \phi)$, however, it gives the finite and very useful quantity $\tilde{p}(\xi; \phi)$.

The spatial frequency content of the reconstructed image is always finite, of course, as is indicated explicitly by the constraint (29.1) on the version of the spectrum that can be computed from the given data. This means that practical filter functions $q(\xi; \phi)$ are obtained by 'windowing' (see §15) the term $|\rho|$ in (33.9). Much effort has been expended on the detailed design of such windows (refer to the references quoted in the introductory comments to this chapter). Each modified projection is conveniently generated from its corresponding given projection either entirely through software, via the first relation in (9.6) followed by computing the one-dimensional Fourier transform of $(|\rho|\,F(\rho; \phi))$, or (equivalently) by effecting the convolution operation on RHS (33.8) with dedicated hardware.

We refer the reader to the third to sixth (inclusive) paragraphs of Example V for a purely descriptive account of the back-projection operation, and for the differences between back-projecting ordinary and modified projections. However, we think it is helpful here to offer a direct interpretation of the formula (33.7), which is the mathematical expression of the connection between the true image and its modified projections. It is instructive to think of the final integral in (33.7) as a prescription for a sequential method of displaying an image. Note that the integrand spans the whole image plane because ξ runs, in principle, from $-\infty$ to ∞. So, for any particular value of ϕ, the amplitude $\tilde{p}(\xi; \phi)$ is recorded at every point on the line parallel to the η-axis and distant ξ from it. This means that the function $\tilde{p}(\xi; \phi)$ spreads like a 'breaker' over the whole image plane, with the 'wavefront' always parallel to the ξ-axis. This operation is known by the graphic name of *back-projection*. The complete reconstructed image is formed by back-projecting all of the given projections after they have been modified. Because the image is defined only for $r < \hat{r}$, as indicated by (25.1), it is a waste of computational effort to back-project outside the circle defined by $(\xi^2 + \eta^2)^{\frac{1}{2}} = \hat{r}$.

Before CT was invented, ingenious analogue methods were devised for back-projecting the given projections themselves. This process is known as *ordinary back-projection* or *ordinary tomography* – it can provide superior resolution to CT because its images can be recorded on film, but its contrast is very much poorer since no attempt is made to compensate for the blurring inherent in the given projections. The procedure described in the previous paragraph is called *modified back-projection* – it is also known as *filtered back-projection* or *convolutional back-projection* because of the form of RHS (33.8). It is possible to perform modified back-projection by analogue means, but this has hardly been looked into at all by the manufacturers, perhaps because so much money has already been invested in doing it digitally. There is no doubt, however, that it is always more convenient and 'cleaner' to do one's processing digitally rather than in analogue fashion. It is, nevertheless, worth keeping in mind that there is usually an analogue process equivalent to any digital computation, and it is sometimes possible for the former to be more efficient and cheaper to realise than the latter.

It is now standard medical imaging practice to invoke crude interpolation (e.g. nearest neighbour or linear) for assigning the amplitudes generated by modified back-projection to the pixels in the reconstructed image.

Although modified back-projection was first suggested by electrical-engineer-astronomers, it is rarely made use of nowadays in astronomical contexts, mainly because the measured data tend to lie on elliptical curves, rather than straight lines, in Fourier space. The interferometric problem (see §25) is generally solved with the aid of the F.F.T. algorithm after sampling Fourier space as finely as possible. It is now almost standard practice to attempt to eradicate the resulting artefacts by employing Högbom's cleaning algorithm (see §17), or an appropriate adaptation thereof (see the final paragraph of §17).

Example V – Various reconstructed images

Projections through the image shown in Fig. Va are here invoked to illustrate how several reconstruction algorithms can be expected to perform under practical constraints. We call Fig. Va the original image.

It is convenient to think of the x- and y-axes (of the Cartesian coordinate system introduced in §3) being horizontal and vertical (i.e. 'across' and 'up' the page) respectively. The left-hand graph in Fig. Vc represents the projection, as defined by (9.4), through the original image in the negative x-direction – i.e. it is the projection at angle $\phi = 90°$ (refer to the first and second paragraphs of §9).

The straight vertical line through the graph is the 'zero' level, with positive values being to the left. It is seen that the projection is wholly positive, as the definition RHS (9.4) requires when the image $f(x, y)$ is itself wholly positive.

Figure Vb is the back-projection (see §33) of the left-hand graph shown in Fig. Vc. Imagine any horizontal line drawn from the left of Fig. Vb to the 'zero' level of the left-hand graph in Fig. Vc. The brightness (or intensity, or density) of the image shown in Fig. Vb is constant along this line and is proportional to the distance from the zero level of the point on the graph intersected by this line. So, the projection is spread evenly back across image-space. The term 'back-projection' is thus satisfactorily descriptive. It is also worth recognizing that an intensity plot along any vertical line through Fig. Vb has the same shape as the left-hand graph in Fig. Vc.

Figure Vd shows the image obtained by back-projecting 8 equispaced projections. These projections were at the angles $\phi = (22.5n)°$, where the integer n ran from 0 to 7. It is remarkable how much of the detail present in Fig. Va can be recognized in Fig. Vd, despite the obscuration caused by the heavy streaking (it is of course only possible to identify the small detail with certainty when one already knows what the original looks like!). Note that the angles of the projections can be estimated from the directions of the streaks. Figure Ve shows the back-projection of 180 equispaced (in angle) projections. The streaking is now imperceptible and all the detail present in Fig. Va can be discerned (if one looks carefully enough!). The contrast is low, however. It must be understood that the faithfulness of Fig. Ve is due in large part to the brightness levels in Fig. Va being binary, in that they are 'bright' or 'dark' (we have done this to make it easier to reproduce some of the other images presented in this example). For instance, the bright spot inside the elliptical feature in Fig. Va would be invisible in Fig. Ve if the spot were, say, one quarter as bright as the ellipse.

The right-hand graph in Fig. Vc is a noisy version of the left-hand graph. The level of the noise (generated from a pseudo-random number routine) is seen to be appreciable. Figure Vf shows the back-projection of 180 equispaced noisy projections (each suffering from the same r.m.s. noise level as that exhibited by the right-hand graph in Fig. Vc). Despite the significant level of noise, Fig. Vf is little different from Fig. Ve.

Figure Vh shows modified projections – see paragraph containing (33.5) through (33.9) and following paragraph – corresponding to the ordinary projections shown in Fig. Vc. Note that the two graphs in Fig. Vh have negative as well as positive parts, because they are convolutions of the ordinary projections shown in Fig. Vc with a filter function – refer to the paragraph containing (33.8). Figure Vg is the magnitude of the back-projection of the left-hand graph in Fig. Vh (the magnitude is displayed because of the difficulty of distinguishing negative from positive parts of an image). Fig. Vi is the image obtained by back-projecting 8 equispaced modified projections. The latter were at the same angles as those invoked to generate Fig. Vd. The detail is much more visible than in Fig. Vd, but the degree of 'streaking' is similar. The image reconstructed by back-projecting 180 equispaced modified projections is almost identical to the original shown in Fig. Va, although there is faint streaking from the sharp corners in the original image (to remove this streaking entirely would require perhaps as many as 300 equispaced projections – provided (29.3) is satisfied, the level of such streaking tends to be imperceptible in real-world reconstructions because naturally occurring images seldom exhibit detail of such angularity – anyway, streaking due to

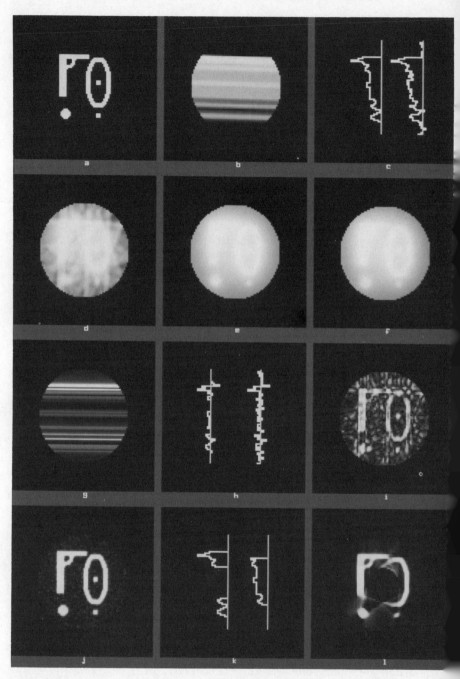

Fig. V.

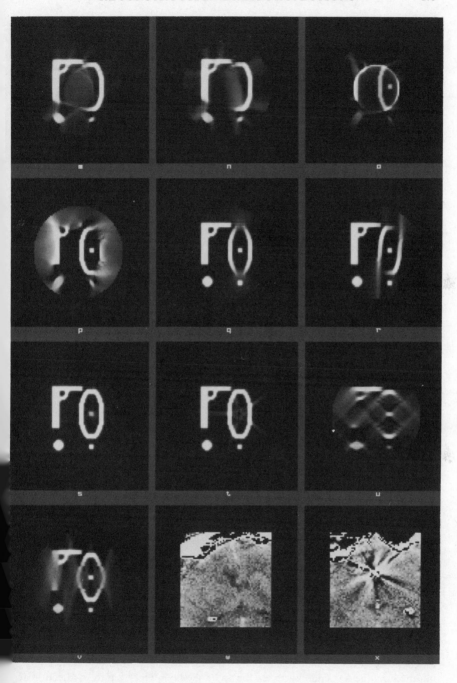

beam-hardening and other imperfections usually swamps streaking of the kind being discussed here).

Figure Vj is the image obtained by back-projecting 180 equispaced modified noisy projections (the right-hand graph in Fig. Vh is a typical modified noisy projection). Note that the differences between Figs. Va and Vj are more noticeable than those between Figs. Ve and Vf. This is because the filtering operation on RHS (33.8) tends to enhance noise. Nevertheless, it must be emphasized that Fig. Vj is much superior to Fig. Vf in that its contrast is very much higher.

The left- and right-hand graphs in Fig. Vk are hollow and truncated, respectively, versions of the projection at angle 90° (i.e. the left-hand graph in Fig. Vc). Figures Vl–Vp are each reconstructions from 180 modified incomplete projections (§30). Figure Vl shows what transpires if no preprocessing is applied to the given hollow projections. Figures Vm and Vn show images reconstructed from 180 preprocessed hollow projections. Preprocessing was effected by joining the 'breaks', for Figs. Vm and Vn respectively, with straight lines and cubic curves (the latter ensure that both the preprocessed projections and their slopes are continuous at the breaks in the original hollow projections). In Fig. Vl the ellipse and gantry, which are clearly separated in Fig. Va, could be considered to be joined. Furthermore, the small square on the right at the bottom of Fig. Va is distorted and faint in Fig. Vl. Note that the quarter-circle 'supporting' the horizontal arm of the gantry is brighter in Figs. Vm and Vn than in Fig. Vl. The separation of the ellipse and the gantry is clearer in Fig. Vn than in Fig. Vm. There is significantly less 'streaking' in Fig. Vn than in Fig. Vm. The reader could be excused for thinking these 'improvements' to be trivial. We emphasize, however, that they tend to be far from trivial for real-world images which exhibit wide ranges of pixel values (remember that Fig. Va is a binary image, in that its pixels are either bright or dark). Niceties of this sort are impossible to make completely clear without expensive plates. We feel that the essential practical points can be got across adequately by interspersing the text with appropriate comments.

Figure Vo shows the image obtained from *un*preprocessed truncated projections. Only the central detail in Fig. Va is reconstructed at all faithfully. Figure Vp was obtained by merely 'completing' each truncated projection at both ends with a straight line – even so, there is a marked improvement. If there was a wide range of brightness levels in the original image, further significant improvement would be got by performing the completion procedure outlined in the penultimate paragraph of §30. Each of the straight lines mentioned above with respect to Fig. Vp meets a given projection where it is truncated and falls to zero at one extremity of the completed projection. The segment of the ξ-axis spanned by each straight line is one-half of the extent of each truncated projection. So, the extent of each completed projection is twice that of a given truncated projection.

Figures Vq and Vr are reconstructions from 180 equispaced modified projections, for which the brightness of the 'ellipse' in Fig. Va varied from projection to projection. The projections were at angles 0° through 179°. For Fig. Vq the brightness varied sinusoidally from the value shown in Fig. Va when $\phi = 0°$ to zero when $\phi = 90°$ and back again to the value shown in Fig. Va when $\phi = 180°$ (of course ϕ never actually reached 180° because the final projection was at 179°, but the brightness would have been the value shown in Fig. Va if the projection at 180° had actually been computed). For Fig. Vr the brightness varied quadratically from zero when $\phi = 0°$ at such a rate that it would have been 1.6 times the value shown in Fig. Va when it would have reached $\phi = 180°$ – but remember that the final projection was at 179°. The inner and outer boundaries of the ellipse are clearly recognizable in Fig. Vq, for which the brightness of the ellipse was almost the same when $\phi = 0°$ and 179°. This is typical of 'smooth' brightness variations. Even when the brightness variations are extremely abrupt (as for Fig. Vr), the shapes of variable brightness regions are often reconstructed quite recognizably. A fair amount of the ellipse is reasonably faithfully reconstructed in Fig. Vr.

Figures Vs and Vt show the effect of a part of the original image (the bright spot inside the ellipse) changing position from projection to projection. The movement was pseudo-random, with a standard deviation of 3 diameters of the moving spot, for Fig. Vs. The degradation of the image is effectively confined to the part of image space where the motion has happened. This is typical of most kinds of irregular or nondescript motion. When the motion is regular, however, as for Fig. Vt (corresponding to a circular movement along a circle of diameter equal to 4 diameters of the moving spot, in a contrary sense to the rotation of the ξ, η-axes with respect to the x, y-axes – refer to the first sentence of the second paragraph of §9), there is often appreciable streaking across much of the reconstructed image (as is apparent in Fig. Vt).

Figure Vu and Vv emphasize how much detail is lost when there are no given projections within particular ranges of ϕ (for $0 < \phi < \Phi$ and $\pi - \Phi < \phi < \pi$ in this example, with $\Phi = 45°$ and 15° for Figs. Vu and Vv respectively). Even when the angle of the 'missing fan' is only 30° (as for Fig. Vv) the degradation is appreciable. In fact, if the angle of the missing fan is (approximately) $(180/m)°$, where m is the integer making this value closest to the actual value, then a reconstruction from m equispaced projections is likely to be as faithful as one incorporating many projections (but none of them within the missing fan). This contention is supported by Fig. Vi being of comparable quality to Fig. Vv.

Figures Vw and Vx are images (of parts of cross-sections through the brain of a hospital patient) formed with a commercial CT scanner. The patient had recently undergone brain surgery, after which a surgical clip had been left (purposely!) inside the skull. Figure Vw shows a cross-section not intersected by the clip (the solid white regions at the top of the image are parts of the skull, while the small bright features near the bottom are ventricles). Figure Vx shows a cross-section containing the clip. The latter is the bright central feature from which 'streaks' radiate like spokes from the centre of a wheel. This streaking is due to beam-hardening (see (i) in §25 and the third paragraph of §32) caused by the density of the metal clip being much higher than that of either brain tissue or the skull. Comparison of Figs. Vw and Vx emphasizes both how severe the effects of beam-hardening can be in practice and how advantageous it would be if an effective, computationally efficient algorithm could be developed for eradicating the streaking.

VI

SPECKLE IMAGING AND INTERFEROMETRY

During the whole of the three centuries since its deleterious effect was first clearly recognized [final paragraph of Prop. VIII of Book I: Newton, B1], up until about fifteen years ago, the Earth's atmosphere has been accepted as the cause of an irreducible limit on the resolution of terrestrial optical astronomical telescopes. This all changed with the introduction of stellar speckle interferometry [Labeyrie, 1], which has sparked a flurry of research into new image processing techniques in observatories, and also in optical and computational laboratories, around the world. Despite current fascination with space telescopes, it is quite likely that the highest resolution will be attained with Earth-based instruments, at least until well into the twenty-first century.

The great majority of the original work bearing on the material discussed in this chapter is referenced in a fairly comprehensive review article [Bates, 10]. Comparatively few other references need therefore be quoted in the remainder of these introductory comments to this chapter. We do, however, recommend [Dainty, B2] as a general reference.

The genesis of the astronomical seeing problem and the formation of speckle images, together with their relation to the problem of forming images from data gathered with radio-astronomical synthesis telescopes, are described in §34 [Ryle, 1; Woolf, 1]. This section also deals with ramifications of the remarkable fact that fixed aberrations of an imaging instrument can be corrected to a significant degree by viewing through a turbulent medium [Cady, 1; Dainty, 1].

The crudest (but nevertheless useful on occasion) of the reported forms of interferometric processing is treated in §35 [§§4.1 and 8.8 of Bates, 10].

The earliest type of speckle processing [Labeyrie, 1, 2], and its close relative 'intensity interferometry' [Hanbury Brown, B1], are analysed and compared in §36. Two immediate extensions of speckle interferometry are also discussed [§§8.2 and 8.3 of Bates, 10].

The techniques examined in §36 make no attempt to recover any of the phase information implicit in the observed data. This 'oversight' is remedied in the methods introduced in §§37 and 38 [§8.1 of Bates, 10].

The two approaches to speckle processing described in §37 [§§8.4 and 12.1 of Bates, 10] can both be traced to Jennison's researches in the early 1950s towards extending radio intensity interferometry in such a way that reliable phase information could be obtained – a readable account of facets of this important development is presented in a book published in the middle 1960s [Jennison, B1].

The newest of the speckle imaging techniques, now generally called 'shift-and-add', is discussed in §38 [§§8.5, 8.6, and 8.7 of Bates, 10]. This approach has also been invoked in ultrasonic imaging contexts [§§12.2 and 12.3 of Bates, 10; see also Bates, 12]. Other pertinent references for this section are [Christou, 1; Milner, T1; Weigelt, 1].

The final section (§39) in this chapter preaches the virtues of attempting to find optimum combinations of the various available processing methods [§8.10 of Bates, 10].

We think it is worth pointing out that speckle imaging is a branch of blind deconvolution. The latter is mentioned in the introductory comments to Chapter III and is discussed in §52.

Example VI illustrates the remarkable restorative powers of speckle imaging. The images presented in this example relate to speckle interferometry (§36) and to shift-and-add (§38). The subtractive deconvolutional technique called Cleaning (see §17) is also illustrated (as promised in the final paragraph of the introductory comments to Chapter III).

34 The seeing problem

Consider an imaging instrument, such as an optical astronomical telescope, viewing an object through a distorting medium (like the Earth's atmosphere). If the interval during which the image is recorded is shorter than any significant fluctuation of the medium, we say that $a(\mathbf{x})$ is recorded *instantaneously* (it is important to recognize that the duration of an *instant*, as here defined, must be at least as long as the response time of whatever is recording the radiation impinging on the focal plane of the instrument). So, $a(\mathbf{x})$ can be thought of as a contaminated convolution of the true image of the object with a psf characterizing the medium at the instant the recording is made. We remind the reader that the true image $f(\mathbf{x})$ is only resolved to the diffraction limit of the particular instrument being used – refer to the paragraph containing (3.10). When the distortion introduced by the medium is severe (which it is in many situations of scientific and technical importance), the apparent resolution of $a(\mathbf{x})$ is well below the diffraction limit. The goal of the kinds of processing discussed in this chapter is to compensate for the distortion, without prior knowledge of the form of the psf.

The psf characterizing a fluctuating medium changes randomly in time. By recording images at a succession of instants, we can discover how separated a pair of instants must be (on the average) for the forms of the psf (at these two instants) to be independent (in the usual statistical sense). We call this average separation the *redistribution time* of the medium. A sequence of M instants, each separated from the previous and succeeding ones by the redistribution time, are conveniently labelled by the integer index m. The forms of the psf at these instants are of course unknown, but it is appropriate to label them by the same index and to write the mth as $h_m(\mathbf{x})$. Since the form of the recorded image varies from each of these instants to the next, it is appropriate to think of a set $\{a_m(\mathbf{x}); m = 1, 2, \ldots, M\}$ of recorded images, each of which is blurred (and contaminated) differently.

In order to give ourselves the best chance of restoring $f(\mathbf{x})$ as faithfully as possible from the set of recorded images, each of them should be preprocessed (see §15). We denote by $a_m(\mathbf{x})$ the form of the mth image, after $a_m(\mathbf{x})$ has been subjected to whatever preprocessing seems most suitable. So, in accordance with the notation introduced in the paragraph containing (14.9), we write

$$a_m(\mathbf{x}) = f(\mathbf{x}) \, \textcircled{0} \, h_m(\mathbf{x}) + c_m(\mathbf{x}), \qquad (34.1)$$

where the mth psf $h_m(\mathbf{x})$ and the mth contamination $c_m(\mathbf{x})$ are ideally independent (statistically) of each other and of their fellows for other values of m. There is, of course, no such thing as complete independence in practice. What this means is, the higher the degree of independence, the more effective are the processing methods which are described in the following sections of this chapter. Another way of looking at it is that, the lesser the degree of independence between the psfs and the contaminations, the larger must M be if the required information is to be recovered to any particular desired accuracy. We emphasize that the $c_m(\mathbf{x})$ account for every imperfection of the preprocessed recorded images, not only the recording noise and any non-linearity of the recording process, but also the non-isoplanatic (i.e. non-point-spread-invariant) part of the distortion introduced by the medium – recall the second paragraph of §4.

The $a_m(\mathbf{x})$ are called *speckle images* or *specklegrams* in optical astronomical contexts (it is worth noting that the blurring depends upon the wavelength of the radiation, so that the wider the field of view covered by each speckle image, the narrower should be the bandwidth of the detected radiation, in order to avoid smearing of the 'speckles' towards the outer parts of the images). Since the ocean is a fluctuating medium as regards the propagation of acoustic waves, the images formed by arrays of sonar transducers can often be usefully thought of as constituting sets of 'acoustic speckle images'. Real-world propagation media do not introduce perfectly *point-spread-invariant* (psi) blurring (characterized by a simple convolutional psf – referred to as isoplanatism in optics – refer to §3), but this does not cast any doubt on the validity of the model (34.1) for a typical preprocessed recorded image because $c_m(\mathbf{x})$ accounts for all parts of $a_m(\mathbf{x})$ that cannot be described by psi blurring.

The techniques discussed in §§35 through 39 rely on the blurring being effectively psi. So, in those situations for which each $c_m(\mathbf{x})$ swamps $(f(\mathbf{x}) \odot h_m(\mathbf{x}))$, none of the techniques are effective; but that is the way of the real-world! In many instances of practical interest, however, the first term on RHS (34.1) seems to model what is measured sufficiently faithfully to permit useful versions of the true image to be reconstructed by remarkably simple processing.

Some propagation media, that introduce severe distortion, do not fluctuate rapidly enough to allow usefully large sets of independently distorted images to be recorded within a single frequency (or wavelength) band in a reasonable length of time. The Earth and biological tissue considered as propagation media for seismic waves and ultrasound, respectively, are in this category. However, their propagation distortions at different frequencies can sometimes be reasonably independent while appearing statistically quite stationary (i.e. seeming to possess the same probability density distributions, etc.).

When coherent radiation (e.g. medical ultrasound, radar, etc.) is used to probe the interior of a body or to search for distant scatterers, it is usually inappropriate to treat the true image as a real non-negative quantity. When it becomes necessary to consider complex images (see §38), the symbol f is replaced by $\hat{f}$ in conformity, for instance, with the early part of §28.

The blurring due to the propagation distortion can make it very difficult to recognize the object being viewed. So, invoking the delightful astronomical usage, there is a 'seeing problem'!

As a preliminary to the imaging techniques described in the remaining sections of this chapter, it is worthwhile briefly examining how speckle images are formed. At any particular frequency, a given propagation medium can be characterized by a *scale length* D_0, the import of which is that images formed with instruments having pupil diameters less than D_0 are negligibly distorted (it is worth remarking that D_0 tends to decrease with increasing frequency). Now consider a 'large' instrument having a pupil diameter D which is appreciably larger than D_0, i.e.

$$D/D_0 \simeq N_0^{\frac{1}{2}}, \tag{34.2}$$

where N_0 is some positive integer. This suggests that the pupil of the large instrument can be partitioned into approximately N_0 *sub-pupils* within each of which there is negligible distortion; but the relative attenuation and propagation delay differ markedly from sub-pupil to sub-pupil. While being a distinctly crude picture, this serves as a very useful approximation which leads to results in conformity with experience.

It is now instructive to reconsider the radiation field, defined by (8.6), existing instantaneously in the pupil plane of the aforesaid large instrument. In the absence of any significant propagation distortion, the field incident upon the pupil can be described by the quantity $S(\mathbf{z}, t)$ defined by

(8.5). When, as always happens in practice, there is sensible propagation distortion and the instrument suffers from aberrations, the field accepted by the pupil can be described by the quantity $G(\mathbf{z}, t)$ defined by (8.6). On invoking the 'sub-pupil concept' introduced in the previous paragraph, the distortion term $\mathfrak{D}(\mathbf{z})$ appearing on RHS (8.6) can be approximated by

$$\mathfrak{D}(\mathbf{z}) \simeq \sum_l \Gamma_{0l} \, \mathrm{perf}_{0l}(\mathbf{z} - \mathbf{z}_{0l}), \qquad (34.3)$$

where the summation is essentially infinite, $\mathrm{perf}_{0l}(\cdot)$ is a 'perfectly' smooth function which is approximately unity over the lth sub-pupil and falls rapidly to zero outside it, $\mathbf{z}_{0l}$ are randomly located position vectors (mean separation of which is D_0), and Γ_{0l} is a complex constant representing the attenuation and phase delay associated with each subpupil. The $\mathrm{perf}_{0l}(\cdot)$ are all different, they are randomly positioned and they can overlap, but their average is a roughly Gaussianly-shaped function of effective width D_0. When $\mathfrak{D}(\mathbf{z})$ is multiplied by an ideal apodization function $\mathfrak{K}_i(\mathbf{z})$, the summation on RHS (34.3) becomes finite:

$$\mathfrak{D}(\mathbf{z})\mathfrak{K}_i(\mathbf{z}) \simeq \sum_{l=1}^{N_o} \Gamma_{0l} \, \mathrm{perf}_{0l}(\mathbf{z} - \mathbf{z}_{0l}). \qquad (34.4)$$

Since $\mathfrak{K}_i(\mathbf{z})$ can be tapered towards the edge of the pupil, the Γ_{0l} and $\mathrm{perf}_{0l}(\cdot)$ in (34.4) differ slightly from their fellows in (34.3); but it is pointless introducing new notation because, first, these quantities are not (and need not be) rigorously defined, and second, it is the general rather than the detailed form of the pupil function which is of interest here.

When the aberrations of the instrument itself are significant, the ideal apodization function $\mathfrak{K}_i(\mathbf{z})$ must be replaced on LHS (34.5) by the actual apodization function $\mathfrak{K}(\mathbf{z})$ introduced in §8. By analogy with the previous treatment of the effects of the propagation distortion, it seems reasonable to characterize $\mathfrak{K}(\mathbf{z})$ by its own sub-pupils of average diameter D_a, i.e.

$$\mathfrak{K}(\mathbf{z}) \sim \sum_{l=1}^{N_a} \Gamma_{al} \, \mathrm{perf}_{al}(\mathbf{z} - \mathbf{z}_{al}), \qquad (34.5)$$

where N_a, Γ_{al}, $\mathrm{perf}_{al}(\cdot)$ and $\mathbf{z}_{al}$ bear the same relation to the new sub-pupils as do the same symbols (with subscript a exchanged for 0) to the previous sub-pupils. At any point $\mathbf{z}$ in the pupil, the distortion due to the medium is superimposed upon the instrument's aberration. However, if either of these is (effectively) a random variable then so is their combination. It still makes sense to partition the pupil field into sub-pupil fields, but the latter are chiefly characterized by either $\mathfrak{K}(\mathbf{z})$ or $\mathfrak{D}(\mathbf{z})$ depending upon which of these two functions represents the severer distortion (note that: the smaller the typical sub-pupil, the worse the distortion). When one of the larger sub-pupils is superimposed upon several of the smaller ones, their combined effect can be described by essentially the same number of smaller sub-pupils. Consequently, $\mathfrak{K}(\mathbf{z})$ dominates $\mathfrak{D}(\mathbf{z})$ when $D_a < D_0$, and vice versa:

$$\mathfrak{D}(\mathbf{z})\mathfrak{K}(\mathbf{z}) \simeq \sum_{l=1}^{N_s} \Gamma_{sl} \, \mathrm{perf}_{sl}(\mathbf{z} - \mathbf{z}_{sl}), \qquad (34.6)$$

where

$$\begin{aligned}
N_s \sim N_0, \quad & \Gamma_{sl} \sim \Gamma_{0l}, \quad \mathbf{z}_{sl} \sim \mathbf{z}_{0l}, \quad \mathrm{perf}_{sl} \sim \mathrm{perf}_{0l} \quad && \text{when } D_0 < D_a \\
N_s \sim N_a, \quad & \Gamma_{sl} \sim \Gamma_{al}, \quad \mathbf{z}_{sl} \sim \mathbf{z}_{al}, \quad \mathrm{perf}_{sl} \sim \mathrm{perf}_{al} \quad && \text{when } D_a < D_0
\end{aligned} \right\} \quad (34.7)$$

where $\mathbf{A} \sim \mathbf{B}$ here implies that, while $\mathbf{A}$ and $\mathbf{B}$ can be different in detail, their important characteristics are the same.

The formula (34.6), although no more than a qualitative statement, is a true reflection of one of the more significant aspects of speckle image processing. This formula expresses the by now well-established result (experimental as well as theoretical – refer to references quoted in the introductory comments to this chapter) that the fixed aberrations of a poor instrument can be compensated, up to the diffraction limit, by superimposing upon its pupil further aberrations. The latter must not only fluctuate, they must also be *more* severe than the fixed aberrations (i.e. $D_0 < D_a$).

Denoting the smaller of D_0 and D_a by D_s, each *sub-pupil image field* (i.e. the instantaneous field in the image plane due solely to the particular sub-pupil) occupies a region (often called the *seeing disc*) of the image plane whose diameter is γ/D_s, where the constant γ is determined by the geometry of the imaging instrument and the frequency of the radiation. The sub-pupil image fields are effectively randomly phased, so that their mutual interference is manifested by randomly distributed bright and dark areas, giving to the image a characteristic mottled, or *speckled*, appearance. On the average, there are N_s bright areas. Since the sub-pupils span the instrument pupil, which is of diameter D, the effective diameters of the bright areas are of the order of γ/D (they can be larger, but they cannot be smaller because γ/D represents the diffraction limit of the instrument – i.e. γ is proportional to the wavelength λ of the radiation).

When $D_s = D_0$, so that the speckling is dominated by the fluctuating propagation distortion, the detailed forms and positions of the speckles change from instant to instant. Suppose the extents (see §7) of the image are less than γ/D_s. Then, averaged over many fluctuation times of the medium, the speckle images form a bright featureless patch, which represents the resolution limit for optical telescopes operated conventionally. It is this patch that has been given the graphically descriptive name 'seeing disc'. To have a chance of unravelling the effects of the propagation distortion, one must operate on individual speckle images because they exhibit detail (albeit all jumbled up) out to the diffraction limit. The latter of course represents far finer detail than the diameter of the seeing disc when the telescope is large (a typical value for D_0 is 0.1 m for $\lambda = 500$ nm, whereas D is close to 5.0 m for many existing giant optical astronomical telescopes).

In speckle imaging contexts, the psfs are unknown *a priori*. They have to be inferred, in effect, from the speckle images themselves. This inference process can be significantly aided by recording a second set of speckle images from an isolated object which, as far as the imaging instrument is concerned, is *unresolvable* (i.e. its angular diameter is less than λ/D). It is convenient to identify such speckle images by surmounting the symbols a, f, h, and c by tildes. An unresolvable object of unit amplitude, positioned at the origin (the position of which can be chosen arbitrarily without loss of generality) of image space, can be represented by

$$\tilde{f}(\mathbf{x}) = \delta(\mathbf{x}). \tag{34.8}$$

Inspection of (34.1) then shows that the mth speckle image of the unresolvable object is described by

$$\tilde{a}_m(\mathbf{x}) = \tilde{h}_m(\mathbf{x}) + \tilde{c}_m(\mathbf{x}). \tag{34.9}$$

In practice, $\tilde{h}_m(\mathbf{x})$ and $\tilde{c}_m(\mathbf{x})$ cannot be the same as $h_m(\mathbf{x})$ and $c_m(\mathbf{x})$, but they can (especially in optical astronomical contexts) be statistically similar, which is all that is required for the techniques described in §§35 through 39.

The *seeing problem* is here posed as: given the sets $\{a_m(\mathbf{x}); \ m = 1, 2, \ldots, M\}$ and $\{\tilde{a}_m(\mathbf{x}); m = 1, 2, \ldots, \tilde{M}\}$ of speckle images, for *usefully large* values of the integers M and $\tilde{M}$, reconstruct the image-form (as defined in §20) of $f(\mathbf{x})$; on the understanding that $a_m(\mathbf{x})$ and $\tilde{a}_m(\mathbf{x})$ are defined by (34.1) and (34.9) respectively, that the term 'usefully large' implies that there are enough speckle images to allow meaningful averages of the kind described in §§35 through 39 to be performed, and that all the $h_m(\mathbf{x})$ and $\tilde{h}_m(\mathbf{x})$ are members of the same statistical ensemble, and that all of the $h_m(\mathbf{x})$, $\tilde{h}_m(\mathbf{x})$, $c_m(\mathbf{x})$ and $\tilde{c}_m(\mathbf{x})$ are independent.

In no branch of applied science which relies on imaging techniques for gathering data (e.g. astronomy, non-destructive testing of materials, medical diagnostics, etc.) is there any sign of satisfaction with the resolving power of existing instruments. There is a continual demand for better resolution. The only way to achieve this is to build instruments with larger apertures. When attempting to construct pupils bigger than the largest which have been built so far, it can be more convenient (it may even be the only practicable approach) to invoke the principles of interferometry (such as are used in the design of radio super-synthesis telescopes) as opposed to those of classical imaging (which describe the functioning of ordinary optical telescopes, microscopes, and eyes(!), for instance).

Interferometers are composed of discrete elements positioned in a plane, which it is convenient to think of as the pupil plane of a generalized imaging instrument. The *effective size* of each element (e.g. the effective area of the receiving aperture of each antenna in a radio interferometer) is usually very much less than that of any of the sub-pupils introduced earlier in this section. It is useful, therefore, to identify a typical element by a single point z in the pupil plane. When observing spatially incoherent sources (see §3) the signals received by the interferometer elements are correlated in pairs. Consider elements at points z' and z''. In most situations of practical importance one can make the *ergodic hypothesis* that the correlator output depends only upon the separation u of the elements, where

$$u = z'' - z'. \qquad (34.10)$$

This assumption can often be deduced directly from (8.10), which is also, of course, an ergodic hypothesis. The conditions for ergodicity are intricate (see references quoted in the introductory comments to this chapter), but experience suggests that there are few instances of interferometric imaging of spatially incoherent sources where it cannot be safely assumed. From an image processing point of view then, interferometers make their measurements in Fourier space. Instead of the speckle images $a_m(x)$ and $\tilde{a}_m(x)$, their spectra $A_m(u)$ and $\tilde{A}_m(u)$, respectively, are observed where

$$A_m(u) = F(u)H_m(u) + C_m(u), \quad \text{and} \quad \tilde{A}_m(u) = \tilde{H}_m(u) + \tilde{C}_m(u), \qquad (34.11)$$

as follows from applying the convolution theorem (7.7) to (34.1) and (34.9) respectively. The term *visibility* is interferometric parlance for what is called *spectrum* in this book – e.g. $F(u)$ is the visibility of the true image $f(x)$.

The contamination of the signal received anywhere in the pupil plane varies, of course, with both z and time (the latter is represented by the integer m – refer to the first two paragraphs of this section). So it can be worth specifying the pair of points (say z_j and z_k) in the pupil plane associated with any given component, say $F(z_k - z_j)$, of the visibility. The j, kth piece of the given data can then be usefully written as

$$A_m(z_j, z_k) = F(u_{j,k})H_{m,j,k} + C_{m,j,k} \quad \text{with} \quad u_{j,k} = z_k - z_j, \qquad (34.12)$$

where $H_{m,j,k}$ characterizes the isoplanatic (or psi) part of the relative distortion, at the mth instant, due to the propagation medium, between the interferometer elements at z_j and z_k.

In certain situations of practical importance (for many modes of operation of radio astronomical interferometers, for instance) it transpires that

$$\text{phase}\{H_{m,j,k}H_{m,k,l}H^*_{m,j,l}\} = 0, \qquad (34.13)$$

and

$$|H_{m,h,j}H_{m,j,k}/H_{m,h,l}H_{m,l,k}| = 1, \tag{34.14}$$

for any integers h, j, k, and l. It is worth emphasizing that no such relations exist for the spectral contaminations $C_{m,j,k}$ which are to be regarded as completely randomly distributed over m, j, and k.

It is worth posing the interferometric seeing problem in two ways. The *simple-minded interferometric seeing problem* is the same as the 'seeing problem' above, except that the spectra $A_m(\mathbf{u})$ and $\tilde{A}_m(\mathbf{u})$ replace the speckle images $a_m(\mathbf{x})$ and $\tilde{a}_m(\mathbf{x})$ respectively; on the understanding that these spectra are defined by (34.11). The *sophisticated interferometric seeing problem*, which is more realistic in many instances, is here posed as: given the set $\{A_m(\mathbf{z}_j, \mathbf{z}_k);\ m = 1, 2, \ldots, M\ \&\ j, k = 1, 2, \ldots, M^+\}$ of visibility components, for usefully large values of the integers M and M^+, reconstruct the image-form of $f(\mathbf{x})$; on the understanding that $A_m(\mathbf{z}_j, \mathbf{z}_k)$ is defined by (34.12).

It is to be understood, for any seeing problem examined in this chapter, that the given data are sampled closely enough to satisfy whatever sampling constraints may be required.

35 Probability imaging – fluctuation interferometry

When the object being viewed is unresolvable, as described by (34.8), a plane wave can be assumed to impinge upon each of the sub-pupils introduced in §34. This implies that each sub-pupil image field can be characterized by a single complex amplitude. The total field existing instantaneously at any point within the seeing disc is thus the sum of N_s random complex amplitudes. Provided N_s is large enough (a value of 50 is certainly satisfactory; even 5 is adequate in some contexts!) this total field must be Rayleigh-distributed. The *probability density function* (pdf), denoted here by $\tilde{p}(\Im)$, for the intensity $\Im$ of a Rayleigh-distributed field is

$$\tilde{p}(\Im) = \exp(-\Im) \qquad \text{for } 0 < \Im < \infty, \tag{35.1}$$

where the average intensity has been normalized for convenience to unity and the symbol p is surmounted by a tilde to emphasize that the object is unresolvable – refer to the paragraph containing (34.8) and (34.9).

With the aid of modern image-capture technology (e.g. vidicon systems, digital-electronic cameras, etc.) the pdf characterizing a set of speckle images can readily be measured. If it is found to closely approximate the form $(b^{-1}\exp(-\Im/b))$ then one can be confident that the object is unresolvable and of intensity, or *brightness*, b.

The image of any spatially incoherent object can be represented as an assemblage of independent unresolvable 'points', i.e.

$$f(\mathbf{x}) = \sum_{j=1}^{J} b_j\, \delta(\mathbf{x} - \mathbf{x}_j), \tag{35.2}$$

where b_j and $\mathbf{x}_j$ are the brightness and position of the jth point. The points closest together must be separated by at least the diffraction limit to ensure that all points can in fact be independent. The pdf characterizing the jth point in isolation is

$$p_j(\mathfrak{I}) = (1/b_j) \exp(-\mathfrak{I}/b_j), \qquad (35.3)$$

whose *characteristic function* $C_j(u)$ is given by

$$C_j(u) = \int_0^\infty p_j(\mathfrak{I}) \exp(\mathrm{i}\, 2\pi u \mathfrak{I})\, d\mathfrak{I} = 1/(1 - \mathrm{i}\, 2\pi b_j u). \qquad (35.4)$$

Since the 'points' are independent, their combined characteristic function is

$$C(u) = \prod_{j=1}^J C_j(u) = \prod_{j=1}^J B_j/(1 - \mathrm{i}\, 2\pi b_j u), \qquad (35.5)$$

where

$$(1/B_j) = \prod_{k=1}^J b_k \prod_{\substack{l=1 \\ l \neq j}}^J (1/b_j - 1/b_l). \qquad (35.6)$$

The inverse Fourier transform of (35.5) shows that the combined psf, characterizing the complete image (35.2), is given by

$$p(\mathfrak{I}) = \sum_{j=1}^J B_j \exp(-\mathfrak{I}/b_j). \qquad (35.7)$$

Given a set of speckle images of an arbitrary object, the pdf $p(\mathfrak{I})$ can be evaluated and modelled numerically by RHS (35.7), thereby permitting the quantities J, B_j, and b_j to be estimated. This procedure is here called *probability imaging*. Quite apart from its inability to form an actual image, because it possesses no means of evaluating the $\mathbf{x}_j$, the estimation from measured data – i.e. from $p(\mathfrak{I})$ – of J and of the b_j is well known to be an extremely error-sensitive numerical problem. One can hardly expect it to be successful in practice if J is greater than, say, 4. If J is given *a priori*, however, there is a much better chance of gaining useful estimates of the b_j and B_j.

Despite its considerable limitations, probability imaging can be useful when one has only a single detector, implying that only one point in the image (i.e. only one picture element or *pixel*) can be measured. The detector must be capable of recording the brightness levels of radiation incident upon it within intervals of duration somewhat less than the characteristic fluctuation time of the medium. A particularly valuable application is the technique here called *fluctuation interferometry*, for which the propagation medium serves as an (effective) instrument pupil. The measurements are made at great distance from the medium with the detector connected to a sensor having low angular resolution. A typical such sensor is a radio antenna with a wide beam which intercepts a large area of the aforesaid medium. The fluctuating signal power picked up by the antenna is directly analogous to the time-varying intensity existing at any particular point in the image plane of the instrument considered in the early paragraphs of §34. It follows that the pdf of this fluctuating power can be represented by the $p(\Im)$ previously discussed in this section.

When the distorting medium is the interplanetary or interstellar propagation medium, and the object is a radio star, fluctuation interferometry can provide useful upper limits on the object's size.

36 Basic speckle processing – intensity interferometry

It is first of all convenient to introduce notation for the effective radius of the seeing disc (denoted by $1/\rho_s$) and for the diffraction limit (denoted by $\hat{\rho}$) of the imaging instrument – i.e. the instrument responds negligibly to spatial frequencies whose magnitudes $|\mathbf{u}|$ exceed $\hat{\rho}$.

The seeing disc, which is introduced in §34, is identical to the average of the second set of speckle images specified in the seeing problem (as posed in §34). So, the seeing disc is described by

$$\langle \tilde{a}_m(\mathbf{x}) \rangle = (1/M) \sum_{m=1}^{M} \tilde{a}_m(\mathbf{x}). \tag{36.1}$$

The spectrum of the seeing disc is, of course, negligible for $|\mathbf{u}| > \rho_s$. The same can be said for the average of the spectra of the first set of speckle images specified in the seeing problem. Therefore

$$\langle \tilde{A}_m(\mathbf{u}) \rangle \simeq 0 \simeq \langle A_m(\mathbf{u}) \rangle \qquad \text{for } |\mathbf{u}| > \rho_s. \tag{36.2}$$

It is remarked in §34 that each speckle image contains information out to the diffraction limit. Because of this Labeyrie recognized that, despite (36.2),

$$\langle |A_m(\mathbf{u})| \rangle \qquad \text{has significant value for } 0 \le |u| < \hat{\rho}. \tag{36.3}$$

It is thus deduced from (34.11) that

$$\langle |A_m(\mathbf{u})|^2 \rangle = |F(\mathbf{u})|^2 \langle |H_m(\mathbf{u})|^2 \rangle + C_{\text{comp}}(\mathbf{u}) \tag{36.4}$$

has significant value for $0 \le |\mathbf{u}| < \hat{\rho}$, where

$$C_{\text{comp}}(\mathbf{u}) = \langle F(\mathbf{u})H_m(\mathbf{u})C_m^*(\mathbf{u}) + F^*(\mathbf{u})H_m^*(\mathbf{u})C_m(\mathbf{u}) + |C_m(\mathbf{u})|^2 \rangle \tag{36.5}$$

is here called the *composite spectral contamination*.

In order to recover $|F(\mathbf{u})|^2$ from $\langle |A_m(\mathbf{u})|^2 \rangle$, an estimate of $\langle |H_m(\mathbf{u})|^2 \rangle$ is needed. Labeyrie further noted, because of the statistical similarity of the psfs associated with both sets of speckle images specified in the seeing problem, that

$$\langle |H_m(\mathbf{u})|^2 \rangle \simeq \langle |\tilde{H}_m(\mathbf{u})|^2 \rangle, \tag{36.6}$$

where $\tilde{H}_m(\mathbf{u})$ is the OTF corresponding to the psf introduced in (34.9). Inspection of (34.9) reveals that the average of the intensities of the spectra of the $\tilde{a}_m(\mathbf{x})$ can be expressed as

$$\langle |\tilde{A}_m(\mathbf{u})|^2 \rangle = \langle |\tilde{H}_m(\mathbf{u})|^2 \rangle + \tilde{C}_{comp}(\mathbf{u}), \tag{36.7}$$

where $\tilde{C}_{comp}(\mathbf{u})$ is given by RHS (36.5) when $F(\mathbf{u}) = 1$ and H is replaced by $\tilde{H}$.

It follows from (36.6) and (36.7) that the average of the spectral intensities of the second set of speckle images specified in the seeing problem can form the basis of a Wiener filter (see §16), here called the *Labeyrie–Wiener filter* $W_L(\mathbf{u})$, for reconstructing $|F(\mathbf{u})|^2$ from $\langle |A_m(\mathbf{u})|^2 \rangle$. Recognizing that LHS (36.7) is necessarily real, it follows from (16.5) that

$$W_L(\mathbf{u}) = \langle |\tilde{A}_m(\mathbf{u})|^2 \rangle / ((\langle |\tilde{A}_m(\mathbf{u})|^2 \rangle)^2 + \Phi_L) \tag{36.8}$$

where the value of the filter constant Φ_L is chosen to accord with the estimated levels of the composite spectral contaminations appearing in (36.4) and (36.7). The intensity of the spectrum of the recoverable true image (see §16) is then given by

$$|\hat{F}(\mathbf{u})|^2 = \langle |A_m(\mathbf{u})|^2 \rangle W_L(\mathbf{u}). \tag{36.9}$$

The processing described by (36.4) and (36.7)–(36.9) is known as *Labeyrie's speckle interferometry*.

The observant reader has no doubt noticed that essentially the same result can be arrived at by averaging the autocorrelations of the speckle images, rather than the intensities of their spectra, because

$$\langle a_m(\mathbf{x}) * a_m(\mathbf{x}) \rangle = \mathsf{F}\{\langle |A_m(\mathbf{u})|^2 \rangle\}, \tag{36.10}$$

as follows from the autocorrelation theorem (7.10). The reason why one usually carries out the averaging in Fourier space is that it is generally far more efficient computationally, owing to the properties of the F.F.T. algorithm (see §12). It is worth commenting, however, that it can be more efficient to average the autocorrelations when the object being viewed is so faint that a typical speckle image registers a relatively small number of photon impacts. The autocorrelation of each speckle image then consists of the vector separations of the points in image space at which the photons impact. Such an autocorrelation is conveniently formed in real time with modern digital technology – see the quoted references for the theoretical and practical implications of this.

When the object is observed with an interferometer, and in those situations for which the simple-minded interferometric seeing problem (see §34) is adequately realistic, one forms the averages on LHSs of (36.4) and (36.7) and constructs the Labeyrie–Wiener filter defined by (36.8). Thus, RHS (36.9) is again the available estimate of $|F(\mathbf{u})|^2$. The latter can be improved, at the points $\mathbf{u}_{j,k} = (\mathbf{z}_k - \mathbf{z}_j)$ specified for the sophisticated interferometric seeing problem (see §34 and refer to (34.12)), by invoking (34.14) which is sometimes called the *closure-amplitude* relation – the type of (iterative) processing that must be employed is discussed further in §39.

All the processing techniques so far examined in this section can be usefully regarded as species of *intensity interferometry*.

While this is not the place for a detailed discussion of interferometric niceties (they are well covered in references quoted in the introductory comments to this chapter), it is worthwhile digressing on the nature of intensity interferometry because it has several widely known connotations which conflict somewhat with each other. The term is sometimes restricted to interferometers (of the kind invented by Hanbury Brown and Twiss) whose individual elements detect only the envelope fluctuations of the radiation incident upon them. Since correlation can then only be effected between signals which have previously been incoherently detected, the signal-to-noise ratio is often very poor. This technique can nevertheless be useful when it is impractical to maintain coherence between signals received at widely separated stations. When such coherence can be maintained satisfactorily, phase-sensitive (or Michelson) interferometry can be performed, usually with a marked improvement in the signal-to-noise ratio. Even though the processing is phase-sensitive it may not give the visibility phase directly. The point is that the visibility phase can only be deduced confidently from 'raw' interferometric data when a number of precise calibrations (accurate surveying of interferometric baselines being among them) have been performed. From a purely image-processing point of view then, Michelson interferometry may not be able to provide any more information than the intensity interferometry of Hanbury Brown and Twiss. A final pertinent comment is that this limitation on Michelson interferometry applies with increasing force the shorter the wavelength of the radiation.

Only $|F(\mathbf{u})|^2$, or equivalently $(f(\mathbf{x}) * f(\mathbf{x}))$, is estimated directly by speckle interferometry. This represents sufficient information for reconstructing the image-form of $f(\mathbf{x})$ in virtually all situations of practical importance (but remember the qualifying comments interspersed through §§22 and 23). However, the algorithm described in §23 involves a fair amount of computational effort, and it is always desirable to obtain results in as simple a manner as possible. There are two types of situation wherein the image-form can be reconstructed straightforwardly.

The first type of situation corresponds to what is called *speckle holography*. There are two forms of this. In the first, the true image must

separate into two parts, as in (21.1), and (21.3) must also apply –
remember that $h(\mathbf{x})$ is here one of the two parts of the image and is not
to be thought of as a psf. If $h(\mathbf{x})$ is unresolvable, as in (21.7), it then
serves as the reference, so that the unknown part of the image can be
reconstructed by the offset holographic techniques described in §21.
Unresolvable references abound in optical astronomical applications because the majority of individual stars are smaller than the diffraction limit
of the largest telescopes. The data can be checked to see if they are
consistent with the form assumed for the true image. It is helpful to refer
to (21.2). Note first that there must be a markedly bright object at the
origin. This is the autocorrelation of $h(\mathbf{x})$ which is effectively a two-
dimensional delta function. Furthermore, the x-extent of Ω_{g*g} must be
twice that of either Ω_{h*g} or Ω_{g*h}. This is easily checked when (21.6)
holds – i.e. when the holography is elementary in the terminology of §21.
When $s < w$, so that the iterative offset holographic procedure has to be
followed, one can examine the relative lengths of the x-extents as the
iterations proceed. The second form of speckle holography arises when
the image is of the form of RHS (21.20), which can be ascertained by
checking whether the autocorrelation consists of unresolvable points.
Triple-autocorrelation holography can then be attempted. One can check
whether all the $|\mathbf{x}_{mn}|$ are different by noting if the autocorrelation of the
reconstructed image equals the autocorrelation computed from the given
$|F(\mathbf{u})|^2$.

The second type of situation referred to above corresponds to what has
been called *large field speckle interferometry*. The true image must be of
the form

$$f(\mathbf{x}) = \delta(\mathbf{x}) + \sum_{j=1}^{J} gj(\mathbf{x}), \qquad (36.11)$$

where the centres of all of the frames Ω_{gj} must be further from the origin
than $1/\rho_s$. Also, the separations of the centres of the Ω_{gj} from each other
must exceed $2/\rho_s$. The average of the first set of given speckle images
(refer to the seeing problem as posed in §34) takes the form of $(J+1)$
separated seeing discs, one corresponding to the unresolvable reference
and the others to the $gj(\mathbf{x})$. It is convenient to imagine a transparency,
here denoted by $t(\mathbf{x})$, formed from a hard-limited version of $\langle a_m(\mathbf{x}) \rangle$. The
term 'hard-limiting' implies that the transparency is completely transpar-
ent at any point in any of the aforementioned separated seeing discs, but
is opaque elsewhere. Note that either Ω_δ or one of the Ω_{gj} corresponds to
the central region of each of the transparent parts of the transparency,
because all of these frames must have the same centres as their corresponding seeing discs. The aforesaid conditions on the separations of the
frames ensure that all cross-correlation terms in the autocorrelation
$(f(\mathbf{x}) * f(\mathbf{x}))$ occupy regions of image space which are separated from each

other and from the region containing all of the $(gj(\mathbf{x}) * gj(\mathbf{x}))$. Now suppose that $t(\mathbf{x})$ is placed over an image-display of the autocorrelation of $f(\mathbf{x})$, with the centre of each of its transparent regions positioned in turn at the origin of image space. For each such positioning, $t(\mathbf{x})$ is rotated about the origin to try and bring image detail into the neighbourhoods of the centres of all of the transparent regions. When this has been done successfully, all of the $gj(\mathbf{x})$ are necessarily displayed because the contents of each transparent region (apart from the one centred at the origin) must represent one of the $(\delta(\mathbf{x}) * gj(\mathbf{x}))$. The autocorrelation of $\delta(\mathbf{x})$ is, of course, positioned at the origin.

37   Extended speckle processing – phase-sensitive interferometry

The basic speckle processing described in §36 discards the phases of the spectra of the given speckle images. The purpose of the kinds of processing discussed in this section is to make as much use as possible of whatever phase information the data actually contain.

The first phase-preservation technique examined in this section is that commonly associated with the names of Knox and Thompson. It is called *KT-processing* here. One begins by choosing a real, two-dimensional vector $\boldsymbol{\alpha}$ satisfying

$$0 \lesssim |\boldsymbol{\alpha}| \lesssim \rho_s, \tag{37.1}$$

where $\lesssim$ means 'is appreciably less than'. The value of $|\boldsymbol{\alpha}|$ is fairly critical in practice, for reasons made clear later. The next step is to Fourier transform the data; just as in Labeyrie's speckle interferometry. However, instead of averaging the spectral intensities, thereby throwing away any phase information that may be present, one computes

$$\langle A_m(\mathbf{u}) A_m(\mathbf{u}+\boldsymbol{\alpha}) \rangle = J(\mathbf{u}, \boldsymbol{\alpha}) H(\mathbf{u}, \boldsymbol{\alpha}) + C_{\text{comp}}(\mathbf{u}, \boldsymbol{\alpha}), \tag{37.2}$$

and

$$\langle \tilde{A}_m^*(\mathbf{u}) \tilde{A}_m(\mathbf{u}+\boldsymbol{\alpha}) \rangle = \tilde{H}(\mathbf{u}, \boldsymbol{\alpha}) + \tilde{C}_{\text{comp}}(\mathbf{u}, \boldsymbol{\alpha}), \tag{37.3}$$

where it is seen from (34.11) that

$$J(\mathbf{u}, \boldsymbol{\alpha}) = F^*(\mathbf{u}) F(\mathbf{u}+\boldsymbol{\alpha}), \tag{37.4}$$

$$H(\mathbf{u}, \boldsymbol{\alpha}) = \langle H_m^*(\mathbf{u}) H_m(\mathbf{u}+\boldsymbol{\alpha}) \rangle, \tag{37.5}$$

$$C_{\text{comp}}(\mathbf{u}, \boldsymbol{\alpha}) = \langle F^*(\mathbf{u}) H_m^*(\mathbf{u}) C_m(\mathbf{u}+\boldsymbol{\alpha}) + F(\mathbf{u}+\boldsymbol{\alpha}) H_m(\mathbf{u}+\boldsymbol{\alpha}) C_m^*(\mathbf{u})$$
$$+ C_m^*(\mathbf{u}) C_m(\mathbf{u}+\boldsymbol{\alpha}) \rangle \tag{37.6}$$

$\tilde{H}(\mathbf{u}, \boldsymbol{\alpha})$ is given by (37.5) with $\tilde{H}_m$ replacing H_m and $\tilde{C}_{\text{comp}}(\mathbf{u}, \boldsymbol{\alpha})$ is given by (37.6) with $\tilde{H}_m$ and $\tilde{C}_m$ replacing H_m and C_m respectively.

It is now assumed that

$$H(\mathbf{u}, \boldsymbol{\alpha}) \simeq \tilde{H}(\mathbf{u}, \boldsymbol{\alpha}), \tag{37.7}$$

for the same reasons that (36.6) is postulated. It follows from (37.3), by analogy with (36.8) and (36.9), that a *KT–Wiener filter* $W_{KT}(\mathbf{u}, \boldsymbol{\alpha})$ can be constructed from $\langle \tilde{A}_m^*(\mathbf{u})\tilde{A}_m(\mathbf{u}+\boldsymbol{\alpha})\rangle$ to recover an estimate of $J(\mathbf{u}, \boldsymbol{\alpha})$ from $\langle A_m^*(\mathbf{u})A_m(\mathbf{u}+\boldsymbol{\alpha})\rangle$; but note that LHS (37.3) is *not* necessarily real. Therefore,

$$W_{KT}(\mathbf{u}, \boldsymbol{\alpha}) = \langle \tilde{A}_m(\mathbf{u})\tilde{A}_m^*(\mathbf{u}+\boldsymbol{\alpha})\rangle / (|\langle \tilde{A}_m^*(\mathbf{u})\tilde{A}_m(\mathbf{u}+\boldsymbol{\alpha})\rangle|^2 + \Phi_{KT}), \quad (37.8)$$

so that an estimate of $J(\mathbf{u}, \boldsymbol{\alpha})$ is given by

$$\hat{J}(\mathbf{u}, \boldsymbol{\alpha}) = \hat{F}^*(\mathbf{u})\hat{F}(\mathbf{u}+\boldsymbol{\alpha}) = \langle A_m^*(\mathbf{u})A_m(\mathbf{u}+\boldsymbol{\alpha})\rangle W_{KT}(\mathbf{u}, \boldsymbol{\alpha}). \quad (37.9)$$

The value of the filter constant Φ_{KT} is chosen to accord with the estimated levels of the composite spectral contaminations appearing in (37.2) and (37.3).

It is to be understood that the given speckle images are also subjected to Labeyrie's processing (see §36) so that point samples of $|\hat{F}(\mathbf{u})|^2$ can be estimated at all desired sample points in Fourier space. Now note from (37.9) that

$$\hat{J}(0, \boldsymbol{\alpha}) = \hat{F}(0)\hat{F}(\boldsymbol{\alpha}), \quad (37.10)$$

since phase$\{\hat{F}(0)\}$ is necessarily zero on account of the image being real (see §8). So, phase$\{\hat{F}(\boldsymbol{\alpha})\}$ is given immediately by (37.10). Note further that

$$\hat{J}(\boldsymbol{\alpha}, \boldsymbol{\alpha}) = \hat{F}^*(\boldsymbol{\alpha})\hat{F}(2\boldsymbol{\alpha}), \quad (37.11)$$

from which phase$\{\hat{F}(2\boldsymbol{\alpha})\}$ is found immediately because all other quantities in (37.11) are already known. Consequently, phase$\{\hat{F}(\mu\boldsymbol{\alpha})\}$ is found recursively for all values of the integer μ for which $|F(\mu\boldsymbol{\alpha})|$ is significant.

It is necessary to repeat the above processing with $\boldsymbol{\alpha}$ replaced by an orthogonal vector $\boldsymbol{\beta}$ of comparable length, i.e.

$$\boldsymbol{\beta} \cdot \boldsymbol{\alpha} = 0 \quad \text{and} \quad 0 \gtrsim |\boldsymbol{\beta}| \gtrsim \rho_s. \quad (37.12)$$

It follows that the values of phase$\{\hat{F}(\nu\boldsymbol{\beta})\}$ are found recursively for all relevant values of the integer ν. One can then return to (37.9) and note that

$$\hat{J}(\mu\boldsymbol{\alpha} + \nu\boldsymbol{\beta}, \boldsymbol{\alpha}) = \hat{F}^*(\mu\boldsymbol{\alpha} + \nu\boldsymbol{\beta})\hat{F}([\mu + 1]\boldsymbol{\alpha} + \nu\boldsymbol{\beta}), \quad (37.13)$$

which demonstrates that phase$\{\hat{F}(\mu\boldsymbol{\alpha} + \nu\boldsymbol{\beta})\}$ can be recursively evaluated at a rectangular grid of points in Fourier space defined by $\boldsymbol{\alpha}, \boldsymbol{\beta}$ for the values of the integers μ and ν for which $|F(\mu\boldsymbol{\alpha} + \nu\boldsymbol{\beta})|$ is significant.

Errors necessarily accumulate with KT-processing because of the recursive phase estimation procedure. The error in a phase evaluated after l recursions can be expected to equal $(l^{\frac{1}{2}}K(\boldsymbol{\alpha}, \boldsymbol{\beta}))$, where $K(\boldsymbol{\alpha}, \boldsymbol{\beta})$ depends upon the contamination level in the given speckle images, the statistics of the medium causing the seeing problem and the lengths of the vectors $\boldsymbol{\alpha}$

and $\boldsymbol{\beta}$. The smaller $|\boldsymbol{\alpha}|$ and $|\boldsymbol{\beta}|$ are, the larger l must be for any particular point in Fourier space. Since $K(\boldsymbol{\alpha}, \boldsymbol{\beta})$ tends to depend mostly on the contamination level when α and β are small, one should avoid taking $|\boldsymbol{\alpha}|$ and $|\boldsymbol{\beta}|$ too close to zero. On the other hand, $A_m(\mathbf{u})$ tends to become increasingly independent (statistically) of $A_m(\mathbf{u}+\alpha)$, or equivalently $A_m(\mathbf{u}+\boldsymbol{\beta})$, implying that $K(\boldsymbol{\alpha}, \boldsymbol{\beta})$, grows appreciably with $|\boldsymbol{\alpha}|$ and $|\boldsymbol{\beta}|$ at the upper ends of the ranges specified in (37.1) and (37.12). Because, in practice, there are always uncertainties concerning details of the statistics of the seeing, the values of $|\boldsymbol{\alpha}|$ and $|\boldsymbol{\beta}|$ are chosen on the basis of experience.

The literature abounds with variants of extended speckle processing (see the quoted references). While several of them are significantly different from KT-processing, they all suffer similarly from the growth with distance from the origin in Fourier space of the errors in the estimated phases. Furthermore, KT-processing is at least as simple as any of them, both to describe and to perform.

It is now instructive to examine the sophisticated interferometric seeing problem (see §34). Consider the special case

$$\mathbf{z}_k - \mathbf{z}_j = \mu\boldsymbol{\gamma} \quad \text{and} \quad \mathbf{z}_l - \mathbf{z}_k = \boldsymbol{\gamma}, \tag{37.14}$$

where μ is any integer and $\boldsymbol{\gamma}$ is a vector satisfying $0 \lesssim |\boldsymbol{\gamma}| \lesssim \rho_s$. It then follows from (34.12) and (34.13) that

$$\langle\text{phase}\{A_m(\mathbf{z}_j, \mathbf{z}_k)A_m(\mathbf{z}_k, \mathbf{z}_l)A_m^*(\mathbf{z}_j, \mathbf{z}_l)\rangle \simeq \text{phase}\{F^*(\mu\boldsymbol{\gamma})F((\mu + 1)\boldsymbol{\gamma})\}, \tag{37.15}$$

provided the averaging is carried out for long enough to ensure the contamination level is below whatever threshold is desired. By taking $\boldsymbol{\gamma}$ equal to $\boldsymbol{\alpha}$ first and then to $\boldsymbol{\beta}$, it is seen that the interferometric seeing problem can be solved by KT-processing. Unfortunately, it is usually impracticable to realize the regular interferometer element spacing implied by (37.14). It is often possible, however, to carry out iterative processing (of the sort discussed in §39) based on (34.13), which is sometimes called the *closure–phase* relation.

It is, of course, mandatory that the interferometric observations are made with a phase-sensitive (Michelson) interferometer, if one wishes to carry out closure–phase processing. Just as in §36, it is worth emphasizing that the phase data obtained directly from such an interferometer are usually very different from the true visibility phases. The latter can only be extracted from the measured data after, first, intricate calibrations of the sort intimated in the brief discussion of intensity interferometry given in §36, and second, extensive computations of the kind outlined in §39. A final point worth making is that KT-processing can be improved by proceeding from the origin of Fourier space by more than one of the possible paths to each point $(\mu\boldsymbol{\alpha} + \nu\boldsymbol{\beta})$ when estimating each of the phase$\{F(\mu\boldsymbol{\alpha} + \nu\boldsymbol{\beta})\}$. This is somewhat the same as invoking least-squares for partial noise compensation of an over-determined set of measurements.

38   Direct speckle imaging

Intuitively one expects a blurred image to be least distorted where it is brightest. This cannot, of course, be generally substantiated, but it often serves as a useful *ad hoc* principle.

Now suppose there exists a set of M independently blurred images, such as the speckle images $a_m(\mathbf{x})$ specified in the seeing problem (see §34). By superimposing their brightest parts, one can expect to form an improved version of the brightest part of the true image. If the blurring is psi (point-spread-invariant or isoplanatic – i.e. can be described by a convolution – see §§4, 5, and 7) then the other parts of the true image must also be superimposed, thereby producing (hopefully) a relatively faithful version of the whole of the true image. This is the *shift-and-add* principle.

It is convenient to re-emphasize here that images must be quantized in amplitude and spatially sampled before they can be subjected to the kinds of processing described in this book. So, any point $\mathbf{x}$ in image space is in effect a discrete pixel (see §5).

The first step in shift-and-add processing is to find the *brightest pixel position* $\mathbf{x}_m$ in each $a_m(\mathbf{x})$, i.e.

$$a_m(\mathbf{x}_m) > a_m(\mathbf{x}) \qquad \text{for all } \mathbf{x} \neq \mathbf{x}_m. \tag{38.1}$$

Even in practice there can be more than one 'brightest point' (because of the aforementioned amplitude quantization), but such a possibility is disregarded for the moment – it is discussed later in this section in the context of 'speckle masks'.

The whole of each speckle image is *shifted* without any rotation until its brightest pixel is at the centre of image space *and* it is then *added* to its fellows which have been so-treated previously. The sum of all M shifted speckle images is finally divided by M to produce what we call the *shift-and-add image* $f_{sa}(\mathbf{x})$, the formula for which is

$$f_{sa}(\mathbf{x}) = \langle a_m(\mathbf{x} + \mathbf{x}_m) \rangle. \tag{38.2}$$

There can be large variations in the brightnesses of the brightest pixels in a set of speckle images. Furthermore, the contamination level of a typical speckle image need not be proportional to the brightness of its brightest pixel. It is sometimes found that the *adjusted shift-and-add image* $f_{asa}(\mathbf{x})$, defined by

$$f_{asa}(\mathbf{x}) = \langle \mathfrak{W}\{a_m(\mathbf{x}_m)\} a_m(\mathbf{x} + \mathbf{x}_m) \rangle, \tag{38.3}$$

is an appreciably more faithful version of $f(\mathbf{x})$ than is $f_{sa}(\mathbf{x})$, where $\mathfrak{W}\{\cdot\}$ denotes a *weighting* related to the brightness of the brightest pixel. Since the theory underlying shift-and-add is little understood at present, the form of $\mathfrak{W}\{\cdot\}$ can only be chosen on the basis of practical experience. As good a choice as any seems to be the simple

$$\mathfrak{W}\{a_m(\mathbf{x}_m)\} = a_m(\mathbf{x}_m). \tag{38.4}$$

The degree of blurring in a typical speckle image can be very large. In optical astronomy, for instance, the effective extents (see §8) of $h_m(\mathbf{x})$ are sometimes 100 times those of $f(\mathbf{x})$. Consequently, many separated distorted replicas of the brightest part of $f(\mathbf{x})$ can be expected in each $a_m(\mathbf{x})$. It is often useful, therefore, to superimpose several of the brighter pixels within a single speckle image. To see how this can be effected it is convenient to think of the particular speckle image as a map of mountainous terrain. The (positive) amplitude of each pixel then represents a height above the image plane, which can itself be regarded as 'sea level'. Now imagine a continually lowering 'cloud cover' with a horizontal upper surface. As it falls, peaks break through. Denote by $\mathbf{x}_{m,n}$ the position of the nth peak to break through (for peaks of equal height, the values of n can be chosen to increase with, say, the x-coordinate of $\mathbf{x}_{m,n}$; for those having the same x-coordinate it could be convenient to require n to increase with increasing y-coordinate – it does not matter what rule is chosen, provided it can be precisely defined). Since the replicas of $f(\mathbf{x})$ contained within $a_m(\mathbf{x})$ tend to be increasingly distorted the fainter they are, it is appropriate to disregard all peaks lower than some pre-set threshold (chosen on the basis of experience – there is no other valid criterion). Suppose there are N_m peaks above the threshold. The mth *speckle mask* $\mathrm{mask}_m(\mathbf{x})$ is defined by

$$\mathrm{mask}_m(\mathbf{x}) = \sum_{n=1}^{N_m} a_m(\mathbf{x}_{m,n})\,\delta(\mathbf{x}-\mathbf{x}_{m,n}). \tag{38.5}$$

The mth *masked speckle image* $ma_m(\mathbf{x})$ is defined by

$$ma_m(\mathbf{x}) = a_m(\mathbf{x}) \circledast \mathrm{mask}_m(\mathbf{x}), \tag{38.6}$$

which is seen to superimpose all N_m of the brighter pixels in $a_m(\mathbf{x})$.

The processing by which each $ma_m(\mathbf{x})$ is formed was actually suggested earlier than shift-and-add, which is perhaps surprising since the latter is considerably simpler, both conceptually and to perform. By averaging all M of the masked speckle images, the *Lynds–Worden–Harvey image* $f_{\mathrm{LWH}}(\mathbf{x})$ (named for the originators of the technique) is obtained:

$$f_{\mathrm{LWH}}(\mathbf{x}) = \langle ma_m(\mathbf{x}) \rangle. \tag{38.7}$$

Inspection of (38.3) through (38.7) reveals that $f_{\mathrm{LWH}}(\mathbf{x})$ reduces to $f_{asa}(\mathbf{x})$ when $N_m \equiv 1$.

Note that $f_{LWH}(\mathbf{x})$ makes use of much more information in each $a_m(\mathbf{x})$ than does $f_{asa}(\mathbf{x})$, so that it tends to exhibit a higher signal-to-noise ratio. However, the brightest replica of $f(\mathbf{x})$ in each speckle image is usually the least distorted, which means that $f_{asa}(\mathbf{x})$ is virtually always inherently more faithful than $f_{LWH}(\mathbf{x})$. There is another point which, until recently, escaped the notice of (presumably) everybody working in this area. [It is a good example of not being able to see the wood from the trees, or vice versa, unless one occasionally takes time off for a calm dispassionate look at what one is doing.] If the aforementioned threshold is set to sea-level, so that N_m assumes its largest possible value for each m, it transpires that $f_{LWH}(\mathbf{x})$ is necessarily closer to $(f(\mathbf{x})*f(\mathbf{x}))$ than to $f(\mathbf{x})$ itself. This, of course, emphasizes the need for an improved theoretical understanding of the shift-and-add principle, in order that optimum threshold levels can be chosen.

In contradistinction to §§36 and 37, there has been no recourse in this section to Fourier space. It therefore makes sense to call the kinds of processing introduced here *direct speckle imaging*. Denoting either $f_{sa}(\mathbf{x})$ or $f_{asa}(\mathbf{x})$ or $f_{LWH}(\mathbf{x})$ by $pf(\mathbf{x})$, it is seen from comparison of (38.2), (38.3), and (38.7) with (34.1) that

$$pf(\mathbf{x}) = f(\mathbf{x}) \, \mathbb{O} \, ph(\mathbf{x}) + pc(\mathbf{x}), \qquad (38.8)$$

where the p placed before f, h, and c stands for 'processed'. Since all images are recognized in this section to consist of discrete pixels, a general expression for $ph(\mathbf{x})$ is

$$ph(\mathbf{x}) = \sum_{k=1}^{K} d_k \, \delta(\mathbf{x} - \mathbf{x}'_k), \qquad (38.9)$$

where the $\mathbf{x}'_k$ are constant position vectors and the d_k are positive constants (normalized so as not to exceed unity, with $\mathbf{x}'_0 = 0$ and $d_0 = 1$). Substituting (38.9) into (38.8) gives

$$pf(\mathbf{x}) = \sum_{k=1}^{K} d_k f(\mathbf{x} - \mathbf{x}'_k) + pc(\mathbf{x}). \qquad (38.10)$$

If all of the d_k except d_0 are small, $pf(\mathbf{x})$ is a faithful version of $f(\mathbf{x})$, provided the level of the processed contamination is reasonably low.

The true image can also be represented as a sum of discrete pixels, i.e.

$$f(\mathbf{x}) = \sum_{j=1}^{J} b_j \, \delta(\mathbf{x} - \mathbf{x}'_j), \qquad (38.11)$$

where the $\mathbf{x}'_j$ are constant position vectors and the b_j are positive constants (normalized so as not to exceed b_0, with $\mathbf{x}'_0 = 0$). While the $h_m(\mathbf{x})$ depend only upon the characteristics of the medium through which the object is viewed, the processed psf $ph(\mathbf{x})$ also depends upon the form of the true image. When

$$b_j \ll b_0 \qquad \text{for all } j > 0, \tag{38.12}$$

it is intuitively clear that $ph(\mathbf{x}) \simeq \delta(\mathbf{x})$. If any of the b_j, for $j > 0$, cease to be negligible, however, some of the d_k, for $k > 0$, must become significant. It follows from (38.10) that $pf(\mathbf{x})$ must then consist of several shifted replicas of $f(\mathbf{x})$. The relative amplitudes of these replicas depend, of course, on the statistics of the seeing and the form of $f(\mathbf{x})$. In such cases, the processed image $pf(\mathbf{x})$ is said to contain *ghosts*.

Experience shows that $pf(\mathbf{x})$ can often be a recognizable version of $f(\mathbf{x})$, although it should be noted that, the larger J is, the smaller the b_j have to be, for $j > 0$, for $ph(\mathbf{x}) \simeq \delta(\mathbf{x})$. Even when the ghosts are so numerous that $f(\mathbf{x})$ is unrecognizable in $pf(\mathbf{x})$, direct speckle imaging can be useful. The point is that one can always estimate $|F(\mathbf{u})|^2$ by processing the speckle images in the manner described in §36. Consequently, provided the image frame Ω_{ph} (see §7) is appreciably smaller than the average of the Ω_{h_m} – this is effectively the same as the seeing disc (see §34) –, $pf(\mathbf{x})$ may serve as a useful initial estimate for $f(\mathbf{x})$ in a composite image processing procedure (see §39). Experience indeed suggests that Ω_{ph} is often much smaller than the seeing disc even when $f(\mathbf{x})$ exhibits little brightness contrast throughout Ω_f.

From the second set of speckle images specified in the seeing problem (see §34) one can compute another processed psf, denoted here by $\tilde{p}h(\mathbf{x})$. Because any processed psf is object-dependent, the Fourier transform of $\tilde{p}h(\mathbf{x})$ only serves as a useful basis for a Wiener filter when $f(\mathbf{x})$ is composed of well separated unresolvable points – i.e. $|\mathbf{x}_j - \mathbf{x}_{j'}|$ exceeds several pixel widths for all $j \neq j'$. In such cases it can be convenient to employ subtractive (see §17) as opposed to multiplicative (see §16) deconvolution. In fact, the simple cleaning algorithm often performs satisfactorily.

Because $f(\mathbf{x})$ and all of the $h_m(\mathbf{x})$ are positive, $f_{sa}(\mathbf{x})$ reveals (a version of) the true image on a bright background – i.e. the contrast tends to be low. The true image gleams fitfully as through a 'fog'. When $f(\mathbf{x})$ is such that deconvolution with $\tilde{p}h(\mathbf{x})$ can be usefully attempted, the 'fog level' can be markedly reduced. It should also be noted here that, under the appropriate conditions, the extensions to 'speckle masking' due to Weigelt and to Milner can be thought of as means for heightening contrast in direct speckle imaging.

When coherent radiation is imaged, the true image can no longer be taken as positive – it must be assumed complex, in fact. It is then appropriate to replace $f(\mathbf{x})$ by $\mathfrak{f}(\mathbf{x})$ in (34.1). Furthermore, the $h_m(\mathbf{x})$ and the $c_m(\mathbf{x})$ are also complex. Consequently, at any particular $\mathbf{x}$ for which $a_m(\mathbf{x} + \mathbf{x}_m)$ varies randomly with m, RHS (38.2) must be small. One sees that RHS (38.2) can only be appreciable at those $\mathbf{x}$ for which phase$\{a_m(\mathbf{x}+\mathbf{x}_m)\}$ depends little upon m. It is indeed found in practice that the fog level tends to be negligible for coherent shift-and-add imaging.

There is an extension of coherent shift-and-add, here called *stochastical imaging*, that requires one to be given only a *single* speckle image. There is even less theoretical justification for this than for the previously introduced forms of direct speckle imaging; but it often works! It is also in many ways the most intriguing of the techniques discussed in this section because it allows *a posteriori* restoration of a single blurred image, even in the absence of detailed *a priori* knowledge of the psf. A worthwhile research problem is the devising of an appropriate theoretical approach for understanding it (and related techniques) better. One starts with

$$a(\mathbf{x}) = \bar{f}(\mathbf{x}) \odot h(\mathbf{x}) + c(\mathbf{x}), \qquad (38.13)$$

where the subscript m appearing in (34.1) has been dropped to emphasize that only a single speckle image is given – the forms of $h(\mathbf{x})$ and $c(\mathbf{x})$ are not given, of course. A set $\{g_m(\mathbf{x}); m = 1, 2, \ldots, M\}$ of pseudo-random psf are generated in the computer and each of them is convolved with $a(\mathbf{x})$ to give a set of pseudo-speckle-images. The latter are here denoted by $\{a_m(\mathbf{x}); m = 1, 2, \ldots, M\}$ in order to avoid multiplication of notation. The coherent shift-and-add image $\bar{f}_{sa}(\mathbf{x})$ is then defined by RHS (38.2), but it should be remembered that the true image is now the complex $\bar{f}(\mathbf{x})$ rather than the positive $f(\mathbf{x})$. Provided $\bar{f}(\mathbf{x})$ possesses a palpably 'brightest' pixel, and provided the distortion introduced by the pseudo-random psf is *more* severe than that characterized by $h(\mathbf{x})$ – refer to the paragraph following that containing (34.7) – , $\bar{f}_{sa}(\mathbf{x})$ is often a distinctly recognizable version of $\bar{f}(\mathbf{x})$. When such is the case, an improved image can usually be obtained by subjecting all given speckle images (if, in fact, there is more than one) to this stochastic processing and, finally, carrying out coherent shift-and-add on all of the stochastical images.

39   Composite image processing

It is always a 'good thing' if more than one non-equivalent way of processing data can be found. The differences between the several results can represent a measure of the inherent noise or contamination level. It is therefore worthwhile developing as many imaging techniques as possible, provided each is viable in its own right.

Another and perhaps more cogent reason for acquiring an armoury of image processing methods is that no single technique is likely to be superior in all circumstances. A third reason, more important still, is that improved results can often be obtained by combining different methods.

The iterative approach to phase restoration introduced in §23 is an example of composite image processing. A number of algorithms are amalgamated to produce results quite beyond any of them if employed singly. This approach can be usefully extended by trying other possibilities (than the two considered in §23) for the initial phase estimate $\Psi(\mathbf{u})$ introduced in (23.12). The phase obtained from KT-processing (see §37) can be used, as can the phase of the Fourier transform of any of the images generated by direct speckle imaging (see §38). A similar comment can be made on the particular one-dimensional restricted phase problem discussed in §24. The $\mathbf{Z}_f$ corresponding to KT-processing and to the various direct speckle images can be computed and compared, thereby helping to identify the members of $\mathbf{Z}_{f*f}$ which belong to $\mathbf{Z}_f$.

It is sometimes impossible to make any headway without resorting to composite image processing. Such is the case for the sophisticated interferometric seeing problem (see §34) when the $\mathbf{z}_j$ are irregularly positioned in the pupil plane. Even if the contamination is neglected, the formula (34.13) – called the closure–phase relation, as noted in §37 – does not permit phase$\{F(\mathbf{u}_{j,k})\}$ to be unambiguously inferred directly. It is necessary to invoke either extra data and/or other *a priori* information and/or a theoretical model to generate an initial estimate of phase$\{F(\mathbf{u})\}$. The latter is then refined iteratively under the constraints

$$\text{phase}\{F(\mathbf{u}_{j,k})\} + \text{phase}\{F(\mathbf{u}_{k,l})\} - \text{phase}\{F(\mathbf{u}_{j,l})\}$$
$$= \text{phase}\{\langle A_m(\mathbf{z}_j, \mathbf{z}_k) A_m(\mathbf{z}_k, \mathbf{z}_l) A_m^*(\mathbf{z}_j, \mathbf{z}_l)\rangle\} \quad (39.1)$$

for all given vectors $\mathbf{u}_{j,k} = \mathbf{z}_k - \mathbf{z}_j$.

Example VI – Illustrations of speckle processing

This example illustrates particular facets of §§34, 36 and 38. We wish to demonstrate to the reader the degree of image degradation typically caused by naturally occurring inhomogeneous propagation media. We have chosen to concentrate mainly on the sort of effects the earth's atmosphere has on optical astronomical images. No actually recorded images are presented here, because we want to illustrate the essence of each point we are making. However, many comparisons, made over a dozen or so years, of computational and experimental results have convinced us that our software for simulating speckle psfs is not only thoroughly realistic, it is also robust in the sense that the results are relatively insensitive to alterations in statistical parameters. The kinds of image-restoration by speckle processing which are illustrated here are similarly robust. All of this is substantiated in references quoted in the introductory comments to this chapter.

Figures VIa through VIt were devised with astronomical contexts in mind. Figures VIu through VIx were inspired by our experiments on ultrasonic imaging of isolated objects embedded in mammalian tissue. Because we wish to emphasize as clearly as possible the differences which can be manifested between coherent and incoherent imaging techniques, we have chosen to base our illustration of these differences on the particularly simple true image shown in Fig. VIu.

Suppose Fig. VIa is the image of a group of stars which would be formed, under ideal seeing conditions, in a focal plane of a large telescope. Suppose further that Fig. VIb is a long-exposure image of the same group of stars formed under typical seeing conditions. Figure VIb is thus equivalent to the seeing disc (see §34). Finally suppose Fig. VIc to be a typical narrow-band short-exposure image of the group of stars (which must of course be bright enough that a recorded image can exhibit all the detail revealed in the image – it is worth remarking that speckle processing can be effective for quite faint celestial objects – see references quoted in the introductory comments to this chapter). So, Fig. VIc is a typical speckle image (refer to §34) of the group of stars. Figure VId is the (logarithm of the) magnitude of the spectrum of Fig. VIc. The log-magnitude is shown in Fig. VId to increase the amount of pictorial information which can be displayed in the printed image.

Labeyrie's speckle interferometry (see §36) involves averaging the squared-magnitudes of the spectra of a number (M say) of speckle images, all having the same statistics but all being statistically independent. Figure VIe shows the (logarithm of the) average of 20 squared-magnitudes (values for M very much larger than 20 are usual in astronomical practice, but we feel it is instructive to demonstrate here that useful results can be got with modest values for M – of course, the reason why values for M in the hundreds, or even hundreds of thousands, are often needed in practice is that interesting celestial objects are usually faint). Figure VIf has been formed in the same way as Fig. VIe but the spectra are of speckle images of an isolated unresolvable object (such an object is needed as a speckle interferometric reference – see §36). Figure VIg shows the result of Wiener filtering (see Example III) Fig. VIe with Fig. VIf (the filter constant was 0.001).

Note that Fig. VIg is somewhat similar to Fig. VIh, which is the squared magnitude of the spectrum of Fig. VIa. Figs. VIi and VIj are the Fourier transforms of the quantities represented by Figs. VIg and VIh respectively. They are gratifyingly similar to each other. Figure VIj is, of course, the autocorrelation of the original image. This simple example emphasizes that the extents of an object can be estimated by speckle interferometry (the extents of an object are necessarily half those of its autocorrelation – see §7).

It is instructive at this juncture to refer to Example IV and note that Fig. VIa, like Fig. IVe, is the kind of image which can be immediately reconstructed by offset Fourier holography (see §21). Figures VIj and IVg should also be compared. The four fainter 'stars' in Fig. VIa stand our clearly on the left in Fig. VIj. Similarly, the part of Fig. IVe consisting of three 'stars' superimposed upon a 'fog' is revealed on the left in Fig. IVg. We emphasize that, whenever the object being viewed possesses an 'offset reference' (like the brighter 'star' on the left in Fig. VIa, or the isolated 'star' in Fig. VIe), speckle interferometry becomes speckle holography so that the image form is immediately obtained – refer to the paragraph before that containing (36.11).

The image shown in Fig. VIa can be restored quite faithfully by straightforward shift-and-add, as defined by (38.2). Before demonstrating shift-and-add imaging it is instructive to illustrate what we mean by speckle images being statistically identical but statistically independent. At first sight, Fig. VIk looks much the same as Fig. VIc. In fact, the psf which blurred Fig. VIa to produce Fig. VIc had identical statistics to that which blurred Fig. VIa to produce Fig. VIk. The two psfs were nevertheless independent, as close inspection of Figs. VIc and VIk confirms. The positions of individual speckles in the two images are different, and randomly so.

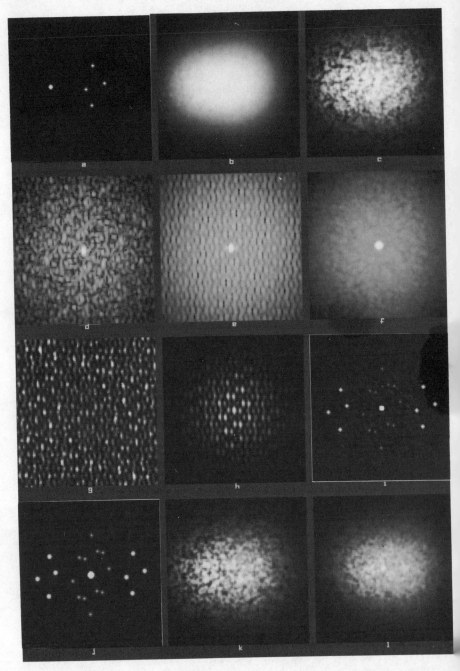

Fig. VI.

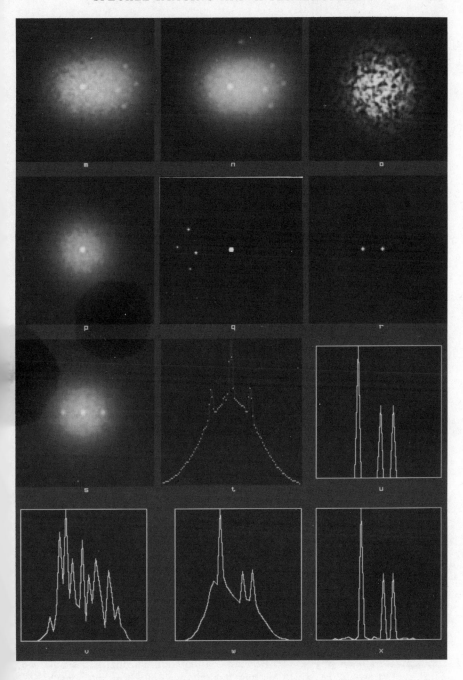

Recall the shift-and-add principle (§38). Each speckle image is shifted, without rotation, until its brightest point is at the centre of image space. All such shifted speckle images are then averaged. Figure VIk can be thought of as the average of a single shifted image. It is impossible to recognize Fig. VIa in Fig. VIk. Figure VIl shows the average of 5 shifted images – the brighter 'star' in Fig. VIa can just be discerned but the other four are still obscured. When 30 speckle images are averaged (Fig. VIm) all the stars in Fig. VIa can be recognized. After 100 shifted images are averaged (Fig. VIn) the form of Fig. VIa is clearly revealed on top of a (more or less) smooth background. It is as though the stars are shining through a fog.

By analogy with speckle interferometry, shift-and-add images can be improved if many statistically independent narrow-band short exposures of an isolated unresolvable star (which again serves as a reference) are observed under statistically similar seeing conditions – refer to the paragraph containing (38.11) and (38.12) and the two following paragraphs. Figure VIo is a typical speckle image of a single unresolvable object. So, Fig. VIo could be one member of the set $\{h_m(\mathbf{x}); m = 1, 2, \ldots\}$ of psfs which characterize the seeing conditions. If Fig. VIa represents $f(\mathbf{x})$ then Figs. VIc and VIk are statistically akin to the set $\{a_m(\mathbf{x}); m = 1, 2, \ldots\}$ of speckle images defined by (34.1). Since $f(\mathbf{x})$ consists of 5 isolated objects (see Fig. VIa), and because of the convolution operation on RHS (34.1), the density of speckles in Figs. VIc and VIk must be significantly greater than that in Fig. VIo. This is consistent with a close examination of these three figures. Figure VIp is the shift-and-add image obtained from 100 statistically independent speckle images all of which are statistically akin to Fig. VIo.

Figure VIq shows the result of Cleaning – see paragraph containing (17.8) through (17.11) – Fig. VIn with Fig. VIp, employing 50 iterations and a loop gain $\tau = 0.1$. Note that the background fog has been removed in Fig. VIq, which is quite a faithful replica of Fig. VIa.

It is instructive to illustrate shift-and-add's propensity for generating 'ghosts' – see paragraph containing (38.11) and (38.12). The shift-and-add image is only a faithful version of the original image if the brightest point is significantly brighter than any other point. If the second brightest point is comparable in brightness to the brightest point then it is quite often mistaken during shift-and-add processing for the brightest point. As a result, the brightest point is sometimes shifted to the wrong point in image space, thereby generating a ghost. When the original image possesses several points of comparable brightness, there can be many ghosts – but remember, from the argument put forward in the paragraph following that containing (38.11) and (38.12), that this does not prevent shift-and-add imaging being useful. Figure VIr represents a pair of unresolvable stars of equal brightness. Each of these stars corresponds equally often to the brightest point in a speckle image, implying that the ghost should be as bright as the true point corresponding to it. Figure VIs is the shift-and-add image of the pair of stars. Figure VIt is a graph of the magnitude of the brightness of Fig. VIs along a horizontal line through the three bright points in Fig. VIs. The ghosting is emphasized by the two outer peaks being of almost equal magnitude.

In fact, the image shining fitfully through the fog in Fig. VIs is very similar to the autocorrelation of Fig. VIr. Because shift-and-add imaging is so simple, we feel that it would be worth putting a great deal of effort into developing algorithms for eradicating the ghosts.

It is also instructive to illustrate the difference between coherent and incoherent shift-and-add (only the latter has been discussed so far in this example). It is convenient to go into this in the context of one-dimensional, rather than as up to now of two-dimensional, images. All such images are here displayed as graphs representing plots of image magnitude – similar, in fact, to Fig. VIt. Figure VIu represents a simple one-dimensional image, which in general possesses an (arbitrary) phase distribution in the coherent case but is wholly positive (i.e. has zero phase) in the incoherent case. Figure VIv shows a typical speckle image – it is only the magnitude of the image in the coherent case but is the complete image in the incoherent case. It is worth noting that a simple average of a large number of coherent speckle images can be expected to be faint, because of destructive interference. A similar average of incoherent speckle images gives a one-dimensional seeing disc.

Figures VIw and VIx shows the results of performing incoherent and coherent shift-and-add, respectively, on sets of speckle images, each consisting of 200 statistically independent images, all of which are statistically akin to Fig. VIv. In Fig. VIw, the three peaks present in Fig. VIu are superimposed upon a (more or less) smooth 'hump', which is equivalent to the background fog previously remarked in Fig. VIn. This hump is also virtually identical to the one-dimensional seeing disc. The hump has disappeared in the coherent shift-and-add image Fig. VIx since only the original image escapes destructive interference. We emphasize that coherent shift-and-add differs from its incoherent counterpart in that, after the brightest point (i.e. the point of greatest magnitude) in any speckle image is shifted to the centre of image space, the phase of the brightest point is subtracted from the phase of the whole image. This means that the phases of all the shifted speckle images are zero at the centre of image space, implying that one replica of the original image in each speckle image is not only aligned with its fellows in the other speckle images, it is also in phase with them. There is, of course, no such 'phasing up' of the remaining parts of the speckle images. So, the only parts which survive the averaging are the aforementioned aligned and phased replicas of the original image.

VII

IMAGE PROCESSING SYSTEM DESIGN

The previous chapters describe particular processing methods which are effective in important practical contexts, as is evidenced by the many studies referred to. For all these techniques to be really useful, as opposed to merely being of general scientific interest, they must be integrated into a comprehensive image processing system. The latter should provide a friendly environment within which the individual techniques can act in harmony. The power of such a system can then be brought to bear on routine or exotic applications of image processing as they arise, without having to develop additional hardware and/or software for each new problem. Whenever specific hardware and software are developed for some purpose or other, they can be absorbed into the system as general tools available for use on future problems.

This chapter and the next are concerned with the design, development and employment of a system such as is alluded to above. These chapters summarize our relevant experience with particular emphasis on image restoration and satellite image processing using the Physics and Engineering Laboratory's image processing system. Extensive documentation on this system is readily available in the form of microfiche as is made clear in section (6) of the References.

The reason for the emphasis on satellite image processing in general image processing system design is the extraordinary amount of digital image data that has been generated by satellites (primarily viewing Earth) during the last decade. The data from these satellites (especially those in the Landsat and NOAA series) has been made widely available, and this, has in turn, given a tremendous thrust to image processing research. The need to handle large images is closely tied to the need for sound image processing system design.

Almost all published work on image processing is concerned with individual techniques and results. When overall system design is discussed it is usually in a very general way that glosses over points which, although they may in some senses be minor, are nevertheless of considerable practical importance. We attempt to cover more detail and be more specific. In §40 we deal with the overall design of an image processing system, and in particular with its hardware design, and in §41 with overall software design. In §42 data structure, which is in itself a vital aspect of software design, is considered, and §43 discusses aspects of data acquisition. Finally, §44 discusses image display. Detailed discussion of individual programme categories is left until Chapter VIII.

Up-to-date information on hardware relevant to §§40, 43, and 44, is best obtained directly from manufacturers. The names of these can be obtained from advertisements placed in image processing journals and in particular the remote sensing journals listed in Section (2) of the References.

References for the discussion of image processing software in §41 are [Castleman, B1; Fraser, 1, 2; Hamlet, 1; IEEE, B1; Lawden, 1; McDonnell, 6; O'Callaghan, 1; Reeves, B1].

With respect to §42, the data structure and header format used in a particular image processing system is normally described in its documentation manual. An example is [McDonnell, 16]. The EROS Data Center has developed a standard format for the distribution of satellite image data [EROS, 1]. Fraser's F.F.T. algorithm is discussed in detail in [IEEE, B1].

Aspects of image display (§44) are dealt with in more detail in [McDonnell, 12; O'Callaghan, 2].

Because it makes more sense to be specific, rather than attempt to keep the discussion as general as possible, when assessing the comparative merits of various approaches to system design and implementation, we refer throughout this chapter to the system we are most familiar with: the PEL (Physics and Engineering Laboratory of the D.S.I.R. – Department of Scientific and Industrial Research) image processing system, located in Lower Hutt, New Zealand. We often find it convenient to mention commercial trade marks for several hardware and software items. The reader should note that UNIX is a trade mark of Bell Laboratories, while VAX, VMS, RSX, and RT-11 are trade marks of Digital Equipment Corporation.

We remind the reader that several widely adopted abbreviations, freely invoked in this and the following chapter, are listed in the Glossary.

The general and particular aspects of system design, which are discussed in detail in this chapter, are illustrated in Example VII with reference to the PEL system.

40 Hardware

The choice of a configuration for an image processing system involves numerous trade-offs. These are discussed in this section and the following ones.

The first decision to be made concerns the purpose of the system. Should it be for research or production? Should it be dedicated to a few particular tasks or should it be general purpose and adaptable? The important point here is that available hardware, operating systems and languages are all evolving rapidly. Many parts of dedicated production systems tend to become obsolete in under five years. This is particularly true of hardware dedicated to a single production task. As experience is gained with a system and the imagery it is producing it is inevitable that the processing requirements will change and increase. This implies that the system should be general purpose and adaptable and also available for use in research and development work as required. Even if the system is intended purely for research, occasional heavy demands will be placed on the system by individual projects. This suggests at least a temporary production mode. Also if a system is used primarily for production it should not need to be interrupted by ongoing research and development activities. All this points to the use of a time-shared operating system which offers reasonable response times for interactive use and which can also operate in batch or background mode for large jobs or sequences of jobs. Several such systems are available for mini and larger computers. The distinction between classes of computers has, of course, become very vague because of the advent of powerful microprocessor systems but our meaning should become clear in the discussion that follows.

The next question concerns the number and sizes of computers that are required. This depends, of course, among other things, on the size of image to be processed and the processing throughput required. The remainder of this section is devoted to topics affecting the choice of computer.

Image processing requires the use of many non-standard peripherals, a number of which are discussed in §§43 and 44. These include interactive television display systems, scanning systems for digitizing images (such as a rotating drum scan system, a CCD (charge coupled device) scanner array or a vidicon scanner), and also hardcopy display devices (such as a rotating drum write system, an electron beam writing system, or devices which plot directly to paper). In our experience it is essential from a logistical point of view that the computer controlling these special purpose devices be an integral part of the image processing system and be readily accessible for use, adjustment and maintenance. In practice, this means that it should not be shared with or controlled by another group.

A minicomputer (or a modern microprocessor equivalent) is ideal for handling image processing peripherals such as a digitizer or an interactive display system. Given access to the full minicomputer when required, installation and maintenance of these devices is relatively straightforward. Interfacing to a large computer system is much more complex and the necessity to locate the peripherals close to the computer can create extreme logistical problems. Furthermore, specialized image devices place a severe I/O (input/output) load on a computer system and a minicomputer can usually handle this load more efficiently. For these reasons we favour a dedicated minicomputer to handle special purpose peripherals.

Disadvantages of a minicomputer are as follows. First, it may lack the required processing power. One way of overcoming this is to run most non-interactive batch jobs on a larger computer in the vicinity. This brings up the question of software transportability which is discussed in the next section. A dedicated larger computer could be used instead of the minicomputer but the cost of this may be prohibitive. Alternatively the processing power of the minicomputer could be enhanced by adding an *array processor*. This is a device capable of carrying out simple computations on each element of a vector array in parallel rather than in series as for a conventional computer. For example, two image lines could be subtracted in one step rather than sequentially subtracting each corresponding pixel pair. Array processors are particularly suited to the neighbourhood processes discussed in §45. To make the best use of an array processor programs should be written with parallel operation in mind rather than simply modified if perchance an array processor should become available.

Replacing frequently used subroutines (e.g. F.F.T. calls – refer to §12) with standard array processor routines can result in a considerable improvement in the speed of programs whose speed is limited by these routines. However, speeding up programs by writing special purpose array processor software would take a considerable investment in development time and would be almost certainly at the expense of program transportability. Specialized software such as this is difficult to write. Such a step would only be worthwhile in a system which is primarily production orientated.

A second disadvantage of many older minicomputers is the lack of total addressable random access memory. This can limit the number of users that can efficiently time-share the system. However, many systems are now becoming available which do not suffer from this limitation. The problem may be partially overcome by using memory resident libraries of common subroutines which are shared by different users rather than included separately in the executable code of each program.

A third and more serious problem is the limited address space available to an individual program. This is typically 64 K (kilo) bytes and is due to the use of a 16-bit address word. This has two main effects. Programs which are too large to run directly need to be overlaid. By this is meant that independent sections of program code are forced to occupy the same region of physical memory when in use. Overlaying is in general undesirable but should nevertheless be achievable directly through the linking process without modifying the program itself. Thus transportability is not affected. All of this assumes that any large program which is to be run on a minicomputer has originally been designed with a calling structure suitable for overlaying. If this has not been done it can be extremely difficult to break a program into overlay segments. The other effect of limited address space is to limit available buffer size. Buffer space is required for the temporary storage of image lines. One way of overcoming this problem is to use virtual memory if the language and operating system allows it. It is important to note here that the term 'virtual memory' is used in two ways. First, we refer here to the use of nonstandard Fortran statements on 16-bit computers, which allow a program access beyond its 64K byte addressable space into the larger (typically 256K byte) memory space which is physically present. This first usage is also called extended memory. Second, it refers to a technique by which a standard program may equivalently access an area of memory which may be much larger than that physically present. This is achieved by swapping blocks of memory to and from disc as required. The technique is used on a number of larger computers such as the 32 bit VAX-11/780. The first type of virtual memory should be avoided where possible, whereas the second is highly desirable. Use of the first type of virtual memory is awkward and harms transportability. Alternatives are to put up with a limited buffer size or to move to a larger word size. The problem of limited address space is more serious for FORTRAN programs than for more modern languages such as PASCAL or C. This is because the former requires that a buffer size be dimensioned to be as large as it will ever need to be, whereas the latter allows a buffer size to be set to whatever is required for a particular run of a program. Some groups have overcome this problem with FORTRAN by using an assembler subroutine to dynamically allocate buffer space as it is required. This is an ingenious solution, but unfortunately again leads to transportability problems.

Choice of computer *word size* is an important consideration. Image acquisition, processing, and storage have traditionally been tied almost exclusively to an 8-bit pixel. For simplicity of software development it is therefore necessary to choose a word length which is a multiple of this. It also makes sense to have a word length which is a power of 2. This makes possible more efficient use of memory space and a more powerful instruction set. For example, a 32-bit computer can include 32-, 16-, and 8-bit instructions. This is of considerable advantage in image processing. The choice is then between 8-, 16-, 32-, and 64-bit computers: 8-bit instructions are normally available as a subset of 16-bit machine instruction sets: 64 bits is unnecessarily big for image processing and may be too expensive. The choice is really then between 16 and 32. The traditional word size of minicomputers is 16 bits, and most image processing systems are based around such a computer. As discussed above the limitations of this size are being felt increasingly. Accompanied with the desired instructions, 32-bit machines are available but are more expensive. Most image processing groups would be too small to afford such a machine at its present cost. Also, a dedicated 32-bit machine would usually be much

more complicated and more expensive to interface special purpose devices to and to write I/O drivers for. The resolution of these trade-offs is
coming quickly in the form of 32-bit microprocessors which combine the
advantages of an accessible dedicated machine and the 32-bit word size.
These should become the standard computers around which to build an
image processing system in the future.

In the meantime a satisfactory compromise is to use a 16-bit minicomputer to do interactive image processing and control special peripheral
devices, and to use a 32-bit computer (if one is available) to run large
image processing jobs. This is the approach adopted on the PEL image
processing system. Large jobs are run on a VAX-11/780 32-bit computer
under the VMS operating system. This computer is part of the DSIR
computer network and its use is shared with a number of other groups.
Large batch jobs are routinely run overnight, and smaller ones during the
day. An LSI-11/23 16-bit computer (this is really a microcomputer, but is
equivalent in power to a minicomputer) with 256K bytes of memory and
a floating point processor operates under the time shared RSX-11M
operating system. This dedicated system is used to control the following
peripherals – an Optronics Colorwrite machine, a 1600 BPI (bytes per
inch) magnetic tape unit, an 80M (mega) byte Winchester disc drive, two
0.5M byte floppy disc units, several VT100 terminals, an RS232 serial
communications line to the VAX-11/780, a line printer, and an Intel
8085 microprocessor system. The Intel 8085, in turn, controls an interactive colour refresh memory display system, a raster scanner, a spectrometer, and a CCD scanner system. This is typical of the range of peripherals
that the dedicated computer needs to control.

Another aspect of hardware design to be considered is *storage space.*
This can be *on-line* or *off-line* to the computer. Image processing makes
heavy demands on disc space which is an on-line storage medium. Lack of
disc space seriously inhibits the effectiveness of a system. There needs to
be enough disc space to store the operating system, program source,
subroutine source, subroutine libraries, linked programs, system
documentation, command file libraries, and test images as well as the
images temporarily loaded for current projects. On our system each band
of a full Landsat image occupies some 7.5M bytes. A disc space of from
150 to 300M bytes on the LSI-11/23 would be ideal. Operating over-
heads increase dramatically as disc space decreases because of the need to
repeatedly load files from and unload files to magnetic tape. If necessary
programs can routinely read or write image files directly from or to
magnetic tape. Note that in this case the magnetic tape is temporarily
used as an on-line storage medium whereas it is normally an off-line
storage medium. This saves disc space but is slower and is not suitable for
a number of algorithms such as the F.F.T. (refer to §12) box filtering
(refer to §46), and rectification (refer to §47), which require random

access by image line, rather than sequential access (these are all discussed in Chapter VIII). Magnetic tape is the standard off-line storage medium for large image files. The aspects to be considered here are the number of units and their density, speed and reliability. One unit is the minimum required. The main advantage of two units is that they can be used to copy tapes easily, and can also be used to process images in a tape-to-tape mode with a conseqent saving in disc storage. Tape densities in common usage are 800, 1600, and 6250 BPI: 1600 BPI has largely replaced 800 BPI, and 6250 BPI is being used increasingly – 1600 BPI needs to be available for reading *foreign* tapes. By foreign here we mean that the tapes have been produced on a different computer system using a non-standard data format. The higher the density used the fewer the tapes required and the less frequently they have to be changed. This is a strong argument in favour of 6250 BPI. As fast a tape unit as possible should be selected as this greatly simplifies the task of loading and unloading image files to and from disc. The main problem with magnetic tape usage is reliability. These problems can be reduced by thorough cleanliness, regular maintenance, and by adopting well thought out procedures to automatically recover from parity errors.

41   Software   *– summarize for yourself*

Overall software design is the most important aspect to be considered in the design of an image processing system. The overriding concern here is with *transportability*. This is the ability to transfer software from one computer to another. The system must be able to evolve steadily despite changes in hardware, operating system and language version. Useful system lifetime is then determined by its design framework which needs to be as general as possible.

The first aspect to consider is the choice of *operating system*. It should be emphasized that this choice is extremely important and will have a strong bearing on the computer chosen. The operating system is the most visible part of a computer to any user and it must be 'user friendly'. It is essential that it supports the use of *command files* (for the execution of a sequence of system commands) and includes a good *screen editor* (i.e. an editor which allows a full VDU screen of text to be viewed and interactively manipulated, rather than just a single line of text). Use of a *line editor* should not even be considered. The main limitation placed on transportability by the operating stystem comes through libraries of command files set up to carry out standard image processing jobs. These command files are an important practical component of an established image processing system. It should be straightforward to modify these with a good editor to run on another computer system. Ideally, the operating system should also offer a sophisticated file structure which supports directories and subdirectories of file names and includes a file protection system. Finally, it should offer a debugger for use with the chosen high-level language(s). This can significantly decrease program development time.

Some designers of image processing systems have gone so far as to provide a comprehensive *image processing command language*. This provides a framework within which menus of available programs are displayed and individual programs are executed. The main advantages are that it usually makes it easier for a user to become familiar with the system, and that the amount of typing required to run a program is reduced. On the other hand the new command language must itself be transportable and this invariably requires significantly more effort than for a normal image processing program. At PEL we have found the system command language which is part of the VMS operating system on our VAX-11/780 to be more than adequate for our requirements, and so we have not developed a special command language for our image processing system. One application in which a simple set of commands is particularly useful is in interactive image enhancement on a television display screen. For this purpose we provide an interactive program which provides and responds to a simple set of commands – but this program is just a component of the overall image processing system.

The next aspect of software design to consider is choice of a *programming language*. This must be widely available and its specifications should vary as little as possible between computers. Some years ago we decided to work with FORTRAN because it seemed the best available language at the time. The original version we worked with was ANSI (American National Standards Institute) FORTRAN 1966. This was soon replaced with FORTRAN IV as improved compilers became readily available for both mini and larger computers. Note that FORTRAN IV is a DEC product and not an ANSI standard. This is what we use at present but we will make the minor modifications necessary to upgrade to ANSI FORTRAN 77 when these compilers become more available on minicomputers. FORTRAN is far from being an ideal language. It does have some advantages, however. It is the most widely used scientific language so that compilers are readily available and are well tested. Also many extensive scientific subroutine packages (e.g. matrix inversion and F.F.T. subroutines) are available in FORTRAN but not in other languages. A large proportion of image processing software is related to I/O transfers. This is the weak point of FORTRAN (and other languages) and so is particularly troublesome for image processing. Our approach has been to do all I/O (except FORTRAN formatted writing of text) through a small number of system dependent subroutines. The latter contain any calls which depend on the FORTRAN compiler version and the hardware configuration. The rest of the software is designed to be system independent. Transferring the software to another computer involves making appropriate modifications to the system dependent subroutines only. A similar approach is taken in most image processing systems. Another difficulty with FORTRAN (and other languages) is that individual manufacturers offer additional non-standard features in their compilers. Use of these makes transportability much more difficult to achieve. This difficulty can be avoided by being very clear about what features are non-standard and scrupulously avoiding them. This depends on experience with a cross-section of compilers. The remaining difficulty with FORTRAN is its awkward overall design or structure. This relates particularly to the manner in which the various types of 'GOTO' statements are used. Although FORTRAN 77 attempts to overcome this difficulty, more modern languages such as PASCAL and the forthcoming ADA are clearly superior. It will, nevertheless, be some time before FORTRAN is displaced as the most widely used language. Once a commitment has been made to FORTRAN and a considerable amount of software developed, it is not easy to convert to a more modern language such as ADA. To rewrite the software completely would require a major and probably excessive effort. The only practicable way of doing this seems to be to write additional software in the new chosen language so that conversion is gradual. This requires that programs in one language can call subroutines in the other. Unfortunately this feature is at present only available on a small number of computers with compilers provided by the manufacturer. This situation is improving however. We are adopting this approach and are writing a proportion of our new software in Pascal.

The next aspect of software design to consider is *processing mode*. This can be either interactive or batch. We have found the use of command files (this is batch mode) to be an efficient way of running large and/or repetitive jobs. Each job is set up by copying and editing a standard command file, or a previous similar command file. This procedure simplifies the scheduling of tasks and reduces the chances of typing errors. This in turn reduces the staffing overhead required to operate the system. For example, we have set up command files to run for several days. These read images from magnetic tape, process them and write them to colour film, pausing when required to ask for a new tape or film to be loaded. If there is an interruption (due to a power failure for example) the command file can be restarted from where it stopped (after inspection of the log file produced by the command file). It is highly desirable to be able to run programs both in batch mode using command files and also interactively from a VDU terminal. This introduces a complication. Every parameter read by a program should be checked for errors as carefully as possible. In the event of an error in batch mode a program should stop and give an appropriate error message. However, when running interactively it is desirable for all errors to cause requests for data input to be repeated. We have resolved this conflict by making programs which can run either interactively or in batch mode always stop on an error. Note that real-time interactive programs (discussed further in §44) require the use of a feedback control device and are not normally run in batch mode.

For a program to be able to run interactively and in batch mode it is important that it can read a file name and open it from within the program itself. This is for simplicity and because the names of files to be used may depend on the interaction. Some large and antiquated systems require that all file names and logical unit numbers be declared outside a program before it is run. Such a system is unsuitable for interactive image processing.

Another aspect to consider is *software documentation*. The usefulness of software is limited by the quality of its documentation. In an evolving system documentation is continually being upgraded and extended. This creates a need for up-to-date documentation to be readily available. The best way of doing this is to have the documentation stored on disc and directly accessible. A complete new manual or any section can then be printed when required using a command file. In some systems a separate documentation file is maintained for each program and subroutine. We have found it simpler to place the documentation for each program or subroutine as comments at the beginning of its source file. When required a manual production program extracts the documentation and formats it as required.

The time taken to develop software can be dramatically reduced by ensuring the software is as modular as possible. This is aided by extensive use of clearly defined subroutines and this leads to short and simple main programs. Once an extensive subroutine library has been built up, a new program can frequently be written simply by copying and editing a previous similar high level program. Program debugging is simplified by insisting that all subroutines, except those crucial for speed, fully check each input parameter and give a unique error message or number for every error.

Most of the CPU time used by a high-level image processing system is taken up in a small number of frequently used subroutines. Rewriting these routines in assembler should result in a significant increase in overall speed. In our experience a speed-up of four is not uncommon. To maintain transportability and to assist debugging we maintain both FORTRAN and assembler versions of these routines.

A further aspect of software design to consider is that of random access memory usage. Our VAX has a large amount of available memory and typically allows each program up to 4M bytes of virtual memory. Use of this virtual memory allows some programs (such as those discussed in §47 for rectification, or interpolation between contours) to run very efficiently.

These programs could usually be run on a minicomputer using appropriately reduced array sizes but would be very inefficient. This consideration can restrict the transportability of some programs to large computers only, or alternatively can require the modification of an array size. But note that even on a larger computer virtual memory may be no substitute for specially designed algorithms that directly access data on disc. Large 2-D transforms (as discussed in §48) are a classic example of this. For example, these can take a few minutes with an optimized scheme for the management of disc and memory data storage, whereas the virtual memory approach may take hours of page thrashing (repeatedly swapping blocks of data between memory and disc).

Some additional problems peculiar to running image processing software on a minicomputer are now discussed. On our LSI-11/23 most programs are limited by I/O speed. This is particularly true when writing to the Optronics Colorwrite display device which writes to film at a maximum rate of 4 lines/sec. Input/output to the Colorwrite was considerably speeded up by buffering the I/O and not waiting after each write request was made. This allows a new line to be calculated while the previous one is being written from the buffer. Minicomputer operating systems are usually more restrictive than those for larger computers. They frequently have bugs or unexpected shortcomings that make implementation more difficult than anticipated. For example, in setting up our software to run on the LSI-11/23 under the RT-11XM operating system we encountered a few FORTRAN compiler bugs, a difficulty over determining the lengths of records read from magnetic tape, and another difficulty with the maximum root overlay segment size for large programs.

42 Data structure

It is now appropriate to introduce the concept of a *generalized image file*. Software development is considerably simplified if, wherever possible, disc data files, and I/O devices to which images may be written or from which they may be read are treated as being *equivalent disc image files* in a single standard format. For example, files containing 1-D histograms, 2-D histograms, graphs, lookup tables, and images can all be treated as images. Devices such as a magnetic tape unit, an interactive display refresh memory and an Optronics Colorwrite machine can be treated as being equivalent disc image files. For these devices I/O routines automatically seek to the required image line and thus mimic random access. The advantage of the generalized image file is that high-level software (such as main programs and important subroutines) can assume that all image files are disc data files. In general, special programs do not have to be written for particular peripheral devices.

New devices can be added to the system simply by modifying the appropriate system dependent I/O subroutines. Unlike real disc files, these devices usually require the specification of additional parameters. We use a system of abbreviations to include all of these as part of the file name. For example, M:RS2FL101FP201NL128NP128 means rewind (R) the magnetic tape unit (M:), skip 2 files (S2), and then read 128 line (NL128) by 128 pixels (NP128) sub-image with first line 101 (FL101) and first pixel 201 (FP201). A more straightforward example is $M:E$ which we interpret as meaning read the next image file on magnetic tape and rewind the tape when the end has been reached (E). Sensible defaults are used for all parameters and as much information as possible is read from the image header (which is discussed later in this section).

A decision must be made as to what size pixels are allowed. At present we allow 1-, 8-, and 16-bit integer pixels, 32-bit real pixels and 64-bit complex pixels. New sizes can easily be added but usually any one program only needs to operate on images with at most two different pixel sizes. We use 16-bit images when 8 bits does not give enough precision (e.g. for an image in which each pixel is a terrain height in metres). Real and complex images are routinely used in restoration work involving the F.F.T. It should be pointed out here that 8-bit pixels are usually viewed as ranging from 0–255. However, some computers (notably DEC) do single byte arithmetic ranging from −128 to 127. It is therefore advisable to avoid 8-bit arithmetic and to do integer arithmetic at 16- or 32-bit precision. We normally use 16-bit precision in doing arithmetic operations on 8-bit images. The 16 bit range is taken to be −32768 to 32767 which is normally adequate for transferring image data between computers. Real (or complex) pixels are machine dependent because of the computer manufacturers' choice regarding the size and position of the exponent and mantissa, and even the base used (e.g. 2 or 10).

This brings us to the question of *headers*. Every image file should have an associated header which contains information about the image. As a good image processing system develops, its capability becomes constrained by its header design. The information retained in the header eliminates the need to repeatedly remember and enter information about the image and thus simplifies system operation and reduces the chances of input errors. In some systems the header is kept in a separate file from the image. As the header must still be written to tape with the image we feel this is an unnecessary complication and so keep the header as an integral part of the image file. The header must be at the beginning of the file so that when it is read from tape the image size can be determined prior to reading the image itself. There is no widely accepted standard format for a header. Adoption of such a standard would lead to problems because as new programs are developed they need to be able to store specialized information in the header. This means the header must be expandable in some way. We have adopted a header format which consists of one or more records of 1024 ASCII characters. The first record contains information about the header size and the image size. The header is in ASCII to make it easy to edit and inspect and to prevent difficulties with converting binary data (e.g. reals) between computers. The record length is long enough that the number of records has to be increased only rarely. This makes it possible to edit the header and/or image in place on disc if required. When it is required to store new data items in the header, new positions within existing header records, or new records can be allocated. We include in our header a history of the processing that has been carried out on an image since its header was first created. This history can be displayed on request and is a valuable aid in running a complicated sequence of operations on an image.

It is worthwhile discussing the contents of the header in more detail. Several types of data should be held. First, are items such as the number of lines and pixels in the image, the number of bits/pixel, the image type (e.g. Landsat MSS), its spectral range (e.g. 700–800 nm), and the Greenwich Mean Time at which it was taken or created. Second, are specific details about the data source. For satellite data these would include the satellite position, velocity and look angle, and the elevation and azimuth angles of the sun. Third, are details of image geometry. These include the latitude and longitude of the image corner points and centre, and information on whether or not the image is rectified (as discussed in Chapter VIII). If the image is rectified the map projection used must be stored as well as the top left and bottom right map coordinates and the sampling distance. Fourth, is calibration information. This includes information on whether or not the image is calibrated, the calibration units, and the calibration range corresponding to the data. Note that an image or pseudo image (such as a histogram) can be calibrated in its horizontal and vertical dimensions, as well as in intensity. Furthermore, a sequence of image frames which make up a movie can be considered to be calibrated in time as well. Finally, it should be noted if the image has been Fourier transformed, and whether the transform was one or two-dimensional (2-D). The reason for the latter is that a 2-D image becomes complex after either a 1-D or a 2-D F.F.T. These cases need to be distinguished. Other items which need to be stored in the header depend on the particular applications for which the system is used. It should be pointed out that there is a programming overhead in maintaining the header information described above. However, this is more than repaid by such advantages as the ability to combine or merge calibrated and rectified data sets automatically.

Another important point to be considered is *disc image file structure*. Disc file structures available depend on the computer operating system to be used. The file structure must allow random access by image line because many algorithms require it. This rules out sequential access file types. Most, if not all, operating systems allow random access files with a fixed record length. The size of the full file must be declared before it is written to. Some systems such as UNIX allow random access files with variable record length. Under UNIX random access is to any byte position in the file. This is a desirable capability but choosing it would restrict transportability as it is not available on most systems. We have already mentioned the desirability of having fixed length (e.g. 1024 byte) header records. This length may be greater or less than the image line length and so suggests the need for variable length records. This runs into the transportability problem stated above. We have resolved this difficulty by writing the header into a block at the start of the image file. The size of the block is the minimum multiple of the image line length required to store the full header. This allows fixed record length random access files to be used to store our images. The fixed record length equals the number of bytes in an image line rounded up to be a multiple of 4 (to give an integral number of 32-bit words). When initially reading the header from an image file the record length may be determined by using an appropriate system call.

It is sometimes important to be aware of the way that image data is physically stored on disc. For example, under the VAX VMS record management system each record of a random access file starts on a physical disc block boundary. Each disc block is of length 512 bytes. This can be very wasteful of disc space. Furthermore, buffered system I/O called through a high-level language such as FORTRAN can be rather slow. Thus, for reasons of space and speed it can sometimes be a worthwhile investment to replace the system I/O calls with an assembler package designed specifically for image files. This is the approach adopted on our system and it has significantly increased I/O speed.

Some algorithms such as rectification of large images require that files be byte-addressable. This can be achieved inefficiently using a FORTRAN subroutine which reads and writes whole image lines and extracts or inserts the required bytes. Alternatively, it can be achieved efficiently using an I/O system such as that of UNIX or by a specially written assembler package. It should be noted that use of a special purpose I/O system does not affect transportability because the I/O system appears unchanged to higher-level programs and subroutines.

Another aspect of disc file structure to consider is disc block contiguity. For most systems, consecutive disc blocks can be read or written more quickly if they are physically contiguous than if they are non-contiguous. It therefore makes sense to request that image files be made as contiguous as possible within the constraints of the operating system.

A final problem with disc image file structure concerns whether the operating system allows the record length of a file to be redefined while keeping the file size constant. Most systems do not allow this to be done.

This capability is used for example in Fraser's program to perform the F.F.T. of a large disc image file in place. We have circumvented the difficulty and maintained transportability by finally copying the result of the F.F.T. to a new file of the correct record length.

In concluding this section it is worthwhile repeating that the main limitations on transportability concern I/O. Thus this aspect of image processing software design warrants careful thought if transportability is to be achieved or approached.

43 Data acquisition

An image processing system may acquire digital image data by several means. The usefulness of the data depends on what is known *a priori* about the image and the information that can be deduced from it. This section summarizes acquisition techniques and discusses aspects of the image that are of interest.

Images are often digitized directly from the scene they represent. Examples of this are images produced from aircraft and satellite multi-spectral scanner systems which digitize an image a line or group of lines at a time. A whole image may be captured in a single exposure using a CCD scanner array. A scan converter system can be used to capture a television image frame and store it in a digital refresh memory. Such images are normally recorded at an accuracy of at most 8 bits per pixel.

Alternatively an image may initially be recorded photographically. The digital image must then be produced from the film or print using a device such as a scan converter, a CCD scanner, a flat-bed scanner, or a rotating drum scanner.

A further possibility is for the image to be made in the computer. For example original data could be obtained from using a graphics digitizer to digitize terrain contours. These would then be written into an initially zero image file. Finally, interpolation between the contours would be carried out (as discussed in §47) to produce a digital terrain model or image.

Frequently, an image is provided as a file on magnetic tape with a foreign format (as discussed in §40). Because of the lack of a universal image format, it is often awkward to read such a tape. If header information is not required, we can read most foreign tapes with a general input program. However, if header information is required from the foreign tape, a separate program is produced for each new format. This seems to be the simplest way of surmounting this difficulty.

This discussion also applies to graphics files. We use a standard graphics format and convert foreign files to this format when they are accepted into our database. The important point here is to ensure that a standard format is rigorously maintained on the system, and that all relevant information about the image or graphics file is available in the header in computer-readable form.

44 Image display

In order to do justice to the quality of the work produced by an image processing system it is important to maintain a high standard of image display. Hardware display devices and related software are discussed in this section.

A display device can be interactive and real-time, or else a hard copy device. The most common interactive display is a colour or monochrome television refresh memory system. Let us consider the case of a colour display system. In general, a multispectral image to be displayed is available in several channels or *bands* each corresponding to a different spectral range or pass-band. Up to three image bands can be displayed at a time by assigning each television colour gun (blue, green, or red) to one of the available bands. The better display systems store the full 8 bits of each band of an image in the refresh memory and display the data through an 8-bit lookup table for each band. The 8-*bit lookup table* is simply a memory chip which in real-time translates each 8-bit pixel value (ranging from 0–255) as it is read from the large refresh memory into an output pixel value (again ranging from 0–255). This output pixel value is then fed through a digital to analogue converter to the appropriate television colour gun. The operator can use the computer to interactively manipulate the lookup tables to give the best display. Some systems offer much more hardware capability (refer to §51). A typical display screen resolution is 512×512 pixels, although more expensive systems with resolutions such as 1024×1024 pixels are becoming available.

Hard copy devices include electron beam and laser recording systems which write direct to rolls of film, electron beam systems which write direct to a print or transparency, and rotating drum systems which write direct to a print or transparency. Our system uses an Optronics C4300 Colorwrite rotating drum machine which can write, one primary colour at a time, to a 25-cm square sheet of colour film using a spot size of 25, 50, or 100 microns. We normally write to Colour positive Kodak Ektachrome film using a rotation speed of four rotations a second and the 50-micron spot size. At the time of our purchase, the Colorwrite was the only colour device in its price range that offered sufficient geometric accuracy for mapping applications. Its writing speed is limited by the rotation speed of its drum, which in turn is limited by the time required to expose the film. The French processing system for use with SPOT satellite imagery achieves a greater throughput by using a much larger drum. Monochrome

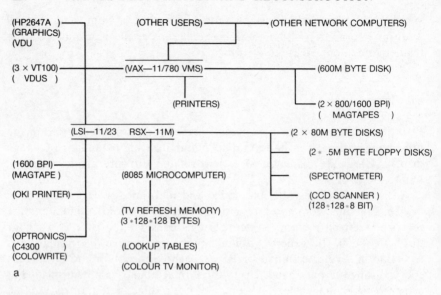

a

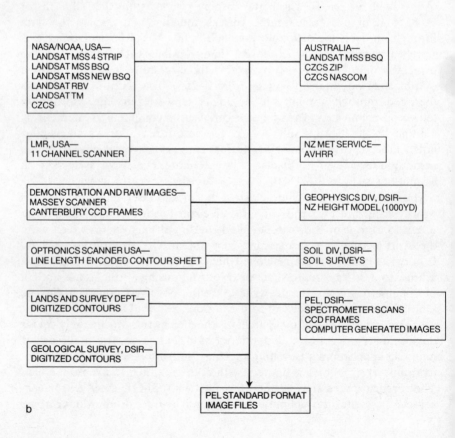

b

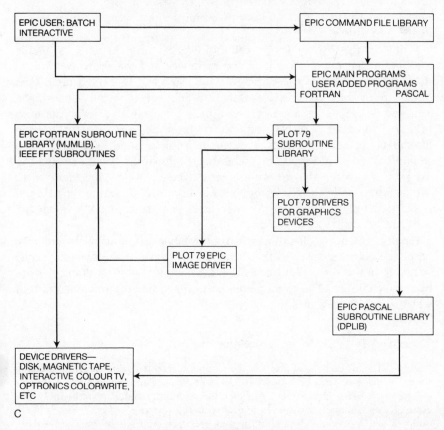

Fig. VII.

and, more recently, colour laser rotating drum systems have become available which produce much more light output and are thus limited by drum rotation speed.

We have carried out several modifications to our Colorwrite machine. In particular, an automatic filter wheel selection mechanism has been installed. This allows, for example, a command file to be used to fill a colour film sheet with images with the Colorwrite running overnight and unattended. Computer controlled stopping and starting of the Colorwrite is used to reduce the wear on the drum bearings. A filter has been added to the power supply to reduce the chance of a power fluctuation causing an error while writing an image. When writing an image line to the Colorwrite, it is sometimes desirable to expand each pixel by a factor such as 2 or 4. This would normally require the use of an extra long buffer array which may be hard to find given the limited memory space available on a minicomputer. We do expansions like this using a simple improvement in the interface hardware. All our photographic processing is carried out in our own facility. This has proved important to the maintenance of quality control.

Whatever the display device, calibration is essential. We have available a program that models a colour display device and the eye's response to it, and then calibrates the display device using a 3-D lookup table as is discussed in §45. This is an elaborate procedure. What is normally required in practice is a set of 1-D 8-bit lookup tables to calibrate each primary colour of a display separately. In setting up these calibration lookup tables several points need to be kept in mind. The calibration should be linear in density, or preferably, in apparent visual density. The dynamic range for the calibration should be carefully chosen so as to avoid saturation in the final products. Furthermore, the calibration should be carried out transparently to high-level programs, so as to keep the software as simple as possible.

Finally, given a well-calibrated display device, it is usual to enhance an image before display so that it makes good use of the dynamic range available in the display device. The most common ways of doing this are by the techniques of histogram equalization and linear stretching, both of which are discussed in §45.

Example VII – PEL image processing system

Figures VIIa through VIIc are block diagrams of facets of the image processing system which has been developed at PEL (the Physics and Engineering Laboratory of the DSIR – Department of Scientific and Industrial Research) in Lower Hutt, New Zealand. This is a typical general purpose system. Figure VIIa outlines the hardware configuration and shows the range of peripherals and computers that have been advantageously incorporated into the system. Like others of its kind, the PEL system has many 'customers' whose separate needs must be accommodated satisfactorily. All of the organizations named in Fig. VIIb have their own data-formatting peculiarities. Each 'foreign' format must be converted to PEL standard format before the data can be processed. Software design is greatly eased by breaking down the total package into a number of functional blocks. This allows an evolving multiple-user software system to be developed (the PEL software system is called EPIC). The design strategy adopted at PEL is summarized in Fig. VIIc.

VIII

PROGRAM CATEGORIES

A general purpose image processing system uses a wide variety of processing techniques. Many of these must be invoked when carrying out a specialized application such as image restoration. This is, first, because the image to be restored may need to be analysed to determine appropriate restoration parameters. Second, the blurred image may need to be preprocessed before restoration is attempted. Third, once restoration has been carried out the image may need to be further processed before it is finally displayed. General image processing techniques are therefore directly relevant to image restoration.

In this chapter we divide image processing techniques into a number of categories and discuss each category in some detail. Many such divisions are possible. For example, programs could be categorized according to their application (restoration, enhancement, rectification, etc.), the type of image being processed (e.g. Landsat, Seasat, electron microscope, astronomical, etc.), the number of bits/pixel of the input image, or the programming language.

We find it most useful to subdivide programs according to their handling of image I/O (input/output) from and to disc or equivalent disc devices (refer to §40). This is firstly because a simple division in this manner is possible, and secondly because the simplest way to write a new image processing program for an existing package is to edit a previous program which has a similar I/O structure to that which the new program requires.

The categories we have chosen provide the headings for §§45 through 51. The first three of these (neighbourhood, random line access, and block) are concerned with processing a single image in a disc-to-disc mode. There is inevitably some overlap between the individual categories. Transformations (in particular the F.F.T.) are considered separately in §48 because of their particular relevance to this book. The multiple-input-image processes introduced in §49 generalize the single-input neighbourhood processes discussed in §45. Useful associated programs, which do not fit neatly into the previous categories, are the subject of §50. Finally, the special requirements of interactive display system software are considered in §51.

General references appertaining to this chapter are [Andrews, B1, B2, B3, 2, 3; Beebe, 1; Bernstein, B1, 1; Foley, B1; Fraser, 1, 2; IEEE, B2; McDonnell, 1 through 18; O'Callaghan, 1, 2; Sawada, 1; Van Wie, 1].

The reference for 'theme processing' in §45 is [Golay, 1]. In §47, the references for PLOT79 and for Voyager I imagery are [Beebe, 1; McDonnell, 5]. As regards §48, Section 1.5 of [IEEE, B2] is particularly relevant.

Example VIII shows how enhancement of different aspects of an image can be effected with the aid of techniques introduced in §§45 through 49. Each technique permits a particular aspect to be displayed to best advantage. It is instructive to compare enhancement, as invoked in this example, with the restoration and reconstruction techniques which are illustrated in Examples III–VI.

As in Chapter VII, we give specific illustrations of our general arguments by making frequent reference throughout this chapter to aspects of the PEL image processing system.

45   Neighbourhood processes

In §§45 through 47 we are concerned with processing a single input image in order to produce a single output image. In each case the image is assumed to be stored on disc, although storage could be effected with an equivalent device of the kind discussed in §42. Each image consists of a number of lines, and each line consists of a number of pixels along the line. Image lines are read from or written to a storage device one at a time. We note that the output image lines must always be written sequentially when the image is first created, even if it is later edited in a *random access mode* (by this we mean that image lines are not accessed sequentially).

By a *neighbourhood process* we mean that the input image is processed sequentially (line by line), and that each output line is calculated from a small neighbourhood of corresponding input lines. These input lines are adjacent and typically number one or three. By a small neighbourhood we mean that the full lines in the neighbourhood can all be stored simultaneously in the program's random access memory space. A neighbourhood process is the most common type of image processing algorithm. It can run efficiently when both the input and output images are equivalent disc files.

It is appropriate here to introduce some new terms. The *pixel value* of a pixel is the amplitude of the image at that pixel. The *histogram* of an image is the frequency distribution (in the conventional statistical sense) of its pixel values. Usually a histogram is derived from an image with integer pixel values (e.g. 1, 8, or 16 bits/pixel) so that a histogram $HIST(J)$ is defined for valid pixel values J. For an 8-bit image J ranges from 0 to 255. The histogram can be plotted as a graph or stored as an image file for later use by other programs.

A *cumulative histogram* $CHIST(J)$ is defined from a histogram $HIST(J)$

by

$$CHIST(J) = \sum_{K=1}^{J} (HIST(K) + HIST(K-1))/2 \qquad (45.1)$$

for $1 \le J \le 255$, and

$$CHIST(0) = 0. \qquad (45.2)$$

A *lookup table* is a function which maps an integer input pixel value into an output pixel value, which is usually an integer with the same range as the input pixel. Consider an 8-bit lookup table $LKUP(J)$. J and $LKUP(J)$ both range from 0 to 255, and the lookup table can be stored as a 1×256 image file or an image line for later use by other programs. We are talking here about a 1-D lookup table because the output is from one input pixel value. 2-D and 3-D lookup tables are discussed in §49.

In the simplest type of neighbourhood process a new image is being created, so no input image lines are required at all. Examples are programs that create a constant image or an image covered by a map grid. The image size could be defined in terms of total image (line, pixel) size, or map corner coordinates and sampling distance. Further examples are the creation of test patterns and lookup tables, or the calculation of images containing calibrated physical data. For example, it is sometimes important to remove the atmospheric radiance contributions from a recorded satellite image of the sea in order to obtain an image of the sea's surface. The component of atmospheric radiance due to *Rayleigh scattering* of sunlight by gas molecules in the atmosphere can be calculated given the wavelength, the satellite look-angle geometry, and the sun position (or equivalently the Greenwich Mean Time and the map coordinates of the image corners).

In the most common type of neighbourhood process each output image line corresponds to a single input line. An image may be copied from one equivalent disc file to another. It may be converted from one number of bits/pixel (refer to §42) to another, with or without scaling the pixel values. A rectangular map grid may be added to an ungridded image, or removed from a gridded image. A subscene may be extracted from an image by reference to either image or map coordinates. An image may be absolutely calibrated (in radiance or height for example) by means of a calibration lookup table, or a calibrated image with a particular calibration range (e.g. 0–20 °C) may be produced from a calibrated image with a wider range.

Each pixel value then has an associated calibration unit such as microwatts/cm^2/steradian for radiance, metres for height, and °C for temperature.

Many simple enhancement algorithms are implemented by means of a lookup table. Each output pixel is a function of the corresponding input pixel. Such algorithms include linear stretching, histogram equalization, and colour coding. It is worthwhile discussing these in turn. An 8-bit lookup table to *linearly stretch* an image from pixel value A to pixel value B is given by

$$LKUP(J) = 255(J - A)/(B - A). \qquad (45.3)$$

$LKUP(J)$ is calculated using real arithmetic, rounded to the nearest integer and then clipped to range between 0 and 255. In other words if $LKUP(J)$ is less than 0 it is set to 0, and if it is greater than 255 it is set to 255.

Histogram equalization is an enhancement technique in which an image is processed by a lookup table to produce an image with an approximately uniform histogram. In general, the lookup table is non-linear and the resultant histogram cannot be exactly uniform, because the pixel values are quantized. In practice, the way the lookup table is derived is to force the cumulative histogram of the resultant image to be as linear as possible. For an 8-bit image, let $CHIST$ be the cumulative histogram of the original image. Then the lookup table is given by

$$LKUP(J) = 255 CHIST(J)/CHIST(255). \qquad (45.4)$$

$LKUP(J)$ is again calculated using real arithmetic and is rounded to the nearest integer. The advantage of histogram equalization is that it leads to a good overall display of the information in an image and is carried out automatically. Linear stretching can lead to an improved display of information for a particular application, but the appropriate stretch values are normally chosen interactively with a television display screen (refer to §44).

Colour coding is an enhancement technique by which each pixel of an image is assigned a specific colour according to its pixel value. The colour image has three primary colour components (blue, green, and red) and each of these is produced by a separate lookup table from the original image.

A frequently needed type of program converts images from a foreign format magnetic tape (refer to §42) to the standard format for the system. The header for each image on the foreign tape (if indeed it has one) will need to be decoded, and the input image lines extracted and reformatted as appropriate from the tape data blocks. A single foreign data file may contain several image bands with the pixel values interleaved in some manner and thus produce several output files. Institutions distributing image data seem unable to keep to a single unchanging data format even for one type of data produced by a single group. For example, Landsat satellite data has been distributed in several differing formats by the EROS Data Center in South Dakota. A different program is required for each format.

Another type of neighbourhood process consists of statistical programs which read a single image, calculate statistical information about it, and possibly write this to an equivalent disc image (e.g. histogram or lookup table) file. Typical statistics include mean and range of the pixel values, saturation level of the histogram, and image areas corresponding to different pixel values in the histogram. The histogram may be used to calculate lookup tables for later use in enhancement. For example, a lookup table could be produced which clips 3 per cent off each end of the histogram and carries out a linear stretch on the rest of the histogram. In order to speed up the statistical programs discussed here, it is sometimes desirable to sample only selected lines and pixels. In this case these programs properly belong under random line access processes (§46). Statistics can also be gathered on image striping (a characteristic of satellite scanner systems discussed in the next paragraph), and used to prepare lookup tables to compensate for the striping.

The *striping* usually arises because the image lines are digitized in blocks using several detectors rather than sequentially using a single detector. The N lines within a block (6 in total for Landsat images) are digitized simultaneously but each line is digitized by a similar but different detector. Mismatches between the N detectors give rise to the striping effects. A histogram is produced from the pixels corresponding to each detector, and $(N-1)$ lookup tables are calculated which match the $(N-1)$ histograms (apart from the first) to the first histogram.

Neighbourhood processes can be used to carry out simple geometric corrections. For example, an input image can be increased or decreased in size to have any desired rectangular output size. A variety of interpolation techniques such as nearest neighbour or bilinear are available (refer to §§11 and 33). By *bilinear interpolation* we mean linear interpolation in two perpendicular directions. For example, a rectangular image may be resampled by interpolating in the pixel direction along each image line, and then in the line direction between the image lines. Other examples of simple geometric corrections are reversing an image left-to-right, or applying deskewing corrections to correct for the effect of the earth's rotation. The latter simply involves translating the pixels in each image line by an appropriate amount to the left or the right.

Many common image processing algorithms involve neighbourhood processes in which each output pixel is a function of a small number of pixels in the neighbourhood of the corresponding input pixel. Typically, a 3×3 pixel neighbourhood is used. Examples of such algorithms include high and low pass filtering directly in image space with a non-recursive filter array (refer to §18). High- and low-pass filters accentuate respectively high and low spatial frequency information in an image. Another example involves images which are *digital terrain models* so that each pixel value represents the height of the terrain at the map reference corresponding to that pixel. The *slope of* such an image can be estimated at each pixel by calculating the first-order finite differences in the x- and y-directions. From such data, maps of slope and also 'aspect' (i.e. the direction in which a sloping piece of ground is facing) can be generated. Furthermore, because the light radiance R received at a detector from a point on the ground is a function of the irradiance L at that point and the relative geometry of the point, the detector and light source (usually the sun), slope data can be converted via a lookup table into the form of a pseudo-relief map, in which 'relief' is simulated by shading. Such automatic *relief-shading* is so attractive from a cartographical point of view that it remains a useful display technique even when one has no particular concern with the variations in height of terrain. Note that, if θ is the angle between the source and the normal to the ground (at the particular point in question), and if ϕ is the angle between this normal and the line from the point to the detector, then a typical model for the relationship between R and L is

$$R = (1 + (\cos(\phi))/\cos(\theta))^{-1} KL, \tag{45.5}$$

where K is a constant characterizing the terrain.

An *image mask* is a 1-bit image (i.e. an image whose pixel values are either 0 or 1) which is used to perform logical (as distinct from arithmetic) operations on other images. For example, a mask may be used to set the value of an image pixel to zero if and only if the corresponding mask pixel is zero. Such masks can be categorized as multi-input-image processes (see §49). However, the (often useful) operation of *growing* (or, in the opposite sense, *shrinking*) the non-zero portion of a mask *is* a neighbourhood process. An efficient growing technique, which is known as *Golay theme processing*, works as follows. Consider the 3×3 pixel neighbourhood which is centred on each pixel of a 1-bit image. This neighbourhood may be ordered as a 9-bit sequence in the form of a 9-bit number. On comparing the latter with a previously prepared lookup table, the pixel values of the 1-bit output image are generated. A typical Golay theme process would be to leave unchanged all pixels whose values are 1, but to set to 1 the value of any pixel whose value is 0 but which is adjacent to at least one pixel whose value is 1. This particular process tends to cause a mask to grow rapidly. It is often appropriate to apply it repeatedly, the number of times being determined according to whatever criterion seems most appropriate for the application which the user has in mind. Note that Golay theme processing must be implemented non-recursively (because if it was implemented recursively, each pixel would necessarily be set to unity!).

Satellite images are usually recorded simultaneously in several spectral bands (conventionally referred to as *image bands*). A set of satellite images is said to be *classified* when each pixel is assigned to a particular class (or category) according to the variation of its amplitude across the image bands.

Examples of classification categories are: water, cloud, forest, pasture, urban, etc. There is a neighbourhood process algorithm which attempts to remove ancillary detail from classified images. It replaces each pixel value by the most frequent of those of its nine nearest neighbours (including itself), provided the most frequent value occurs at least four times. Depending upon the application, it can be appropriate for this algorithm to be successively applied a (previously specified) number of times.

46 Random line access processes

A *random line access process* is an image processing algorithm in which the input image lines are accessed non-sequentially. To run efficiently the input image file must be stored on disc, but the output image file can be an equivalent disc file. Simple examples are programs which swap an image from top to bottom (possibly in place), or edit individual image lines.

We now discuss some random line access processes which are of particular interest. Numerous techniques are available for reducing the noise level in, for example, a given blurred image. These depend on making reasonable assumptions about the characteristics of the noise and of the underlying image. The differences between these characteristics can be exploited to reduce the noise level. It is desirable that the noise reduction process does not add to the image blurring. A *median filter* is a filter which replaces each pixel value with the median of those pixel values which lie within a rectangular box of a fixed size which surrounds and is centred on the pixel being processed. This is a useful tool for removing high-frequency noise, spikes, bit dropouts, bad lines, scratches, dust, and other anomalies. Ideally, the size of the box should be chosen small enough to avoid blurring the image, and large enough to ensure that the pixels within the box are not dominated by the blemish being corrected. A fast ordering algorithm, which is related to the F.F.T., is normally used in calculating the median. In this case the filter is called the *Tukey median filter*.

Another useful technique with similar application is *low-pass box filtering* in which each pixel value is replaced by the average pixel value within a surrounding rectangular box of fixed size. Box filtering has a number of advantages over filtering with the F.F.T. In particular it is very fast (its speed being comparable to that of convolution with a 2×2 direct non-recursive filter), it can operate directly on 8- or 16-bit images and it requires little memory space. Furthermore, repeated applications of box filters of reducing size can be used to approximate a more complicated filter such as a Gaussian. We have included options in our box filtering programs such that each pixel value within the box is only used if it lies within a chosen valid data range, and such that the change produced by the filter for a given pixel is only implemented if the change lies within a chosen valid range. These simple modifications make box filtering a much more powerful tool. It can, for example, selectively smooth a noisy image in areas where it should be smooth while leaving it unchanged near edges, thereby introducing very little additional blurring.

It is worthwhile discussing the box-filtering algorithm in some detail in order to show the reason for its speed. In its simplest form low-pass box filtering can be described as follows. Let $f_{m,n}$ be pixels of an $M \times N$ image $f(x, y)$ which is to be box-filtered to give pixels $b_{j,k}$ of an $M \times N$ image $b(x, y)$. Then for a $(2J+1) \times (2K+1)$ box filter, we see that

$$b_{j,k} = (1/(2J+1)(2K+1)) \sum_{m=j-J}^{j+J} \sum_{n=k-K}^{k+K} f_{m,n}. \tag{46.1}$$

In order to understand how the algorithm works, it is first necessary to consider the way in which image data is stored in computer memory. We need to maintain an N word-storage buffer *IBUF* and to utilize a single word variable *ISUM*. *IBUF* is updated before calculating each output image line, and *ISUM* is used in calculating each pixel of the output image line.

For the lth line, the nth word in the buffer *IBUF* is conveniently written as

$$IBUF_l(n) = \sum_{m=l-J}^{l+J} f_{m,n} \text{ for } n \in \{1, 2, \ldots, N\} \tag{46.2}$$

The buffer $IBUF_{l+1}(n)$ is generated from $IBUF_l(n)$ and the input image lines $(l+J+1)$ and $(l-J)$. This requires that these two input lines are randomly accessed. $IBUF_{l+1}(n)$ is obtained from $IBUF_l(n)$ by adding the $(l+J+1, n)$th pixel and subtracting the $(l-J, n)$th pixel:

$$IBUF_{l+1}(n) = IBUF_l(n) + f_{l+J+1,n} - f_{l-J,n}, \tag{46.3}$$

where we emphasize that $f_{m,n}$ is identically zero, by definition, if $m < 1$, $m > M$, $n < 1$ or $n > N$. Note that the initial values $IBUF_l(n)$ must be calculated explicitly from (46.2).

An intermediate value *ISUM* needs to be calculated for each pixel k along each output line j. We call this particular value $ISUM_{j,k}$, and it is given by

$$ISUM_{j,k} = \sum_{n=k-K}^{k+K} IBUF_j(n). \tag{46.4}$$

Note that the formula

$$ISUM_{j+1,k} = ISUM_{j,k} + IBUF_j(k+K+1) - IBUF_j(k-K) \tag{46.5}$$

gives $ISUM_{j+1,k}$ in terms of $ISUM_{j,k}$, so that there is a correspondence between (46.5) and (46.3). An initial value $ISUM_{j,1}$ is calculated for $ISUM$ using (46.4) and noting that $IBUF(n)$ is zero for $n < 1$ or $n > N$. The remaining values of $ISUM$ are calculated using (46.5). After each value $ISUM_{j,k}$ is obtained for $ISUM$, the corresponding $b_{j,k}$ is calculated from

$$b_{j,k} = (ISUM_{j,k})/(2J+1)(2K+1).$$ \hfill (46.6)

While the RHSs of (46.1) and (46.6) are exactly equivalent, the steps represented by (46.2) through (46.5) permit RHS (46.6) to be computed significatly more efficiently than is possible for a direct evaluation of RHS (46.1).

If the line and pixel coordinates (m, n) are evaluated modulo M and N respectively rather than forcing the pixel value to be zero if $m < 1$, $m > M$, $n < 1$ or $n > N$, then (46.1) represents periodic convolution (refer to §§12, 14, and 15) with a low-pass box filter.

Consider the case of periodic convolution of an $M \times M$ box with an $N \times N$ image where N is a power of 2 and $M < N$. Then direct implementation of the box filter would take of the order of $(MN)^2$ multiplications and additions. Using the F.F.T., the corresponding number of operations is $6N^2 ln(N)$. In this case the $M \times M$ box must be embedded in an $N \times N$ image frame and pixels must be stored as real. For box filtering, the numbers of operations are $4N^2$ additions and N^2 multiplications. Thus, for large N the box filtering speed is independent of M and approaches that for a 2×2 direct convolution. For $N = 1024$ and $M = 101$ this is approximately 4000 times faster than direct filtering and 15 times faster than using the F.F.T. This is a substantial improvement in speed.

The above comparisons assume that the image can be stored in random access memory. This is usually impracticable for a large image and so it must be stored on disc. Fourier transformation of a large image on disc is time consuming because of the transposition required (this is discussed in §48). Also a larger disc area is required to store an image as real than if it was an 8- or 16-bit integer image. Box filtering overcomes both of these difficulties because the image is usually stored as integer (8- or 16-bit), and only one image line needs to be read into memory at a time in calculating each output image line. This is an important practical advantage.

Box filters can be used for high-pass filtering simply by subtracting a multiple of the low-pass box filtered image from the original image. This process is called *high-pass box filtering*. The subtraction from the original image line is carried out after each low-pass box filtered image line is calculated. The overall box filtering technique as described above can be applied in one, two, or three dimensions and can be adapted for other related techniques such as median filtering.

A further example of a random line access process is the edge extension technique discussed in §15. It is used to extrapolate a blurred image which is a truncated convolution (because it has been truncated by its recording frame) into a larger image which is a periodic convolution within its image frame. Edge extension is implemented as follows. Let the given 8-bit blurred image be $M \times N$ pixels in size and the psf be enclosed by a frame of minimum size $J \times K$ pixels. Then the given image is placed centrally in an image of size $(J + M - 1) \times (K + N - 1)$, with a border of zeros. A program is then run which uses 2-D linear interpolation (refer to §11) in a periodic sense (refer to §15) to fill in the border. An iterative procedure consisting of box filtering (with a reducing box size) followed by replacement of the central image area with the given blurred image can then be used to smooth the border region as required. This procedure converts a truncated convolution to a periodic convolution and ensures stable restoration results (refer to the discussion of the consistency of periodic convolutions in §14).

47 Block processes

A *block process* is one in which the output image is divided into rectangular blocks each of which is calculated separately. To each output block there corresponds a rectangular subimage of the input image which must be loaded entirely into random access memory and which is used to calculate the output block. This technique is used for algorithms which require too much random access memory to be run as a neighbourhood process. The input and output images must both be disc image files.

The main technique in this category is rectification, which is used to geometrically distort an image so that it matches a chosen map projection or another image. In the case of image restoration, rectification is frequently required to remove distortions and thus to make a psf approximately spatially invariant. For example, blurred Voyager I images of Jupiter's moon Io needed to be rectified to remove camera lens distortions before they could be restored. The usual procedure is to identify a number of ground control points (GCPs) in the image (these could be camera reseau marks) and to note their coordinates in the given image and their desired coordinates in the rectified image. A least squares fit is then used to derive a polynomial warping function of appropriate order (e.g. cubic) that decribes the mapping. If the Cartesian coordinates of the input image are (x', y') and the map coordinates are (x, y) then the cubic polynomials which describe the mapping are

$$x' = Ax^3 + Bx^2y + Cxy^2 + Dy^3 + Ex^2 + Fxy + Gy^2 + Hx + Iy + J,$$
$$y' = Kx^3 + Lx^2y + Mxy^2 + Ny^3 + Ox^2 + Pxy + Qy^2 + Rx + Sy + T. \tag{47.1}$$

The rectification is carried out calculating the individual rectangular blocks of the output image one at a time. The corner points of each output image block are mapped by the polynomial to the input image. For each pixel within the output image block the corresponding position in the input image is determined using linear interpolation between the mapped coordinates of the block corner points. The value of each pixel in the output image is calculated by interpolating between the values of pixels near its corresponding position in the input image. Common interpolation techniques used are nearest neighbour (see §33), bilinear (see §45), and *cubic* (as discussed above in this paragraph) listing them in order of increasing accuracy and execution time (note that they make use of the 1, 4, and 16 nearest pixels, respectively). We have found bilinear interpolation to be a satisfactory compromise in many applications. The number of output image blocks must be large enough to achieve the desired accuracy.

Associated with block processes, and therefore included here, are *virtual processes*. By the latter we mean processes which, to run efficiently, need to access input and/or output images (however large) as if they were in random access memory. This is achieved by always maintaining a sub-image block (or window) of an appropriate size in memory. This window is moved around an image as required. The main application of this technique is in converting vectors to raster or image format. Digitized graphical information (such as line drawings and contour sheets) is usually stored as a sequence of vectors which make up a vector or graphics data file. Each vector may be tagged with an associated value such as the height corresponding to a contour on which the vector lies. *Vector to raster conversion* is the process by which a vector data file is written into an image (or equivalently raster format) file. It involves taking a graphics data file containing such things as contours and text information and writing it into an image file as if it was a vector plotting device. Many extensive software packages are available for writing graphics data files to *vector-driven devices*, such as plotting tables, which respond to commands requiring them to plot vectorial quantities. Rather than writing special software to write graphics data files into image files, it is preferable to write a device driver (this needs to be a virtual process) which makes the image look like a vector driven device. (A *device driver* is a package of I/O subroutines that can be called by a program to control a device of some sort.) We have written such a device driver for our image processing system which can be called by the PLOT79 package of graphics software which we have obtained from the University of Utah.

Another type of virtual process which is worth a mention can be used to interpolate between digitized contours which have been written into an image. The pixel value along each contour in the image is set equal to the contour value (such as its height). This process allows, for example, digitized height contours to be converted to a full digital terrain model sampled on a rectangular image grid.

48 Transformations

In this section we consider implementations of the F.F.T. – refer to §12. Our remarks also apply in some measure to other algorithms such as the Haar transform and Hadamard (or Walsh) transform. It should be noted that the convolution theorem (refer to §7) only holds for the F.F.T., however. This is an overriding advantage. The main uses of the Haar and Hadamard transforms are in image coding. Our interest is in taking the F.F.T. of a real image whose extent in both the x- and y-directions is a power of 2. It should be noted that this size constraint is not a serious restriction in practice. In the case of image restoration for example, an image which is a full convolution (i.e. the convolution has not been truncated – refer to §15) can be extended by packing with zeros (refer to §12) until its extent is a power of 2 in each direction. More generally, a truncated convolution may be scaled with the aid of an interpolation routine (refer to §45) until its dimensions are powers of 2. Restoration should of course be followed by rescaling. We have found this procedure to have little adverse effect on restoration accuracy provided the resulting quantization noise is significantly less than the original noise level. An F.F.T. program can be written for dimensions which are not a power of 2, but this is unnecessary for the reason given above.

Many good F.F.T. programs and subroutines are available. The best source that we know of is the software and documentation made available by the IEEE Digital Signal Processing Committee. Subroutines are available for doing a 2-D F.F.T. totally in random access memory, or the 2-D F.F.T. of a large image on disc. We use the very efficient routines of Fraser to do the forward and inverse F.F.T. of a large image on disc in our standard format. The F.F.T. of a real $N \times N$ image occupies the same total space as the original image and is packed, in the manner described below, until it contains $((N/2) \times N)$ complex pixels. We have adopted Fraser's packing convention as our standard. Note that the full F.F.T. of an $N \times N$ real image contains $N \times N$ complex pixels. However, because of conjugate symmetry about the zero frequency term, $((N/2) - 1)$ of the lines are redundant because they have a corresponding complex conjugate line, and two of the lines have internal complex conjugate symmetry. These last two lines can be packed into a single line to give a non-redundant image containing $((N/2) \times N)$ complex pixels. For general programs acting on complex data, this packed format is undesirable because it leads to untidy programs. The point is that each program dealing with complex images must treat the packed lines as exceptions because they are not in the standard format for complex data. For this reason we use a separate program to convert to or from a standard unpacked 2-D complex image which contains $((1+(N/2))\times N)$ complex pixels.

If $f_{m,n}$ denotes the m, nth pixel of a $M\times N$ pixel image, and $F_{j,k}$ denotes the j, kth pixel of its F.F.T., then

$$F_{j,k} = \sum_{m=0}^{M-1} \sum_{n=0}^{N-1} f_{m,n} \exp(i\, 2\pi(jm/M + kn/N)) \tag{48.1}$$

for $j \in \{0, 1, \ldots, M-1\}$ and $k \in \{0, 1, \ldots, N-1\}$. the inverse of (48.1) is

$$f_{m,n} = (1/MN) \sum_{j=0}^{M-1} \sum_{k=0}^{N-1} F_{j,k} \exp(-i\, 2\pi(jm/M + kn/N)). \tag{48.2}$$

Note that the 2-D transform is implemented as a 1-D transform of each line, followed by a transposition, followed by a final 1-D transform of each line. The routines that Fraser provides for the transposition can be used separately from the F.F.T. This provides an easy way of transposing large images whose dimensions are powers of 2.

A program that we find useful in conjunction with the F.F.T. converts a complex image to an 8-bit image. The logarithm of either the modulus, or the positive real or positive imaginary parts of the F.F.T. can be selected. The amplitude of each of these images is scaled to have a maximum of 255. The logarithm of the positive real part is particularly useful for analysing a blurred image and identifying any psf which possesses a particularly simple form, such as that characterizing uniform linear blur or out-of-focus blur. Positions of phase reversals in the F.F.T. are clearly revealed by where this logarithm exhibits apparent discontinuities. When the latter describe sufficiently simple paths in Fourier space – e.g. straight lines, circles, or even somewhat irregular ovals – they often correspond to the (real) zeros of the OTF (it is, of course, impossible to identify complex zeros of the OTF by any such elementary kinds of processing). This relates to the techniques discussed in §52.

49 Multi-input-image processes

In this section we consider algorithms which generate a single output image from several input images. We distinguish between these algorithms on the basis of their I/O structures. The processing is sequential, as for neighbourhood processes (see §45), in the sense that corresponding lines of the input images are read in and operated on, to produce a single line of the output image, before the next line of each input image is dealt with. Because there are often quite a few of them, it is advisable to store the input images on disc.

One class of multi-input-image algorithm calculates each output pixel as a linear combination of the corresponding pixels of the input images. A typical example is *principal components analysis* which involves computing the correlation (or covariance) matrix of the input images. This matrix is diagonalized by operating on the input images on a pixel-by-pixel basis to generate a set of uncorrelated images, which we call the principal components. Only the first few components tend to possess much in the way of information content. It often transpires that a comparatively large number (e.g. 11) of input images can be transformed, with negligible loss of information, into a significantly smaller number (e.g. 3) of principal component images. One can also usually get away with less subsequent processing.

Other algorithms calculate the maximum, minimum, mean or median of corresponding pixels of the input images. This can be useful when processing satellite imagery. After correcting for the additive radiance effect of atmospheric haze, for instance, corresponding pixels in different image bands (see §45) can be ratioed to remove the shading effects due to terrain relief. This *band ratioing* technique produces a flat-looking image except for areas which are in shadow when the input images are recorded.

A small image may be inserted into a larger one, or mosaiced into it according to some merging or matching algorithm. Sequences of image frames may be joined together to build up a movie sequence.

Images may be calibrated by multiplying or dividing by an image which contains a spatially varying correction function. The multiplication or division is done on a pixel by pixel basis. This is useful for example in correcting aircraft photographic survey images for *falloff* (a decrease in average image intensity with distance from the image centre which is due to the camera lens). To obtain a satellite image of the sea surface sophisticated corrections are usually required to remove the effect of the intervening atmosphere (refer to §45). The total radiance due to the atmosphere has two main components. These are due to Rayleigh scattering (refer to §45) and *aerosol scattering* (from aerosol particles in the atmosphere). The Rayleigh scattering component is calculated for each image band. At near infra-red wavelengths the radiance contribution from the sea is negligible, so subtracting the Rayleigh scattering component from the recorded image gives the aerosol scattering component. This can be used to estimate the aerosol scattering at visible wavelengths, so that the visible bands can be corrected for the atmosphere by subtracting out the atmospheric contributions.

If an image is read in conjunction with a corresponding digital terrain model or height image file (refer to §45) then an image with a parallel or point perspective view can be calculated. The term *perspective* refers to the manner in which the image of a 3-D surface is projected onto a 2-D image. This can be by means of parallel rays, or rays emanating from a point. Given an image of a 3-D surface taken from one position (e.g. from vertically above) it can be useful to use the surface height information to calculate the image that corresponds to a different viewing position or angle. A parallel perspective view of the surface from the left or right is very easy to calculate as each input image line can be processed separately. The original image, and the new perspective image, may then be used as a stereo image pair. A general point perspective view is best calculated as a block process (refer to §47).

A pair of 8-bit image bands may be read together and used to calculate a 256×256 2-D histogram distribution. Each pixel $f_{j,k}$ of the latter, where j and k both range from 0 to 255, equals the number of times a pixel has value j in the first band and k in the second band (i.e. it is a scatter plot of the occurrences of the first band against the second). Such histograms can be calculated for a whole image, or (by the use of masks) for different classes of regions present within the image. These 2-D histograms can be displayed in colour by taking the logarithm of each pixel value (except zero), scaling to a maximum amplitude of 255 and introducing colour coding.

A 2-D lookup table can be defined in a similar way to the (1-D) lookup table defined in §45. A 2-D lookup table is a function which produces an output pixel value as a function of a pair of input pixel values, which are normally integers with the same range as the input pixel value. Consider pixel values J, K and an 8-bit 2-D lookup table $LKUP(J, K)$. J, K and $LKUP(J, K)$ all range from 0 to 255, and the 2-D lookup table can be stored as a 256×256 8-bit image file for later use by other programs.

Normalized 2-D histograms derived for several classes of images can be used to calculate a 2-D lookup table which identifies, at each pixel, the particular image whose histogram is largest for that pixel value. The 2-D lookup table provides a method for quickly classifying large images. The same principle can be applied in 3-D provided sufficient memory space is available.

Many classification techniques (refer to §45) are available for use with multiple image bands. In parallelepiped classification, for example, classes are specified in terms of what we call valid data ranges. For each class, there is a valid range for each image band. For instance, 'water' might be assigned pixel values in the range 10–20 in band 1, the range 0–15 in band 2, and the range 0–10 in band 3. We say that a pixel belongs to a particular class if, for each image band, the pixel value lies within the valid data range for that class. So, a pixel having the value 10 in all three bands would be classified as 'water' (in the above instance). It can happen that a pixel possesses values belonging to more than one valid set of data ranges. It is then assigned to the first class which is found to be compatible with its pixel values in the different image bands. A pixel is left unclassified when its values do not coincide with the set of data ranges valid for any specified class. A popular but much more time-consuming technique is called *maximum likelihood classification* (this is not the same as the maximum likelihood technique introduced in §18), for which distribution functions specify the probability that pixels belong to particular classes. For each pixel a probability is calculated for each class, and the pixel is assigned to that class for which the probability is highest.

Wiener filtering (refer to §16) falls into the multi-input-image process category because the F.F.T.s of the blurred image and the psf are processed together. We read corresponding lines of each F.F.T. and process them to produce one line of the F.F.T. of the restored image. We then read the next pair of lines (of the input F.F.T.s) and process them to get the next line of the restored F.F.T., and so on until all lines are processed. Once this has been accomplished, the inverse F.F.T. is evaluated to form the restored image itself, which is typically displayed as pixels whose values span an 8-bit range.

50   Utilities

In this section we describe programs which do not fall neatly into any of the other categories and which are often vital to the smooth operation of an image processing system.

We have already mentioned in §41 the need for a program to automatically produce documentation from text files and program source files. On the VAX-11/780 under VMS an extensive HELP file facility is available. This provides hierarchical information on system topics. Our software documentation could have been incorporated into this system. However, we decided against this because it would be difficult to transfer to other systems. Instead we have provided equivalent command procedures for obtaining documentation on a visual display unit (VDU) screen.

Special programs are required for handling image headers (refer to §42). For example, it is useful to be able to print out the contents of a header, extract it from an image, edit it, and put it back in the image.

It is often necessary to send small images from one computer to another through a computer network. The data are usually required to be coded in ASCII (or EBCDIC) rather than binary, so programs are required to transform standard images to and from an equivalent ASCII format. It is also frequently necessary to print selected portions of an image to a line printer or VDU screen, so a general purpose utility program is required for this. Similarly, a program is required to summarize the contents of a foreign magnetic tape (refer to §45), so that it can be quickly decoded and reformatted.

A versatile program is required to manipulate and analyse a ground control point (GCP) file (refer to §47) in order to calculate and store a satisfactory warping function for use in rectification.

Another type of utility program that we have found useful translates image processing command files (refer to §41) from one computer and/or operating system to another. Command files cannot in general be transported directly from one computer to another. However, this type of program can greatly simplify the task.

51 Interactive processes

Interactive television display devices offer a wide variety of special hardware features, the use of which requires system dependent special purpose software. The more the software takes advantage of these special features, the less transportable it becomes. The advantages these features offer are real-time interaction, and great processing speed. Algorithms can be specially written to take advantage of the available hardware options. Typical features are a trackball and special terminal keys which enable an operator to move a cursor around the display screen, draw overlaid graphics on it, move the centre of the viewed image (this is called *panning*) and change its magnification either continuously or in integer steps (this is called *zooming*).

In a colour display device each image channel or band is stored in a separate refresh memory bank the contents of which are copied to the display screen every 1/25 or 1/30 seconds. Usually a minimum of three channels are stored with a typical size of 512×512 8-bit pixels. The 8-bit output of each memory bank channel is processed by a selected 1×256 8-bit lookup table which in turn is mapped through a D/A converter to a blue, green, or red display screen colour gun. In more expensive display devices the output of the memory bank or lookup table can optionally be sent to a hardware arithmetic or logical processing unit and then written back into the same or a different memory bank. The processing is done in real-time. Possible processing capabilities include arithmetic and logical operations as well as direct convolution with small (e.g. 7×7) non-recursive filter arrays (refer to §18). Histograms can also be hardware generated from a refresh memory bank in real-time.

Only very simple algorithms can be implemented in hardware, as compared to what is possible with software. This is because it is much easier to write software for a particular task than to design and build special-purpose hardware. It is worth keeping in mind, though, that the simplest algorithms are often the most useful, and the increase in speed of a hardware implementation can be considerable. Furthermore, the capability of interactive real-time processing allows some otherwise difficult tasks to be accomplished straightforwardly. For example, images can be interactively stretched, matched, classified, or ratioed in ways which, while they tend to be virtually impossible to justify really objectively, appear optimal to individual users for their particular applications.

Software to control an interactive processing system requires the availability of numerous separate options and is therefore suitable for writing with an overlay structure (refer to §41). This allows large interactive programs to be run on a minicomputer. The restriction of the display screen total pixel size (e.g. 512×512) may be circumvented by writing a program which treats the display screen as a window (movable and with variable magnification) into an disk image file of arbitrary size.

A display screen can be used to show time sequential movies or movie loops. The movie frames can be read sequentially from a single disc file and written to the display screen a frame at a time (rather than line by line). This method is suitable when the frames change slowly and rates of several frames/second are acceptable. Alternatively, all the refresh memory banks available can be used to store frames. These can be cycled through at full video rate under software control. For example, eight 512×512 8-bit memory banks could be used for 32 frames of a 256×256 8-bit movie loop.

An interactive display system is very useful for finding image ground control points (GCPs) – refer to §47. For each GCP its image coordinates and corresponding map coordinates must be found and stored in a GCP file for later use in setting up the rectification warping function (such as a cubic polynomial – refer to §47). A convenient way to find the GCPs on a routine basis for satellite imagery is to prepare a national or regional GCP file containing the map coordinates for the selected GCPs. Each GCP for a particular image is found by moving the display screen cursor to the point on the image and entering its corresponding number in the master GCP file. This gives the GCP coordinates on both the image and the map and these are stored in the GCP file for the image. In the case of a cubic polynomial warping function at least 20 GCPs are normally required. Alternatively, the GCPs can be found automatically for an image whose location and orientation is already known approximately. For each master GCP a surrounding sub-image from a previous rectified image is stored. This is adjusted for scale and orientation and correlated with the new image in the region where the GCP is estimated to be. The location of the correlation peak gives the GCP location in the image to be rectified.

Example VIII – Illustrations of image enhancement

We show how images formed from data gathered by LANDSAT satellites can be enhanced with the aid of techniques described in this chapter. Figures VIIIa, VIIIc and Ia are typical Landsat images. They are of the Tarawera region (in the North Island of New Zealand) and were formed from data in the 600–700 nm, 700–800 nm, and 800–1100 nm bands respectively. It has become conventional to call these *bands* 2, 3, and 4 respectively.

Figure VIIIa was obtained by destriping (refer to §45) the original data for band 2, rectifying it (refer to §47) to the New Zealand Map Grid projection with a sampling distance of 50 m, and finally extracting the sub-scene. Figure VIIIb shows the result of histogram equalizing (refer to §45) Fig. VIIIa. Figures VIIIc and VIIIe correspond, for band 3, to Figs. VIIIa and VIIIb for band 2.

The enhancement of detail, apparent in Figs. VIIIb and VIIIe, as compared with Figs. VIIIa and VIIIc, respectively, graphically illustrates the virtues of histogram equalization.

The information implicit in an image can sometimes be quite well revealed without resorting to non-linear stretching of the pixel values. It is often adequate to apply a linear stretch via a lookup table (refer to §45) to the image histogram. Figure VIIIg is the histogram (scaled to a maximum amplitude of 100) of Fig. VIIIc. Figure VIIIh is the lookup table corresponding to the linear stretch (from pixel values 4 to 156, clipping some 2 per cent off each end of the histogram) which produced Fig. VIIId from Fig. VIIIc. That Fig. VIIId reveals the implicit detail much more clearly than Fig. VIIIc is readily understood when one recognizes from Fig. VIIIg that Fig. VIIIc contains only about half of the 256 available pixel values. The distribution of pixels in the histogram (i.e. Fig. VIIIi) of Fig. VIIId is significantly more uniform than in the histogram (i.e. Fig. VIIIg) of Fig. VIIIc.

The histogram shown in Fig. VIIIi is of course some way from being fully equalized. Figure VIIIj is the non-linear lookup table which was applied to the histogram (i.e. Fig. VIIIg) of Fig. VIIIc to generate the pixel values for Fig. VIIIe. Figure VIIIk is the histogram for the latter image. The average density of pixel values in Fig. VIIIk is roughly uniform, which is equivalent to stating that the cumulative histogram – refer to (45.1) – corresponding to Fig. VIIIk is approximately linear. Note, however, that it is far from obvious that Fig. VIIIe is preferable in any significant particular to Fig. VIIId. Both are unquestionably big improvements on Fig. VIIIc, however.

Enhancement of image detail by histogram equalization is degraded by the presence of contamination. The level of the latter can be lowered by low-pass filtering, although spatial resolution must then be sacrificed. We illustrate this through Fig. VIIIf, which is a low-pass filtered version of Fig. VIIIc obtained by applying a 15×15 pixel box filter (refer to §46) to Fig. VIIId. The histogram (i.e. Fig. VIIIl) of Fig. VIIIf has much more closely spaced (in amplitude) pixel values than the histogram (i.e. Fig. VIIIi) of Fig. VIIId. The resolution of the image is, however, reduced accordingly.

Images in a pair of spectral bands can on occasion be simultaneously enhanced to display information which is hardly evident when either band is processed singly. We now illustrate this point with reference to two-dimensional histogram equalization (see §49). Figure VIIIm shows the logarithm of the 2-D histogram of Figs. VIIIa and VIIIc. Pixel values, in the range 0 to 255, are plotted horizontally and vertically downwards for bands 2 and 3, respectively, with the origin – i.e. zero pixel value in both hands – at the top left-hand corner.

The value of each pixel in the 2-D histogram shown in Fig. VIIIm is proportional to how often the corresponding combination of band 2 and band 3 pixel values occurs in Figs. VIIIa and VIIIc respectively. We can think of a 2-D histogram as consisting of several overlapping distributions of pixel values characterizing different classes (using this term as introduced in the final two paragraphs of §45) – e.g. various types of ground cover, such as water, forest, pasture, bare ground, etc. The need for histogram equalization is revealed far more dramatically in 2-D (see Fig. VIIIm) than in 1-D (see Fig. VIIIg) because such a small

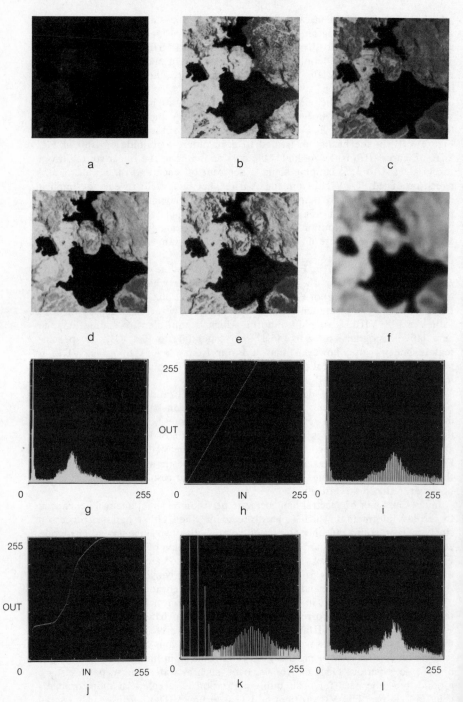

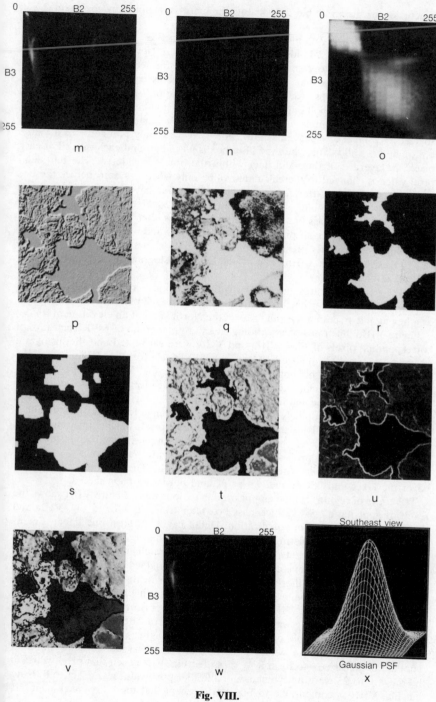

Fig. VIII.

proportion of the available 2-D histogram space is spanned by the band 2 and band 3 pixel values.

Figure VIIIn is the 2-D histogram generated from Figs. VIIIb and VIIIe. Much more of the available 2-D histogram space is covered. Nevertheless, Fig. VIIIn is not an optimum histogram because it has been formed from a pair of 1-D histograms. Note also that these two 1-D histograms consist of comparatively large quantization steps. An even more efficient use of 2-D histogram space can be achieved by interpolating in 2-D between the non-zero pixel values in Fig. VIIIn to produce a much smoother histogram. We begin with five passes of a special 5×5 pixel low-pass box filter. For each pass, only non-zero pixel values are treated as valid data, when the filter box is placed around each pixel. For each pixel, the average of the valid data within the box is calculated. The particular pixel value is changed to the calculated value only when it is zero before the box is placed around it. After five passes with this filter have been completed, we perform further smoothing with a single pass of a 5×5 standard low-pass filter, followed by a single pass of a 3×3 standard low-pass filter (by a standard filter we mean one which treats all pixel values as valid data and which changes all pixels according to the filter's prescription). Figure VIIIo shows the result of applying this processing to Fig. VIIIn. The available 2-D histogram space is much better utilized in Fig. VIIIo.

Figure VIIIp illustrates the relief-shading technique – refer to the paragraph containing (45.5). It has been generated by treating each pixel value in Fig. VIIId as a height and assuming the sun to be in the North West at an elevation of 45°.

Figure VIIIq illustrates band ratioing (see third paragraph of §49). The ratios of corresponding pixels of Figs. VIIIa and VIIIc were calculated and the histogram of these ratios was equalized to generate the pixels for Fig. VIIIq. Note the apparent 'flatness' of this image compared to either Fig. VIIIa or Fig. VIIIc.

The classification technique (see §45) makes it easy for us to generate images which only display specified information. For instance, Fig. VIIIr shows the result of printing as white all pixels within the valid data range for water (in the band 3 image shown in Fig. VIIIc) and printing as black all pixels outside this range. This emphasizes the shapes of the lakes. A different emphasis can be obtained by invoking Golay theme processing (refer to §45). Figure VIIIs shows the effect of growing by 5 pixels the white regions of Fig. VIIIr. The resulting white areas cover the lakes and 250 m wide strips of land bordering their shores.

The edges of regions can be emphasized in various ways. Figure VIIIt shows the result of applying a 15×15 high-pass box filter (refer to §46) to Fig. VIIIc and then employing histogram equalization. Detail having at least one linear dimension of 15 or more pixels in extent is markedly enhanced. One consequence of this is that the borders between land and water are clearly identified (as black contours in this case). White contours between land and water can be generated as follows. The maximum difference between the values of the nine pixels surrounding each pixel in an image are calculated and the resulting array of pixel values is linearly stretched. Figure VIIIu shows the result of performing this operation on Fig. VIIIc. The stretching was between pixel values of 1 (black) and 127 (white) Figure VIIIv is an example of an image, which highlights gradually varying detail in the original scene, but whose borders (between areas of markedly different type) are emphasized – to form it we classified Fig. VIIIu such that pixel values in the range 0 to 25 were unchanged, and all other pixel values were set to 0. Note that Fig. VIIIv is comparable to Fig. VIIIe except that there are black contours around the edges of the lakes.

One of the major triumphs of satellite imagery is that it has facilitated the location of particular agricultural, silvicultural, mineral, etc., features all over the earth's surface. Such location is effected most readily when histograms are palpably 'clumpy' – i.e. they separate into clearly recognizable parts. each corresponding (hopefully) to a single feature of interest. The histogram shown in Fig. VIIIm is more diffuse than we would wish. Figure VIIIw shows the result of de-emphasizing those pixel values in Fig. VIIIm which are relatively faint. Although Fig. VIIIw contains little detail, it displays three distinct clumps (it just so happens, incidentally, that the bright spot at the top left corresponds to water).

Relief plots can sometimes be more informative than grey-scale images (i.e. images whose pixel values are indicated by their brightnesses). This tends to be truer the simpler the image. To illustrate this we show a perspective view (Fig. VIIIx) of the Gaussian psf depicted in Fig. IIi (reference to Example II reminds us that the effective width of the Gaussian is 15 pixels). The perspective is from the 'south-east' with pixel values treated as 'heights'. Figure VIIIx was produced with the aid of the PLOT79 graphics package.

IX

TECHNICAL PRACTICALITIES

Image restoration relies for its success not only on the use of sound techniques, but also on the manner in which a number of practicalities are handled. These are appropriately discussed in this final chapter, because their treatment is conveniently based on much of what we have discussed previously. Particularly relevant are §§5, 15, and 25 and Chapters III, VII, and VIII.

The determination of the psf, which is itself a species of restoration problem, is discussed in §52. The techniques which we describe are edge recognition [Huang, 2; McDonnell, 8, 14, T1; Smith, 2; Tescher, 1], blind deconvolution [Huang, 2; Stockham, 1, 2], zero recognition [Gennery, 1; McDonnell, 11], and calibration [Andrews, B3; Ekstrom, 1; McDonnell, 5]. Advantage can sometimes be taken of the existence of constraints that may apply to the form and symmetry of the psf [McDonnell, 5, 9, T1].

Given the psf, the accuracy of restoration is then largely determined by the contamination level. It is important to know as much as possible about the statistics of the contamination, so as to maximise one's chances of minimizing the contamination level before restoration is attempted. We examine this in §53, for which relevant references are [Andrews, B3; Dainty, B1; Huang, 1]. Finally, in §54 we discuss how restoration accuracy can be assessed, both subjectively and objectively [Andrews, B3; Cannon, 1].

Example IX illustrates aspects of practical problems associated with estimating the form of the psf from the actually recorded image (refer as well to §§4 and 14). While the main emphasis is upon techniques described in §52, we also recall facets of the general treatment of deconvolution presented in Chapter III and illustrated in Example III.

52 Determining the point spread function

Before an image can be restored, the psf must be determined. Four techniques are considered here: edge recognition, blind deconvolution, zero recognition, and calibration. In choosing and implementing any technique it is important to make use of whatever *a priori* information is available concerning the psf and the true image. This includes anything which is known about the form, extent and symmetry of $h(\mathbf{x})$ as well as the likelihood of $f(\mathbf{x})$ containing sharp edges. An estimate $\hat{h}(\mathbf{x})$ of $h(\mathbf{x})$ may be improved by applying constraints, symmetry arguments, enhancement techniques (such as those introduced in §15 and in Chapter VIII), and finally by iteratively varying $\hat{h}(\mathbf{x})$ and assessing each successive version of $\hat{f}(\mathbf{x})$ with the aid of a quantitative measure of restoration accuracy. A spatially varying psf is best determined by partitioning the given blurred image into sub-images, so that the most appropriate spatially invariant psf can be estimated for each sub-image – this is, of course, intimately related to the method of sectioning described in §15. If so desired, a model can then be developed to characterize the variation of $h(\mathbf{x})$ throughout $a(\mathbf{x})$.

Suppose it is known that the true image contains lines across which the intensity varies extremely rapidly – i.e. there are 'sharp edges' in $f(\mathbf{x})$. The technique known as *edge recognition* is based on the realization that inspection of the recorded image usually permits the locations of such edges to be identified. This is always subjective, of course, but rarely objectionably so. We now show that, by scanning across as many seemingly blurred edges as we can find, the form of $\hat{h}(\mathbf{x})$ can be reconstructed by invoking projection theory (see §25). It is convenient to consider a straight edge perpendicular to the Cartesian q-axis, located at $q = q_0$. In the neighbourhood of this edge, the true image must take some such form as

$$f(\mathbf{x}) = e(\mathbf{x})\, \text{step}(q - q_0), \tag{52.1}$$

where $e(\mathbf{x})$ is the image amplitude which would be observed if the edge could be somehow removed, and the *unit step function* is defined by

$$\text{step}(x) = 0 \quad \text{for } x < 0,$$
$$= 1 \quad \text{for } x > 0. \tag{52.2}$$

The recorded image in the neighbourhood of the edge is the convolution of $f(\mathbf{x})$ and $h(\mathbf{x})$, which it is here useful to write out in full – refer to (7.5):

$$a(\mathbf{x}) = a(x, y) = f(x, y) \,\textcircled{0}\, h(x, y)$$

$$= \int\!\!\int_{-\infty}^{\infty} e(x - x', y - y')\, \text{step}(q - q_0 - q')h(x', y')\, dx'\, dy', \tag{52.3}$$

where, if ψ is the angle between the q and x axes,

$$q = x \cos(\psi) + y \sin(\psi). \tag{52.4}$$

We now postulate that $\partial e(\mathbf{x})/\partial q$ is small throughout the neighbourhood of the edge. Since

$$d\, \text{step}(x)/dx = \delta(x) \tag{52.5}$$

we see that the partial derivative of (52.3) with respect to q reduces approximately to

$$\partial a(\mathbf{x})/\partial q \simeq \int_{-\infty}^{\infty} e(x, y) h(x, y)\, dp, \qquad (52.6)$$

where the p-axis is perpendicular to the q-axis. Provided $e(\mathbf{x})$ varies appreciably more slowly than $h(\mathbf{x})$ near the edge, we see from §9 that RHS (52.6) is proportional to the projection, at angle ψ, of $h(\mathbf{x})$. By scanning as many different edges, and different parts of the same edges (if they seem long enough), as possible we can average out at least a fair amount of any effects due to variations in $e(\mathbf{x})$. We thereby obtain a set of approximate projections of $h(\mathbf{x})$ at different angles. By invoking the theory developed in §25, we can then reconstruct an estimate $\hat{h}(\mathbf{x})$ of the psf.

It is worth remembering that the form of the psf is very simple in many practical applications. It is often one-dimensional or circularly symmetric, in which cases a single projection suffices to fully characterize $\hat{h}(\mathbf{x})$. In fact, for a one-dimensional psf varying in the ξ-direction, we see that

$$\partial a(\mathbf{x})/\partial q \simeq \bar{e}\hat{h}(\xi/\cos(\beta)), \qquad (52.7)$$

where $\bar{e}$ is the average of $e(\mathbf{x})$ over all the scanned edges and β is the angle between the ξ- and q-axes.

The technique called *blind deconvolution* can be used when a sequence $\{a_m(\mathbf{x}); m = 1, 2, \ldots, M\}$ of recorded images is available, where each image can be usefully expressed in the form

$$a_m(\mathbf{x}) = f_m(\mathbf{x}) \circledast h(\mathbf{x}) + c_m(\mathbf{x}). \qquad (52.8)$$

The true images in the set $\{f_m(\mathbf{x}); m = 1, 2, \ldots, M\}$ are understood to belong to an ensemble, all of whose members, while they must be different, have significant characteristics in common. For instance, we may be given a single large photograph, containing somewhat similar features (houses, say, or trees dotted around in paddocks) throughout the whole scene, taken with a camera having a defective imaging/focusing system, whose psf seems to spread out over an area Ω_h much smaller than that of the photograph. The $a_m(\mathbf{x})$ might be constructed by subdividing the photograph into may pieces, each of which is several times the size of Ω_h. To avoid generating artefacts, the subdivision should be effected by edge extension or windowing (see §15) rather than by mere truncation. There must be available for comparison another set $\{\tilde{f}_m(\mathbf{x}); m = 1, 2, \ldots, \tilde{M}\}$ of true images, all of which are members of the same ensemble that the $f_m(\mathbf{x})$ belong to. For instance, the $\tilde{f}_m(\mathbf{x})$ might be obtained by subdividing another photograph of a similar scene taken with a properly functioning camera. The houses, or the trees and paddocks, need not be the same ones as before. They are only required to be of similar type. In other words, all of the true images mentioned so far must be different, but their statistics should (ideally) be identical. There will, in practice, always be differences between the statistics of the two sets of true images; but this merely emphasizes that the effectiveness of the deconvolution procedure described in the next paragraph depends upon such differences being, in some useful sense, small.

By analogy with the processing described in §36 (and invoking the notation established there), we see that an estimate $|\hat{H}(\mathbf{u})|$ of $|H(\mathbf{u})|$ is given by

$$|\hat{H}(\mathbf{u})| = (\langle |A_m(\mathbf{u})|^2 \rangle / \langle |\tilde{F}_m(\mathbf{u})|^2 \rangle)^{\frac{1}{2}}, \qquad (52.9)$$

because (and this is the essence of blind deconvolution) the conditions set down in the preceding paragraph permit us to assume that

$$\langle |F_m(\mathbf{u})|^2 \rangle \cong \langle |\tilde{F}_m(\mathbf{u})|^2 \rangle, \qquad (52.10)$$

where $\cong$ is here taken to mean 'is effectively equal to'. While the recovery of phase$\{\hat{H}(\mathbf{u})\}$ from $|\hat{H}(\mathbf{u})|$ in general requires techniques of the kind introduced in Chapter IV, it can sometimes (e.g. when the form of $h(\mathbf{x})$ is very simple) be accomplished by inspection.

It is instructive to note the similarities between blind deconvolution and several of the speckle processing methods described in Chapter VI.

There exist important technical applications for which the form of $h(\mathbf{x})$ is known to be simple. Good examples are images blurred by relative motion between camera and object or by out-of-focus lens systems. Knowing that $h(\mathbf{x})$ has a simple form is not enough to permit the true image to be immediately recovered. Further details of the form of the psf must be obtained. In the above examples, we must estimate the parameters of the motion or the amount by which the lenses are defocused.

When the form of the psf is simple, the OTF tends to possess pronounced real zeros, which can be readily recognized in the spectrum of a recorded blurred image. Since $H(\mathbf{u})$ is often fully characterized by its zeros (refer to §13), this *zero recognition* approach can sometimes serve as a convenient technique for estimating the detailed form of $h(\mathbf{x})$, provided enough of the zeros of $H(\mathbf{u})$ can indeed be recognized by inspection.

Blurring due to linear motion is characterized by $H(\mathbf{u})$ being zero on parallel straight lines in Fourier space. The motion has to be quite complicated before the imaginary parts of (some of) the zeros are large enough that the lines where these zeros are located are no longer apparent. Provided the imaginary parts are small enough, their neglect need not overly distort the form of $h(\mathbf{x})$ reconstructed with the aid of the numerical procedures outlined in §13. When the psf exhibits elliptical symmetry (as for out-of-focus blurring, which of course exhibits circular symmetry if the effects of the defocusing are independent of direction in the image plane) the spectrum tends to possess zeros on quasi-ellipses in

Fourier space. Recognition of such zeros allows the symmetry of the psf to be specified and its detailed form to be estimated by an appropriate model-fitting procedure. When the blurring is due to the simplest type of defocusing, which is characterized by a psf that is constant over a disc in image space and zero outside it, the zeros of the OTF allow immediate estimation of the diameter $\bar{d}$ of the aforesaid disc, because the ideal form of the OTF is proportional to $J_1(\pi\rho\bar{d})$, where ρ is the radial coordinate in Fourier space and $J_1(\cdot)$ denotes the Bessel function of the first kind of order unity (whose zeros are tabulated in many reference works).

It is sometimes more convenient to put up with the trouble and expense of *a posteriori* image processing than to attempt to collimate an imaging instrument before using it. This is now standard practice when trying to obtain the ultimate resolution from the best electron microscopes (one just has to endure the irreducible aberrations of the electron lenses and make all possible ancillary measurements which permit one to estimate the parameters characterizing the aberrations). It is the only thing one can do when the aberrations vary with time, but slowly enough that the psf can be estimated by viewing a reference object either before or after the object one wishes to image. We give the name *calibration* to this approach to psf determination, because it relies on being able to obtain a *reference recorded image* $a_R(\mathbf{x})$, for which the true image $f_R(\mathbf{x})$, here called the *reference true image*, is known. The calibration technique thus reduces to a Deconvolution Problem: given $a_R(\mathbf{x})$ and $f_R(\mathbf{x})$, reconstruct $h(\mathbf{x})$; on the understanding that these quantities are related by

$$a_R(\mathbf{x}) = f_R(\mathbf{x}) \, \textcircled{0} \, h(\mathbf{x}) + c(\mathbf{x}). \qquad (52.11)$$

Since $h(\mathbf{x})$ and $f_R(\mathbf{x})$ play the respective roles of $f(\mathbf{x})$ and $h(\mathbf{x})$ in Chapter III, we can recover an estimate of $h(\mathbf{x})$ using whichever of that chapter's methods seem most appropriate. When $h(\mathbf{x})$ is known to be one-dimensional, $h(\mathbf{x})$ can be independently estimated along as many parallel lines as practicable in the recorded image, permitting the final estimate to be improved by averaging. Such a procedure is appropriate for removing linear motion blur from satellite images of planets and their moons. Satisfactory results can be obtained even when $a_R(\mathbf{x})$ and $f_R(\mathbf{x})$ are recorded under somewhat different conditions.

Of the above four techniques, the most straightforward is calibration. When it cannot be used, one looks to zero recognition. This is only applicable, however, for certain simple types of psf. Edge recognition would be our next choice, but if it is unsatisfactory we would turn to blind deconvolution. Even when none of these techniques are applicable, all is not lost because there may be sufficient *a priori* information available for it to make sense to try a succession of different types of psf, finally adopting the one which seems to give the 'best' results.

Our restoration of a 'star spot' on the supergiant Betelgeuse provides a useful example of the kind of reasoning one can employ when one's *a priori* knowledge of the psf is palpably imperfect. For restoring a poorly resolved optical image of the star, we had available $a(\mathbf{x})$ and a dubious estimate, here denoted by $h'(\mathbf{x})$, of the psf obtained from another nearby star known to be unresolvable by the telescope used to gather all the image data. Neither $a(\mathbf{x})$ nor $h'(\mathbf{x})$ were circularly symmetric, although we had sound physical reasons for believing that the genuine psf, $h(\mathbf{x})$ itself, should be circularly symmetric. It was therefore appropriate to form circularly averaged versions, $a_c(\mathbf{x})$ and $h'_c(\mathbf{x})$, of $a(\mathbf{x})$ and $h'(\mathbf{x})$ respectively. The centres of $a_c(\mathbf{x})$ and $h'_c(\mathbf{x})$ were first estimated by finding the best fit (to within a threshold specified in terms of the prevailing apparent contamination level) of circles to the perimeters of each of them.

Deconvolving $h'_c(\mathbf{x})$ from $a_c(\mathbf{x})$ multiplicatively (see §16) provided a crude circularly symmetric version $f'_c(\mathbf{x})$ of the true image. From an astrophysical point of view it seemed unreasonable to be able to deduce more from $f'_c(\mathbf{x})$ than an estimate of Betelgeuse's apparent diameter and the parameters characterizing its degree of limb-darkening (i.e. how its brightness falls off towards its edge). So, we fitted to $f'_c(\mathbf{x})$ as best we could a model of a circularly symmetric limb-darkened disc, which we here denote by $\hat{f}_c(\mathbf{x})$. On deconvolving $\hat{f}_c(\mathbf{x})$ from $a_c(\mathbf{x})$ we obtained what we held to be an improved estimate $\hat{h}(\mathbf{x})$ of the psf. Finally, we revealed asymmetric detail on the disc of Betelgeuse by deconvolving $\hat{h}(\mathbf{x})$ from $a(\mathbf{x})$. The important aspect of all this is the power of our appeal to circular symmetry. This exemplifies the value of ingenious simplicity in image processing.

In almost any application, an initial estimate of the psf can be improved by enhancement (see §15 and the techniques discussed in Chapter VIII) and by invoking appropriate constraints (e.g. positivity). Other possible constraints are the normalization of, and limiting the size and/or range of, the psf. Spatial filtering of $\hat{h}(\mathbf{x})$ is often useful to smooth or sharpen it as required.

53   Determining the contamination level

Important aspects of image data are their sampling distance (i.e. sample spacing, or pixel size), the spectral band-pass of the imaging/recording apparatus, the level of any contaminatory noise and its statistics, the amplitude quantization range (i.e. the number of bits to which the data are resolved in amplitude), particular parameters of the recording (e.g. degree of uniformity, any geometric distortions, dynamic range, and its associated coefficients of linearity and saturation) and the overall psf (due to the imaging/recording instrumentation together with any effects extraneous to the apparatus). Some of these may have to be deduced from the image itself (e.g. the psf – refer to §52). Certain degradations can result from imperfect recording devices and others from coarse digitization (i.e. under-sampling spatially and in amplitude). The more that is known about such effects, the better they can be compensated, thereby usually restoring images significantly more faithfully.

The restoration algorithms introduced in this book are based on the blurred image being a linear record of whatever radiation impinged on the imaging instrument. We further assume that there is no saturation and that the image is geometrically correct, in the sense that the psf is spatially invariant. Many recorded images require much preprocessing before they can be usefully operated on by such algorithms. Certain crucial preprocessing has already been discussed in some detail, in §15 and in Chapter VIII. We now mention other image-massaging techniques which can make all the difference to the quality of a restored image.

Non-linearities due to instrumental calibrations should be corrected in the reverse order to that in which they occur, if at all possible. Whenever feasible, such non-linearities should be measured, so that they can be straightforwardly compensated once the data are stored in the computer. Spatial non-uniformities, like those caused by the responses of lens systems (wide-angle ones in particular), are usually measured easily enough. They are conveniently removed by using a correction image as a reference. One of the most serious difficulties arises when an image is recorded on film with an unknown gamma. We always try and discover what gamma is commonly obtained when processing the particular film. If this leads to results which appear unsatisfactory, we vary the gamma about the initially assumed value until the results improve, in some (hopefully) objective sense. In our experience, the faithfulness of image restoration is not overly sensitive to errors of up to 0.1 D in estimation of film gamma. We emphasize that far worse effects are caused by improperly handling the truncation of the image by the recording frame (which is of course why we give so much attention to it in §15).

The contamination is treated as additive throughout this book. Even though there is no loss of generality in doing this (as explained in §4), it is worth remembering that all of the techniques examined in detail in Chapter III are implicitly predicated upon $c(\mathbf{x})$ being 'independent' of $b(\mathbf{x})$ – these two quantities are introduced in (4.1). Although such a model is attractively simple, and is the most commonly employed in practice, it is perhaps less 'realistic' than other feasible approaches.

Some types of contamination, film grain noise in particular, are more accurately modelled as being multiplicative. Indeed, some computer simulations suggest that small improvements in the quality of restored images might be achievable if contamination was to be modelled multiplicatively rather than additively. However, we find that restoration quality is limited in practice much more by edge effects, by the choice of deconvolution technique and by other aspects of the contamination than by any inadequacies of the additive noise model. In any case, when deconvolving by the multiplicative method (see §16), one can always adjust the form of $\Phi(\mathbf{u})$ to compensate for uncertainties in the noise model.

Regions of an image can become saturated when the spatial variations of the integrated radiation intensity exceed the dynamic range of the recording apparatus (e.g. a conventional camera employing film or an electronic camera incorporating an array of photo-diodes). The fraction of the image which is saturated can be estimated by inspecting the *histogram* of $a(\mathbf{x})$ – this is the histogram of amplitude levels in $a(\mathbf{x})$. Frequent inspection and assessment of image histograms is necessary if composite image processing procedures, consisting of several individual stages, are to be successful. It is desirable that successive stages should not introduce further saturation, because the quality of the restored image falls off rapidly once the saturated fraction of the image rises above some value set mainly by the size of the frame $\Omega_{h^{-1}}$ of the inverse psf. When multiplicative deconvolution (see §16) is used, saturation effects can be localized by increasing $\Phi(\mathbf{u})$, thereby reducing the effective size of $\Omega_{h^{-1}}$ – refer to the discussion before (15.31) – it is worth recognizing that this sets an irreducible limit on the faithfulness of the restored image.

The formula (16.5) for the Wiener filter implies that the noise-to-signal ratio should be known *a priori*. Since it is, in fact, rarely known in practice, it can only be guessed on the basis of previous experience of similar applications. We usually vary the filter constant Φ about its initially guessed value and choose the one which gives the most 'satisfactory' restoration – refer to the discussion following (16.8). We feel that this is actually characterizing $\Phi(\mathbf{u})$ at high spatial frequencies. We are unclear at present whether it is really worthwhile attempting to modify the filter constant at low frequencies. The point is that low-frequency contamination is less objectionable to the eye. Furthermore, our experience is that the accuracy of the estimate of noise-to-signal ratio is not all that crucial to the success of image restoration. For example, if the true level is 3 per cent, any value between 2 and 4 per cent usually gives effectively the same restoration accuracy.

The faithfulness of restoration is particularly sensitive to errors in estimating the size of Ω_h (recall that techniques for effecting such an estimation are discussed in §52). For example, when the psf exhibits a sharp edge (e.g. for blurring due to uniform linear motion or to defocused imaging), the restored image tends to be seriously degraded when the estimate of any linear extent of Ω_h is in error by more than 10 per cent.

54 Assessing restoration accuracy

Having restored an image, we would like to know how well we have done. This is only possible if we can devise a useful measure of restoration accuracy. It is easy enough to generate subjective measures, and on occasion these may be the only feasible ones. Objective measures are of course much more desirable. Apart from anything else, they permit one to speculate usefully on whether there is any point in trying to improve the image further, by employing some iterative scheme for instance.

There are two categories of objective measures: those for use in computer simulations, on the one hand, and in practice, on the other. In a computer simulation, all the quantities – i.e. the true image, the psf and the contamination, as well as the recorded image – are available for inspection at any time. There is a terrible temptation to cheat, of course! While this might be considered a disadvantage, it has a positive aspect, in that it allows unequivocal criteria to be developed for evaluating the imperfections of restored images. We have found that the most useful accuracy measures in all cases are appropriate norms of the differences between quantities, i.e.

$$\varepsilon_I = \|\bar{\mathfrak{f}} - \hat{\mathfrak{f}}\| \quad \text{and} \quad \varepsilon_F = \|\bar{\mathfrak{F}} - \hat{\mathfrak{F}}\|, \tag{54.1}$$

where the notation introduced in (17.22) has been adopted. We call ε_I and ε_F, respectively the *image space norm* and the *Fourier space norm*. The quantities inside the 'norm signs' are defined differently for computer simulations and for practical cases. Denoting either of these norms by ε, we define it by

$$\varepsilon = \iint\limits_{\Xi} |\bar{\mathbf{f}}(\mathbf{a}) - \hat{\mathbf{f}}(\mathbf{a})|^\tau \, d\Xi(\mathbf{a}), \tag{54.2}$$

where $\mathbf{a}$ represents either $\mathbf{x}$ or $\mathbf{u}$, the region of (image or Fourier) space which is integrated over is Ξ and $d\Xi(\mathbf{a})$ is the element of area in the space. The symbol $\mathbf{f}$ in (54.2) represents either of the symbols $\mathfrak{f}$ or $\mathfrak{F}$ appearing in (54.1). The exponent τ is positive.

The choice between ε_I and ε_F depends upon in which space one's data are more reliable. In most applications we choose ε_I; but remember that data recording is sometimes done entirely in Fourier space (recall the relevant discussions in Chapters IV and VI), in which case it is much more objective to adopt ε_F.

In computer simulations it is convenient to take $\bar{\mathbf{f}}(\mathbf{a})$ to be the true image or its spectrum. Neither of these quantities is of course available in practical applications, in which case it is appropriate for $\bar{\mathbf{f}}(\mathbf{a})$ and $\hat{\mathbf{f}}(\mathbf{a})$ to be successive versions (of the image or its spectrum) given by an iterative restoration scheme. The quantity $\hat{\mathbf{f}}(\mathbf{a})$ can of course be the first restoration of either the image or its spectrum, in a computer simulation. The value of τ is largely a matter of taste, but it is usually taken to be either 1 or 2. We tend to favour the former value.

The choice of Ξ is important from a practical point of view. It is appropriate to take Ξ to cover the whole of the region of image or Fourier space occupied by the significant part of the quantity one wishes to restore, in applications for which nothing is precisely known *a priori* concerning this quantity. Conversely, if the true form of this quantity is given over a restricted region Ξ_r of the space, best results tend to be obtained by setting Ξ equal to Ξ_r. For instance, suppose the recorded image is of a bank robber photographed by a hidden camera, whose resolution is insufficient (for practical reasons, such as the need for the camera to be inconspicuous and to have a wide field of view) for the robber's face to be unambiguously identifiable. After the robbery, the part of the bank where the robber was photographed can itself be photographed with high resolution. The region of this second photograph not occupied by the robber is appropriately taken to be Ξ_r. Another pertinent example is a blurred image of a scene known to contain very sharp edges (compare with the edge-recognition technique introduced in §52). It is then appropriate to take Ξ_r to be the regions of image space spanning the immediate neighbourhoods of all apparent sharp edges in the recorded image.

Much of the practical success of image restoration depends on simple techniques like the one described in §53 for estimating film gamma. Similar simple-minded iterative procedures can refine other important parameters (e.g. levels of various 'noises', extents of the psf when one is only given its general form *a priori* – see §52). We cannot emphasize too strongly the virtues of 'simplicity', when it comes to the devising of image processing techniques. We are perhaps better pleased with the performance of our edge extension (§15) and shift-and-add (§38) methods than with any of our other contributions to the image processing business, and yet it would be difficult to think up simpler techniques.

While the accuracy measures defined by (54.1) and (54.2) are more than adequate from a strictly computational viewpoint, they are somewhat inappropriate perceptually. The eye introduces its own nonlinearities into images on transferring them to the brain (or at least so it seems). Consequently, we conjecture that $\tilde{f}(a)$ and $\hat{f}(a)$ should be subjected to some sort of transformation to match the display device to the visual cortex's varying sensitivity to the amplitude and spatial frequency spectra of the image. Nobody seems as yet to have any very good idea of how to go about doing this, and so we leave it as a final worthwhile unsolved problem for the ingenious reader.

Example IX – Determining the psf

This example illustrates several of the techniques introduced in §52. We examine two blurred versions of the first figure in Example I (i.e. Fig. Ia) and show how the

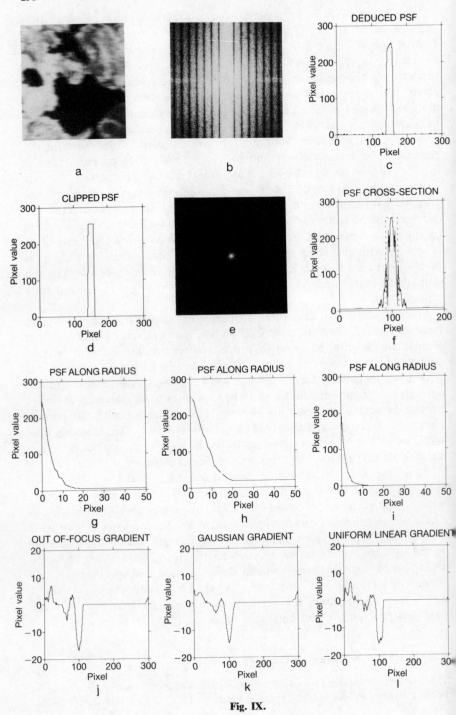

Fig. IX.

forms of the psfs can be determined. Besides illustrating ways (e.g. calibration and radial averaging) of making use of available *a priori* information, we give examples of techniques (e.g. zero recognition and edge recognition) which estimate the psfs directly from blurred images. There is of course always some *a priori* information, but it is less constraining for the recognition techniques than for calibration or radial averaging. The blurred versions of Fig. Ia which are examined here are Figs. If and IXa.

Figure IXa is a truncated (to 256×242 pixels) version of the convolution of Fig. Ia with the particular psf $h(\mathbf{x})$ which caused the blurring. The convolution theorem (7.7) thus ensures that the spectrum (i.e. Fig. IXb, which is the histogram equalized – see §45 – form of the logarithm of the magnitude of the F.F.T. of an overlapped edge-extended – see §15 – version of Fig. IXa) of Fig. IXa is ideally the product of the OTF $H(\mathbf{u})$ and of the spectrum of Fig. Ia. Any zero apparent in Fig. IXb must thus be either a zero of $H(\mathbf{u})$ or a zero of the spectrum of Fig. Ia. [Since edge extension can never be perfect, Fig. IXb must exhibit a significant contamination level. The latter implies that Fig. IXb is not in actual fact a simple product of two spectra. This is unlikely to prevent the recognition of zeros, however. Its most probable effect is to shift their positions slightly.]

It is, of course, conceivable that the parallel, dark, vertical lines in Fig. IXb could be manifestations of the zeros of the spectrum of the unblurred image, but that would be extremely unlikely in practice, especially when (as is the case here) the blurred image seems to be a picture of a natural landscape. When zeros advertise their positions 'blatantly', as they do in Fig. IXb, we can usually feel confident that they characterize the OTF. So, we must assume that $H(\mathbf{u}) = 0$ along each of the dark lines in Fig. IXb. Since the dark lines are parallel, it is a reasonable deduction that both the OTF and the psf are functions of single Cartesian coordinates, i.e. $H(\mathbf{u}) = H(u, v) = H(u)$ and $h(\mathbf{x}) = h(x, y) = h(x)$ with the u- and x-directions being across and the v- and y-directions being up the paper. So, the loss of resolution apparent in Fig. IXb must be due to a linear blur – refer to Table I.1. The separations of the dark lines determine the extent of $h(x)$, which graphically illustrates the power of the zero recognition technique. Furthermore, the comparatively high amplitudes of each of the bright vertical bands in Fig. IXb suggest that $h(x)$ must cut off pretty abruptly at both ends of its x-extent. It is of course impossible to deduce from Fig. IXb the detailed variation of $h(x)$ with x (the psf was actually a uniform blur of 15 pixels extent), but knowing its extent and recognizing that it begins and ends abruptly, one could try deconvolving (refer to Chapter III and the illustrations of deconvolution presented in Example III) Fig. IXa with a variety of what seem like reasonable forms for $h(x)$.

We next put a different interpretation on Figs. Ia and IXa in order to illustrate the calibration technique. We suppose that Fig. IXa is formed by an imperfect imaging system whose psf can be expected to remain the same while further images are recorded. Provided we can make temporary use of a well-collimated system to form a faithful image (e.g. Fig. Ia) of the same scene, the latter can serve to calibrate the imperfect system. The psf of the latter is found by deconvolving the faithful image from the blurred image. When, as is usual, the blurred image is truncated, we can invoke overlapped edge extension (see §15) and Wiener filtering (see §16) to restore $h(x, y)$. If its form is found to vary only slightly with y, we may think it proper to assume the y-variations to be due solely to contamination and processing errors. We can therefore generate what we think is an improved estimate of the psf by forming the average psf $h(x)$, which is

$$(1/M) \sum_{m=0}^{M} h(x, y_m),$$

where $h(x, y_m)$ is the version of $h(x, y)$ restored from the mth horizontal line (with $m = 1$ corresponding to the bottom line) of Fig. IXa, and M is the total number of lines (256 in this example, because Fig. IXa contains 256×242 pixels). Figure IXc shows the result (using a filter constant $\Phi = 0.0009$, corresponding to an effective contamination level of 3 per cent) of this sequence of operations when Figs. Ia and IXa are the faithful and blurred images respectively. Figure IXc is an amplitude plot of $h(x)$. [Note that all subsequent figures in this example, apart from Fig. IXe, are amplitude plots.] The maximum amplitude of $h(x)$ has been scaled to 255.

In a particular application there could be *a priori* information available suggesting that $h(x)$ should be more symmetrical than the curve shown in Fig. IXc. We might then employ linear stretching (refer to §46) to obtain a preferred form for the psf (e.g. Fig. IXd, which was generated from Fig. IXc by stretching between the pixel values 40 and 216).

Figure IXe shows the result of calibrating Fig. If with Fig. Ia. The solid curve in Fig. IXf is an amplitude plot along the horizontal line in Fig. IXe intersecting its brightest point. Reference to Example I reminds us that the blurring of Fig. If was due to a psf having the form of a disc (i.e. out-of-focus blur – refer to Table I.1) of radius 15 pixels. The dashed line in Fig. IXf is an amplitude plot along any straight line in Fig. IIa (which depicts a disc of radius 15 pixels) passing through the centre of the disc. So, Fig. IXe is a reasonable approximation to the actual psf. A better estimate of the latter can be generated if, for instance, the available *a priori* information suggests the psf should be a disc. It is clear from Fig. IXf that, by averaging many solid curves, each being an amplitude plot along a different straight line passing through the brightest point in Fig. IXe, and making the best fit of this average to a rectangular curve, we would arrive at an estimate of close to 15 pixels for the disc's diameter. This is an example of the radial averaging technique, which can be invoked to improve the estimate of any psf thought to be circularly symmetric.

To get the most out of averaging of the kind outlined in the previous paragraph (or any similar technique, for that matter), it is necessary to match the Wiener filter constant Φ (invoked when calibrating the blurred image by deconvolving the faithful image from it) to the contamination level. In fact, it is advisable to perform several deconvolutions, each for a different value of Φ, to obtain a set of estimates for the psf. Amplitude plots of the latter tend to reveal the form of the actual psf (or an approximation thereof) sitting on a pedestal. The optimum estimate is obtained when the height of the pedestal equals $\Phi^{1/2}$. Figures IXg–IXi show the results of radial averaging preceded by calibration (of Fig. If by Fig. Ia), for $\Phi = 0.0009$, 0.0001, and 0.01 respectively. The pedestal in Fig. IXh matches most closely the calibration filter constant.

Image restoration using a psf, whose form has been estimated by calibration, tends to be more faithful when the pedestal is removed. To do this optimally requires a previously obtained estimate for the psf's extent. Accordingly, best results are sometimes only generated after several calibration and averaging iterations (see references quoted in the introductory comments to this chapter).

Figure IXj shows the result of taking first-order finite differences between successive pixels along the 160th horizontal line of Fig. If. Even though Fig. If is appreciably blurred, we can confidently assert that this line intersects the shore of a lake, in the neighbourhood of the 100th pixel (counting from the left). The resolution of the picture shown in Fig. If is such that we would expect the land–water interface to be 'sharp' (i.e. of width no greater than a single pixel) in an image captured by a properly collimated imaging system. It is reasonable to assume, therefore, that the large negative-going peak in Fig. IXj is proportional to the projection (at angle $0°$ – i.e. the horizontal direction) of the psf which has caused the blurring apparent in Fig. If (see second, third, and fourth paragraphs of §52). Figure IXk shows the form that Fig. IXj would have if the blurring of Fig. If had been due to the Gaussian psf shown in Fig. IIi. Figure IXl is the amplitude plot of the first-order finite differences along the 160th line of Fig. IXa. The reason for the large peaks (each in the neighbourhood of the 100th pixel) in Figs. IXj through IXl being negative-going is that we represent water by lower pixel values than land.

Note that the large peaks in Figs. IXj and IXk have very similar shapes, implying that edge-recognition needs to be done with great care if one is to distinguish confidently between different types of psf. On the other hand, the large peak in Fig. IXl is obviously not a projection through either a circular disc or a Gaussian. The reason we can recognize this so easily is that Fig. IXa has been blurred by a one-dimensional psf. In 1-D the forms of the projection of a psf and the psf itself are the same, although one is stretched with respect to the other unless the projection is perpendicular to the blurring direction (which it of course is for Fig. IXl). The presence of the 'wiggles' at the bottom of the large negative-going peak in Fig. IXl would lead us to suppose that the psf to be associated with Fig. IXa should be a non-uniform linear blur (as depicted in Table I.1, for instance). Of course, if we calculated finite difference plots across as many land–water interfaces as we can confidently recognize in Fig. IXa, and then averaged them, the resulting shape of the large negative-going peak would be much closer to rectangular (i.e. the amplitude of the aforementioned wiggles in Fig. IXl would be appreciably reduced).

It is worth belabouring the point raised in the latter half of the previous paragraph. A projection through a psf can be regarded as a blurring of that psf. Consequently, the forms of the projections must be accurately estimated if we are to have any chance of faithfully reconstructing the form of a psf by invoking the techniques described in Chapter V. This implies that we should calculate as many versions of the projections as we can, and at all feasible angles, so as to generate the best possible set of projections of the psf.

REFERENCES

For the reasons given in §1, we have listed each of our references under one of the seven headings: (1) Books, (2) Journals, (3) Special issues of journals, (4) Conference proceedings, (5) Technical report series, (6) Papers, (7) Theses. A number, n say, precedes each reference listed under the first, sixth, and seventh headings. This number is prefixed by B or T for the first or seventh headings, respectively, and it is enclosed in square brackets for the sixth heading. The number n identifies the nth reference to the author (or editor) whose name is immediately to the right of n. Two of the items listed under the third heading are preceded by S followed by either 1 or 2. The way in which individual references are identified in the introductory comments to Chapters II through IX (and in §1) is explained in the third and fourth paragraphs of §1.

(1) Books

B1   Abramowitz, M. and Stegun, I.A. (1965). *Handbook of mathematical functions.* Dover, New York.

B1   Andrews, H.C. (ed.) (1978). *Tutorial and selected papers in digital image processing.* IEEE Computer Society.

B2   Andrews, H.C. (1970). *Computer techniques for image processing.* Academic Press, New York.

B3   Andrews, H.C. and Hunt, B.R. (1977). *Digital image restoration.* Prentice-Hall, Englewood Cliffs, NJ.

B1   Berstein, R. (ed.) (1978). *Digital image processing for remote sensing.* IEEE Press.

B1   Beurger, M.J. (1959). *Vector space.* John Wiley, New York.

B1   Born, M. and Wolf, E. (1976). *Principles of Optics.* Pergamon, London.

B1   Bracewell, R.N. (1978). *The Fourier transform and its applications* (2nd edn) McGraw-Hill, New York.

B1   Brigham, E.O. (1974). *The fast Fourier transform.* Prentice-Hall, Englewood Cliffs, NJ.

B1   Campbell, G.A. and Foster, R.M. (1948). *Fourier integrals for practical applications.* Von Nostrand, New York.

B1   Castleman, K.R. (1979). *Digital image processing.* Prentice-Hall, Englewood Cliffs, NJ.

B1   Cathey, W.T. (1974) *Optical information processing and holography.* John Wiley, New York.

B1   Chandrasekhar, S. (1960). *Radiative transfer.* Dover, New York.

B1   Dainty, J. C. and Shaw, R. (1974). *Image science.* Academic Press, London.

B2   Dainty, J.C. (ed) (1984). *Laser speckle and related phenomena* (2nd edn). Springer-Verlag, Berlin.

B1 Davis, P.J. (1963). *Interpolation and Approximation*. Blaisdell Publ. Co., New York.

B1 Erf, R.K. (1978) *Speckle metrology*. Academic Press, London.

B1 Foley, J.D. and Van Dam, A. (1982). *Fundamentals of interactive computer graphics*. Addison-Wesley, Reading, Mass.

B1 Gonzalez, R.C. and Wintz, P. (1977). *Digital image processing*. Addison-Wesley, Reading, Mass.

B1 Goodman, J.W. (1968). *Introduction to Fourier optics*. McGraw-Hill, New York.

B1 Hanbury Brown, R. (1974). *Intensity interferometer*. Taylor & Francis, London.

B1 Hawkes, P.W. (1980). *Computer processing of electron microscope images*. Springer-Verlag, Berlin.

B1 Herman, G.T. (ed.) (1979). *Image reconstruction from projections: implementation and applications*. Springer-Verlag, Berlin.

B2 Herman, G.T. (1980). *Image reconstruction from projections: the fundamentals of computerized tomography*. Academic Press, New York.

B3 Herman, G.T. and Matterer, F. (eds) (1981). *Mathematical aspects of computerized tomography*. Springer-Verlag, Berlin.

B1 Huang, T.S. (ed.) (1975). *Picture processing and digital filtering*. Springer-Verlag, Berlin.

B1 IEEE Digital Signal Processing Committee (ed.) (1975). *Selected papers in digital signal processing II*, IEEE Press.

B2 IEEE Digital Signal Processing Committee (ed.) (1979). *Programs for digital signal processing*, IEEE Press.

B1 Isaacson, E., and Keller, H.B. (1966). *Analysis of Numerical Methods*, John Wiley, New York.

B1 Jennison, R.C. (1966). *Radio astronomy*. Newnes, London.

B1 Misell, D.L. (1978). *Image analysis, enhancement and interpretation*. North-Holland, Amsterdam.

B1 Newton, I. (1730). *Opticks*. Fourth Edition, corrected, Printed for William Innys at the West-End of St. Paul's, London.

B1 Pratt, W.K. (1978). *Digital image processing*. John Wiley, New York.

B1 Prenter, P.M. (1975). *Splines and variation methods*. John Wiley, New York.

B1 Rabiner, L.R. and Rader, C.M. (eds) (1972). *Digital signal processing*. IEEE Press.

B1 Ramachandran, G. and Srinivasan, R. (1970). *Fourier methods in crystallography*. John Wiley, New York.

B1 Reeves, R.G. (ed.) (1975). *Manual of remote sensing*, Vols 1 and 2. The American Society of Photogrammetry, Falls Church, Virginia.

B1 Rhodes, W.T., Fienup, J.R., and Saleh, B.E.A. (eds) (1984). *Transformations in optical signal processing*. Society of Photo-Optical Instrumentation Engineers (SPIE Vol. 373), Bellingham, WA 98227, USA.

B1 Roberts, J.A. (ed) (1984). *Indirect imaging*, Cambridge University Press.

B1 Rosenfeld, A. (1969) *Picture processing by computer*. Academic Press, New York.

B2 Rosenfeld, A. and Kak, A.C. (1982). *Digital picture processing* (2nd edn). Academic Press, New York.

B1 Sage, A.P. and Melsa, J.L. (1971). *System identification*. Academic Press, New York.

B1 Saxton, W.O. (1978) *Computer techniques for image processing in electron microscope.* Academic Press, New York.

B1 Scaife, B.K.P. (1974). *Studies in numerical analysis.* Academic Press, London.

B1 Sedláček, B. (1975). *Lectures: Vols. A, B & C.* International Union of Crystallography (Commission on crystallographic computing), International Summer School Institute of Macromolecular Chemistry (Academy of Sciences), Prague.

B1 Smith, H.M. (1975). *Principles of holography* (2nd edn). John Wiley, New York.

B1 Van Schooneveld, C. (1979). *Image formation from coherence functions in astronomy* (IAU Colloquium #49). Reidel, Dordrecht, Holland.

B1 Walkup, J.F. and Krile, T.F. (eds) (1981). *Workshop on future directions for optical information processing.* E.E. Dept., Texas Tech. University, Lubbock, TX 79049, USA.

B1 Watson, G.N. (1966). *A treatise on the theory of Bessel functions* (2nd edn). Cambridge University Press.

B1 Woolfson, M.M. (1961). *Direct methods in crystallography.* Oxford University Press.

(2) Journals

Advances in Computer Vision & Image Processing
Applied Optics
Astronomy & Astrophysics
Astrophysical Journal
Computer
Computer Vision, Graphics & Image Processing
IEEE Spectrum
IEEE Transactions on ASSP, AE, AP, C, GRS, MI, & NS
Image & Vision Computing
Journal of the British Interplanetary Society
Journal of the Optical Society of America
Monthly Notices of the Royal Astronomical Society
Optica Acta
Optical Engineering
Optics Communications
Optics Letters
Optik
Physics Reports
Proceedings of the IEE parts A, F & H
Proceedings of the IEEE
Progress in Optics

(3) Special issues of journals

S1 *Computer* **7,** May 1974.
IEEE Transactions on Geoscience & Remote Sensing **GE-18,** April 1980.
Optical Engineering May–June 1974.

Proc. IEEE **59,** July 1972.
Proc. IEEE **69,** May 1981.
S2 *Proc. IEEE* **71,** March 1983.

(4) Conference proceedings

Proceedings of the Image Understanding Workshop. (Sponsored by US Defence Advanced Research Projects Agency.)
Proceedings of the International Conference on Pattern Recognition.
Proceedings of International Symposium on Machine Processing of Remotely Sensed Data.
Proceedings of the Society of Photo-optical Instrumentation Engineers.

(5) Technical report series

Computer Science Dept, University of Maryland, College Park, Maryland.
Dept of Electrical Engineering, University of Canterbury, Christchurch, New Zealand.
Digital Image Analysis Laboratory, Dept of Electrical Eng., University of Arizona, Tucson.
Image Processing Institute, University of Southern California, Los Angeles.
Los Alamos Scientific Laboratory, Los Alamos, New Mexico.
Medical Image Processing Group, Hospital of the University of Pennsylvania at Philadelphia.
Physics and Engineering Laboratory, DSIR, Lower Hutt, New Zealand. (Documentation related to the PEL image processing system is available on request in the form of microfiche.)

(6) Papers

[1] Ables, J. (1974). Maximum entropy spectral analysis, *Astronomy & Astrophysics* Supplement Series **15,** 383.

[1] Adler, M. and Andrews, H.C. (1974). Space variant point spread function pseudoinversion, in *Image processing research,* USCIPI Report 530 (ed. W.K. Pratt).

[1] Anderson, G.B. and Huang, T.S. (1971). Frequency-domain image errors, *Pattern Recognition* **3,** 185.

[1] Andrews, H.C. (1972). N topics in search of an editorial: Heuristics, superresolution and bibliography', *Proc. IEEE* **60,** 891.

[2] Andrews, H.C. (1974). Digital image restoration – a survey, *Computer* **7**(5), 36.

[3] Andrews, H.C., Tescher, A.G., and Kruger, R.P. (1972). Image processing by digital computer, *IEEE Spectrum* **9,** 20.

[1] Arguello, R.J., Sellner, H.R., and Stuller, J.A. (1972). Transfer function compensation of sampled imagery, *IEEE Transactions on Computing* **C-21,** 812.

[1] Backus, G. and Gilbert, G. (1970). Uniqueness in the inversion of inaccurate gross earth data, *Phil. Trans. R. Soc. London* A **266,** 123.

[1] Barnes, C.W. (1966). Object restoration in a diffraction-limited imaging system, *J. Opt. Soc. Am.* **56,** 575.

[1] Barrett, H.H., Hawkins, W.G., and Joy, M.L. (1983). Historical note on computed tomography, *Radiology* **147,** 172.

[1] Bates, J.H.T., McKinnon, A.E., and Bates, R.H.T. (1981). Subtractive image restoration. I: Basic theory, *Optik* **61,** 349.

[2] Bates, J.H.T., McKinnon, A.E., and Bates, R.H.T. (1982). Subtractive image restoration. II: Comparison with multiplication deconvolution, *Optik* **62,** 1.

[3] Bates, J.H.T., Fright, W.R., Millane, R.P., Seagar, A.D., Bates, G.T.H., Norton, W.A., McKinnon, A.E., and Bates, R.H.T. (1982). Subtractive image restoration III. Some practical applications, *Optik* **62,** 333.

[4] Bates, J.H.T., Fright, W.R., and Bates, R.H.T. (1984). Wiener filtering and cleaning in a general image processing context, *Mon. Not. R. Astr. Soc.*, **211,** 1.

[1] Bates, R.H.T. (1969). Contributions to the theory of intensity interferometry, *Mon. Not. R. Astr. Soc.* **142,** 413.

[2] Bates, R.H.T., Kennedy, W.K., and McDonnell, M.J. (1974). Efficient digital restoration of images blurred by linear motion, *Letters in Applied & Engineering Sciences* **2,** 133.

[3] Bates, R.H.T. and Napier, P.J. (1972). Identification and removal of phase errors in interferometry, *Mon. Not. R. Astr. Soc.* **158,** 405.

[4] Bates, R.H.T. and Gough, P.T. (1975). New otulook on processing radiation from objects viewed through randomly fluctuating media, *IEEE Transactions on Computers* **C-24,** 449.

[5] Bates, R.H.T., McDonnell, M.J., and Gough, P.T. (1977). Imaging through randomly fluctuating media, *Proc. IEEE* **65,** 138.

[6] Bates, R.H.T. (1978). On phase problems. I, *Optik* **51,** 161.

[7] Bates, R.H.T. (1978). On phase problems. II, *Optik* **51,** 223.

[8] Bates, R.H.T., Lewitt, R.M., McDonnell, M.J., Milner, M.O., and Peters, T.M. (1978). Practical Image processing, *Physics in Technology* **9,** 101.

[9] Bates, R.H.T. and Fright, W.R. (1982) Towards imaging with a speckle interferometric optical synthesis telescope, *Mon. Not. R. Astr. Soc.* **198,** 1017.

[10] Bates, R.H.T. (1982). Astronomical speckle imaging, *Physics Reports* **90,** 203.

[11] Bates, R.H.T. (1982). Fourier phase problems are uniquely solvable in more than one dimension: I: underlying theory, *Optik* **61,** 247.

[12] Bates, R.H.T. and Robinson, B.S. (1982). A stochastical imaging procedure", in Acoustical Imaging (eds E.A. Ash and C.R. Hill) **12,** 185.

[13] Bates, R.H.T. and Fright, W.R. (1983). Composite two-dimensional phase-restoration procedure, *J. Opt. Soc. Am.* **73,** 358.

[14] Bates, R.H.T. (1984). Uniqueness of solutions to two-dimensional Fourier phase problems for localized and positive images, *Computer Graphics, Vision & Image Processing* **25,** 205.

[15] Bates, R.H.T., Fright, W.R. (1984). Reconstructing images from their Fourier intensities, *Advances in Computer Vision & Image Processing* (ed. T.S. Huang) **1,** 227.

[16] Bates, R.H.T., Garden, K.L., and Peters, T.M. (1983). Overview of computerized tomography with emphasis on future developments, *Proc. IEEE* **71,** 356.

[17] Bates, R.H.T., Napier, P.J., McKinnon, A.E., and McDonnell, M.J. (1976). Self-consistent deconvolution: I – theory, *Optik* **44**, 183.

[18] Bates, R.H.T. and Milner, M.O. (1977). Langrange polynomials of a complex variable for two-dimensional interpolation, *International Journal for Numerical Methods in Engineering* **11**, 1801.

[19] Bates, R.H.T. and Wall, D.J.N. (1977). Null field approach to scalar diffraction, *Phil. Trans. R. Soc. London* A **287**, 45.

[20] Bates, R.H.T. (1980). General introduction to the extended boundary conditon, in *Acoustic, Electromagnetic and Elastic Wave Scattering – Focus on the T-Matrix Approach*, (eds V.K. and V.V. Varadan). Pergamon Press, Oxford, 21.

[1]   Beebe, N.H.F. (1979). A user's guide to PLOT79. Departments of Physics and Chemistry, University of Utah, Salt Lake City, Utah.

[1]   Bergland, G.D. (1969). A guided tour of the fast Fourier transform, *IEEE Spectrum* **6**, 41.

[1]   Bernstein, R. and Ferneyhough, D.G. (1975). Digital image processing, *Photogrammetric Engineering & Remote Sensing* **41**, 12, 1465.

[1]   Billingsley, F.C. (1973). Digital image processing for information extraction, *International Journal of Man–machine Studies* **5**, 203.

[1]   Biraud, Y. (1969). A new approach for increasing the resolving power by data processing, *Astronomy & Astrophysics* **1**, 124.

[2]   Bracewell, R.N. and Riddle, A.C. (1967). Inversion of fan-beam scans in radio astronomy, *Astrophysical Journal* **150**, 427.

[2]   Bracewell, R.N. and Riddle, A.C. (1967). Inversion of fan-beam scans in radio astronomy, *Astrophysical Journal* **150**, 427.

[1]   Breckinridge, J.B. (1975). Coherence interferometer and astronomical applications, *Applied Optics* **42**, 25.

[1]   Bruck, Y.M. and Sodin, L.G. (1979). On the ambiguity of the image reconstruction problem, *Optics Communications* **30**, 304.

[1]   Bryan, R.K. and Skilling, J. (1980). Deconvolution by maximum entropy, as illustrated by application to the jet of M87, *Mon. Not. R. Astr. Soc.* **191**, 69.

[1]   Bryngdahl, D. and Lohmann, A.W. (1968). Holographic penetration of turbulence, in *Proceedings of a NASA/ERC seminar* (ed. M. Nagel) Cambridge, Mass.

[1]   Cady, F. M. and Bates, R.H.T. (1980) Speckle processing gives diffraction-limited true images from severely aberrated instruments, *Optics Letters* **5**, 438.

[1]   Campbell, K., Wecksung, G.W., and Mansfield, C.R. (1974). Spatial filtering by digital holography, *Optical Engineering* **13**, 175.

[1]   Cannon, T.M., Trussel, W.J., and Hunt, B.R. (1978). Comparison of image restoration methods, *Applied Optics* **17**, 3384.

[1]   Cochran, W.T., Cooley, J.W., Favin, D.L., Helms, H.D., Kaenel, R.A., Long, W.W., Maling, G.C., Nelson, D.E., Rader, C.M., and Welch, P.D. (1967). What is the Fast Fourier Transform?, *IEEE Transactions on Audio & Electroacoustics* **AU-15**, 2, 45.

[1]   Carson, E.R, and Jones, E.A. (1979). Use of kinetic analysis and mathematical modelling in the study of metabolic pathways *in vivo*, *New England Journal of Medicine* **300**, 1016.

[1]   Christou, J. (1984). Private communication.

[1] Cooley, J.W. and Tukey, J.W. (1965). An algorithm for the machine calculation of complex Fourier series, *Mathematics of Computation* **19**, 297.

[1] Cormack, A.M. (1963 and 1964). Representation of a function by its line integrals, with some radiological applications, I & II, *Journal of Applied Physics* **34**, 2722, and **35**, 2908.

[1] Cornwell, T.J. (1983). *Astronomy & Astrophysics* **121**, 281.

[1] Crayford, H.T. and McDonnell, M.J. (1982). New Zealand ground control points, *PEL DSIR Manual* No. 269.

[1] Cresswell, A.F. and McDonnell, M.J. (1982). S bend correction and calibration of LRM aircraft scanner imagery, *PEL DSIR Report* No. 743.

[1] Dainty, J.C. (1973). Diffraction-limited imaging of stellar objects using telescopes of low optical quality, *Optics Communications* **7**, 129.

[1] Davenhall, A.C., Bunclarke, P.S., Fraser, C.W., McLean, B.J., Stapelton, J.R., and Stewart, G.C. (1981). An astronomical image processing system written in FORTH, *Journal of the British Interplanetary Society* **34**, 145.

[1] Davison, M.E. (1983). The ill-conditioned nature of the limited angle tomography problem, *SIAM Journal of Applied Mathematics* **43**, 428.

[1] De Santis, P. and Gori, F. (1975). On an interative method for superresolution, *Optica Acta* **22**, 691.

[1] Ekstrom, M.P. (1973). A numerical algorithm for identifying spread functions of shift invariant imaging systems, *IEEE Transactions on Computing* **C-22**, 322.

[2] Ekstrom, M.P. (1973). A spectral characterization of the ill-conditioning in numerical deconvolution, *IEEE Transactions on Audio & Electroacoustics* **AU-21**, 345.

[3] Ekstrom, M.P. (1973). On the numerical feasibility of digital image restoration, *Proc. IEEE* **61**, 1155.

[1] Ennos, A.E. (1978). Speckle interferometry, *Progress in Optics* (ed. E. Wolf) **16**, 235.

[1] Erickson, H.P. (1973). The Fourier transform of an electron micrograph – first order and second order of image formation, *Advances in Optical & Electron Microscopy* **5**, 163.

[2] Erickson, H.P. and Klug, A. (1971). Measurement and compensation of defocussing and aberrations by Fourier processing of electron micrographs, *Phil. Trans. R. Soc. London* B **261**, 105.

[1] EROS Data Center (1982). Landsat multispectral scanner computer-compatible tape format, Version 1.0, EROS Data Center, Sioux Falls, South Dakota (available from EROS).

[1] Fiddy, M.A., Brames, B.J., and Dainty, J.C. (1983). Enforcing irreducibility for phase retrieval in two dimensions, *Optics Letters* **8**, 96.

[1] Fienup, J.R. (1978). Reconstruction of an object from the modulus of its Fourier transform, *Optics Letters* **3**, 27.

[2] Fienup, J.R. (1979). Space object imaging through the turbulent atmosphere, *Optical Engineering* **18**, 529.

[3] Fienup, J.R. (1982). Phase retrieval algorithms: a comparison, *Applied Optics* **21**, 2758.

[4] Fienup, J.R. (1983). Reconstruction of objects having latent reference points, *J. Opt. Soc. Am.* **73**, 1861.

[5] Fienup, J.R., Crimmins, T.R., and Holsztynski, W. (1982). Reconstruction

of the support of an object from the support of its autocorrelation, *J. Opt. Soc. Am.* **72,** 610.

[6] Fienup, J.R. (1984). Autocorrelation unfolding, in [Rhodes, B1], 203.

[1] Franke, R. (1982). Scattered data interpolation: tests of some methods, *Mathematics of Computation* **38,** 181.

[1] Fraser, D. (1977). Fast random access I/O for PDP11 FORTRAN, *Proceedings of the DECUS Conference*, Townsville, Queensland, August 1977, 1507.

[2] Fraser, D. and O'Brien, E. (1979). Fast image rotation techniques using a colour image display, *Proceedings of the DECUS Conference*, Christchurch, New Zealand, August 1979, 1601.

[1] Frieden, B. R. (1967). Band-unlimited reconstruction of optical objects and spectra, *J. Opt. Soc. Am.* **57,** 1013.

[2] Frieden, B.R. (1968). Optimum nonlinear processing of noisy images, *J. Opt. Soc. Am.* **58,** 1272.

[3] Frieden, B.R. (1972). Restoring with maximum likelihood and maximum entropy, *J. Opt. Soc. Am.* **62,** 511.

[4] Frieden, B.R. (1973). Restoration of pictures by Monte Carlo allocation of 'grains', in *Proceedings of the August Meeting of the Optical Society of America*, Washington, DC.

[5] Frieden, B.R. (1974). Image restoration by discrete convolution of minimal length, *J. Opt. Soc. Am.* **64,** 682.

[6] Frieden, B.R. (1981) Maximum-information data processing: application to optical signals, *J. Opt. Soc. Am.* **71,** 294.

[7] Frieden, B.R. and Burke, J.J. (1972) Restoring with maximum entropy II: superresolution of photographs of diffraction-blurred images, *J. Opt. Soc. Am.* **62,** 1207.

[8] Frieden, B.R. and Swindell, W. (1976). Restored pictures of Ganymede, moon of Jupiter, *Science* **191,** 4233.

[1] Garden, K.L., Bates, R.H.T., Won, M.C., and Chikwanda, H. (1984). Computerized tomographic imaging is insensitive to density variation during scanning, *Image & Vision Computing* **2,** 76.

[1] Gennery, D.B. (1973). Determination of optical transfer function by inspection of frequency domain Plot, *J. Opt. Soc. Am.* **63,** 1571.

[1] Gerchberg, R.W. (1974). Superresolution through error energy reduction, *Optica Acta* **21,** 709.

[2] Gerchberg, R.W. and Saxton, W.O. (1972). A practical algorithm for the determination of phase from image and diffraction plane pictures, *Optik* **35,** 237.

[1] Golay, M.J.E. (1969). Hexagonal parallel pattern transformations, *IEEE Transactions on Computing* **C-1,** 733.

[1] Gough, P.T. and Bates, R.H.T. (1972). Computer generated holograms for processing radiographic data, *Computers & Biomedical Research* **5,** 700.

[1] Granger, E.M. (1968). Restoration of images degraded by spatially varying smear, in *Proceedings of a NASA/ERC Seminar* (ed. M. Nagel), Cambridge, Mass.

[1] Greenleaf, J.F. (1983). Computerized tomography with ultrasound, *Proc. IEEE* **71,** 330.

[2] Greenleaf, J.F. (1983). Computerized tomography with ultrasound, *Proc. IEEE* **71,** 330.

[1] Gull, S.F. and Daniell, G.J. (1978). Image reconstruction from incomplete and noisy data, *Nature, Lond.* **272,** 686.

[1] Hall, E.L. (1972) A comparison of computations for spatial frequency filtering, *Proc. IEEE* **59,** 887.

[2] Hall, E.L., Kruger, R.P., Dwyer, S.J., Hall, D.L., McLaren, R.W., and Lodwick, G.S. (1971). A survey of preprocessing and feature extraction techniques for radiographic images, *IEEE Transactions on Computing* **C-20,** 1032.

[1] Hamlet, R.G. and Haralick, R.M. (1978). Transportable package software, University of Maryland, Computer Science Technical Report TR-706.

[1] Harris, F.J. (1976). On the use of windows for harmonic analysis with the discrete Fourier transform, *Proc. IEEE* **66,** 51.

[2] Harris, J.L. (1964). Diffraction and resolving power, *J. Opt. Soc. Am.* **54,** 931.

[3] Harris, J.L. Sr. (1966). Image evaluation and restoration, *J. Opt. Soc. Am.* **56,** 569.

[4] Harris, J.L. Sr. (1968). Potential and limitations of techniques for processing linear motion-degraded imagery, in *Proceedings of a NASA/ERC Seminar* (ed. M. Nagel), Cambridge, Mass.

[1] Hayes, M.H. (1982). The reconstruction of a multidimensional sequence from the phase or magnitude of its Fourier transform, *IEEE Transactions on Acoustics, Speech & Signal Processing* **ASSP-30,** 140.

[1] Helstrom, C.W. (1967). Image restoration by the method of least squares, *J. Opt. Soc. Am.* **57,** 297.

[1] Hill, N.R. and Ioup, G.E. (1976). Convergence of the van Cittert iterative method of deconvolution, *J. Opt. Soc. Am.* **66,** 487.

[1] Högbom, J.A. (1974). Aperture synthesis with a non-regular distribution of interferometer baselines, *Astronomy & Astrophysics* Supplement Series **15,** 417.

[1] Honda, T. and Tsujiuchi, J. (1975). Restoration of linear-motion blurred pictures by image scanning method (Effect of total width of the scanning aperture), *Optica Acta* **22,** 537.

[2] Honda, T., Tsujiuchi, J., and Ishiquro, S. (1974). Image processing by multiple aperture scanning, *Optica Acta* **21,** 653.

[1] Hoppe, W. (1970). Principles of electron structure research at atomic resolution using conventional electron microscopes for the measurement of amplitudes and phases, *Acta Crystallographica* **A26,** 414.

[1] Horner, J.L. (1969) Optical spatial filtering with the least-mean square error filter, *J. Opt. Soc. Am.* **59,** 297.

[1] Hounsfield, G.N. (1980). Computer medical imaging, Nobel lecture, *Journal of Computer-Assisted Tomography* **4,** 665.

[1] Huang, T.S. (1966). Some notes on film grain noise, in *NSF Summer Study Report* (ed. S. Morgan), Woods Hole, Mass.

[2] Huang, T.S., Schreiber, W.F., and Tretiak, O.J. (1971). Image processing, *Proc. IEEE* **59,** 1586.

[1] Hunt, B.R. (1972). Data structures and computational organization in digital image enhancement, *Proc. IEEE* **60,** 884.

[2] Hunt, B.R. (1975). Digital image processing, *Proc. IEEE* **63,** 693.

[3] Hunt, B.R. (1983). Digital image processing, *Advances in Electronics & Electronic Physics* (ed. P.W. Hawkes) **60,** 161.

[4] Hunt, B.R. (1983). Information theory and visual displays, *Reports on Progress in Physics* **46**, 515.

[5] Hunt, B.R. and Janney, D.H. (1974). Digital image processing at Los Alamos scientific laboratory, *IEEE Computer* **7**(5), 57.

[1] Inuiya, M. and Ichioka, Y. (1973). Image restoration of blurring due to uniform linear motion by TV techniques, *Optics Communications* **8**, 382.

[1] Jansson, P.A. (1970). Method for determining the response function of a high-resolution infrared spectrometer, *J. Opt. Soc. Am.* **60**, 184.

[2] Jansson, P.A., Hunt, R.M., and Plyler, E.K. (1970). Resolution enhancement of spectra, *J. Opt. Soc. Am.* **60**, 596.

[1] Kak, A.C. (1979). Computerized tomography with X-ray, emission and ultrasound sources, *Proc. IEEE* **67**, 1245.

[1] Knoll, G.F. (1983). Emission computed tomography, *Proc. IEEE* **71**, 320.

[1] Kohler, D. and Mandel, L. (1973). Source reconstruction from the modulus of the correlation function: a practical approach to the phase problem of optical coherence theory, *J. Opt. Soc. Am.* **63**, 126.

[1] Labeyrie, A. (1970). Attainment of diffraction limited resolution in large telescopes by Fourier analysing speckle patterns in star images, *Astronomy & Astrophysics* **6**, 85.

[2] Labeyrie, A. (1977). High resolution techniques in optical astronomy, *Progress in Optics* (ed. E. Wolf) **14**, 49.

[1] Lahart, M.J. (1974). Maximum-likelihood restoration of nonstationary imagery, *J. Opt. Soc. Am.* **64**, 17.

[1] Lawden, M.D. and Pearce, D. (1980). Making the VICAR image processing system portable, *Journal of the British Interplanetary Society* **33**, 369.

[1] Lee, S.H. (1974). "Mathematical operations by optical processing, *Optical Engineering* **13**, 198.

[2] Lee, W.H. (1970). Sampled Fourier transform hologram generated by computer, *Applied Optics* **9**, 639.

[1] Lewis, B.L. and Sakrison, D.J. (1975). Computer enhancement of scanning electron micrographs, *IEEE Transactions on Circuits & Systems* **CAS-22**, 267.

[1] Lewitt, R.M. (1983). Reconstruction algorithms: transform methods, *Proc. IEEE* **71**, 390.

[2] Lewitt, R.M. and Bates, R.H.T. (1978). Image reconstruction from projections: III & IV: projection completion methods, *Optik* **50**, 189 & 269.

[3] Lewitt, R.M., Peters, T.M., and Bates, R.H.T. (1978). Image reconstruction from projections. II: modified backprojection methods, *Optik* **50**, 85.

[1] Lohmann, A.W. and Werlich, H.W. (1967). Holographic production of spatial filters for code translation and image restoration, *Physics Letters* **25A**, 570.

[1] Lui, B. and Gallagher, N.C. (1974). Optimum Fourier transform division filters with magnitude constraint, *J. Opt. Soc. Am.* **64**, 1227.

[1] MacAdam, D.P. (1970). Digital image restoration by constrained deconvolution, *J. Opt. Soc. Am.* **60**, 1617.

[1] Macovski, A. (1983). Physical problems of computerized tomography, *Proc. IEEE* **71**, 373.

[1] Mascarenhas, N.D.A. and Pratt, W.K. (1975). Digital image restoration under a regression model, *IEEE Transactions on Circuits and Systems* **CAS-22**, 252.

[1] McDonnell, M.J. (1975). Nonrecursive image restoration using a finite filter array, *Optik* **43**, 159.

[2] McDonnell, M.J. (1976). A sampling function appropriate for deconvolution, *IEEE Transactions on Information Theory* **22**, 617.

[3] McDonnell, M.J. (1980). Image restoration parameter choice – a quantitative guide, TR-965, Computer Vision Laboratory, University of Maryland (available from University of Maryland).

[4] McDonnell, M.J. (1981). Box filtering techniques, *Computer Graphics & Image Processing* **17**, 65.

[5] McDonnell, M.J. (1981). Restoration of VOYAGER 1 images of Io, *Computer Graphics & Image Processing* **15**, 79.

[6] McDonnell, M.J. (1982). Automatic relief shading, *PEL DSIR Report* No. 764.

[7] McDonnell, M.J. and Bates, R.H.T. (1975). Preprocessing of degraded images to augment existing restoration methods, *Computer Graphics & Image Processing* **4**, 25.

[8] McDonnell, M.J. and Bates, R.H.T. (1975). Restoring parts of scenes from blurred photographs, *Optics Communications* **13**, 347.

[9] McDonnell, M.J. and Bates, R.H.T. (1976). Digital restoration of an image of Betelguese, *Astrophysical Journal* **208**, 443.

[10] McDonnell, M.J. and Cresswell, A.F. (1982). PEL image processing capability – LRM scanner demonstration, *PEL DSIR Report* No. 752.

[11] McDonnell, M.J. and Fowler, A.D.W. (1979). Computer analysis of suspected Clarence River 'UFO images' filmed by TV1, *PEL DSIR Report* No. 632.

[12] McDonnell, M. J. and Fowler, A.D.W. (1981). Colour image calibration, in *Proceedings of the Landsat '81 Conference*, Canberra, Australia.

[13] McDonnell, M.J., Fowler, A.D.W., and Pairman, D. (1982). North Island Landsat computer mosaic, *PEL DSIR Report* No. 767.

[14] McDonnell, M.J., Kennedy, W.K., and Bates, R.H.T. (1976). Identifying and overcoming practical problems of digital image restoration, *New Zealand Journal of Science* **19**, 127.

[15] McDonnell, M.J. and Pairman D. (1982). Physics and Engineering Laboratory image processing system documentation manual, *PEL DSIR Manual* No. 264.

[16] McDonnell, M.J., Pairman, D., Fowler, A.D.W., and Timmins, S.M. (1982). PEL image processing capability – Landsat demonstration, *PEL DSIR Report* No. 751.

[17] McDonnell M.J., Pairman, D., and Fowler, A.D.W. (1983). PEL image processing capability – sea demonstration, *PEL DSIR Report* No. 817.

[18] McDonnell, M.J., Pairman, D., and Fowler, A.D.W. (1983). PEL image processing capability, *PEL DSIR Report* No. 827.

[1] McGlamery, B.L. (1967). Restoration of turbulence-degraded images, *J. Opt. Soc. Am.* **57**, 293.

[1] McKinnon, A.E., McDonnell, M.J., Napier, P.J., and Bates, R.H.T. (1976). Self-consistent deconvolution; II-applications, *Optik* **44**, 253.

[1] McKinnon, G.C. and Bates, R.H.T. (1981). Towards imaging the beating heart usefully with a conventional CT scanner, *IEEE Transactions on Biomedical Engineering* **BME-28**, 123.

272 REFERENCES

[1] Mersereau, R.M. and Oppenheim, A.V. (1974). Digital reconstruction of multidimensional signals from their projections, *Proc. IEEE* **62**, 1319.

[1] Meuller, D.F. and Reynolds, G.O. (1967). Image restoration by the removal of random media degradations, *J. Opt. Soc. Am.* **57**, 11.

[1] Misel, D.L. (1978). The phase problem in electron microscopy, *Advances in Optical & Electron Microscopy* (eds R. Barer and V.E. Coslett) **7**, 185.

[1] Nagel, M. (ed.) (1968). Evaluation of motion-degraded images, in *Proceedings of a NASA/ERC Seminar*, Cambridge, Mass.

[1] Napier, P.J. and Bates, R.H.T. (1974) Inferring phase information from modulus information in two-dimensional aperture synthesis, *Astronomy & Astrophysics* Supplement Series **15**, 427.

[2] Napier, P.J., Thompson, A.R., and Ekers, R.D. (1983). The very large array: design and performance of a modern synthesis radio telescope, *Proc. IEEE* **71**, 1295.

[1] Nathan, R. (1971). Image processing for electron microscopy, I: Enhancement procedures, *Advances in Optical & Electron Microscopy* **4**, 85.

[1] Nawab, S.H., Quatieri, T.F., and Lim, J.S. (1983). Signal reconstruction from short-time Fourier transform magnitude, *IEEE Transactions on Acoustics, Speech & Signal Processing* **ASSP-31**, 986.

[1] Nityananda, R. and Narayan, R. (1982). Maximum entropy image reconstruction – a practical non-information-theoretic approach, *Journal of Astrophysics & Astronomy* **3**, 419.

[1] O'Callaghan, J.F. and Fraser, D. (1980). The DCR image processing laboratory brief description of facilities for Landsat-oriented applications, *Proceedings of the DECUS Conference*, Sydney, July 1980.

[2] O'Callaghan, J.F., Robertson, P.K., and Fraser, D. (1981). Colour image display – it's not that simple, *Proc. Landsat '81 Conference*, Canberra, Australia, September 1981.

[1] Oppenheim, A.V., Schafer, R.W., and Stockham, T.G. (1968) Nonlinear filtering of multiplied and convolved signals, *Proc. IEEE* **56**, 1264.

[1] Patterson, C.L. and Buechler, G. (1974). Digital image processing at the aerospace corporation, *IEEE Computer* **7**(5), 5.

[1] Philip, J. (1963). Reconstruction from measurements of positive quantities by the maximum-likelihood method, *Journal of Mathematical Analysis & Applications* **7**, 327.

[2] Philip, J. (1973). Restoration of pictures by quadratic programming and by Fourier transformation in the complex domain, *Proc. IEEE* **61**, 468.

[1] Phillips, D.L. (1962). A technique for the numerical solution of certain integral equations of the first kind, *J. Ass. Comput. Mach.* **9**, 84.

[1] Rabiner, R.L. (1971). Techniques for designing finite-duration impulse-response digital filters, *IEEE Transactions on Communications & Technology* **COM-19**. 188.

[2] Rabiner, R.L. and Rader, C.M. (eds) (1972). Digital signal processing, *IEEE Press Selected Reprint Series*.

[1] Radon, J. (1917). Über die bestimming von funktionen durch ihre integralwerte längs gewisser mannigffaltigkeiten, *Sächsiche Berichte Akademie der Wissenschaften* **69**, 262.

[1] Richardson, W.H. (1972). Bayesian-based iterative method of image restoration, *J. Opt Soc. Am.* **62**, 55.

REFERENCES

273

[1] Riemer, T.E. and McGillem, C.D. (1973). Constrained optimization of image restoration, *Applied Optics* **12**, 2027.

[1] Rino, C.L. (1969). Bandlimited image restoration by linear mean-square estimation, *J. Opt. Soc. Am.* **59**, 547.

[1] Robbins, G.M. (1970). Image restoration for a class of linear spatially-variant degradations, *Pattern Recognition* **2**, 91.

[2] Robbins, G.M. and Huang, T.S. (1972). Inverse filtering for linear shift-variant imaging systems, *Proc. IEEE* **60**, 862.

[1] Robinson, G.S. (1972). Logical convolution and discrete Walsh and Fourier power spectra, *IEEE Transactions on Audio & Electroacoustics* **AU-20**, 271.

[1] Roetling, P.G., Haas, R.C., and Kinzly, R.E. (1968). Some practical aspects of measurement and restoration of motion-degraded images, in *Proceedings of a NASA/ERC Seminar* (ed. M. Nagel), Cambridge, Mass.

[1] Rosenbaum, M., Hancil, V., and Korners R. (1980). Resolution of overlapping chromatographic peaks by parameter estimation of their model, *Journal of Chromatography* **191**, 157.

[1] Rosenfeld, A. (1972). Picture processing, *Computer Graphics & Image Processing* **1**, 394.

[2] Rosenfeld, A., (1979). Picture processing: 1978, *Computer Graphics & Image Processing* **9**, 354.

[1] Rushforth, C.K. and Harris, R.W. (1968). Restoration, resolution and noise, *J. Opt. Soc. Am.* **58**, 539.

[1] Ryle, M. (1975). Radio telescopes of large resolving power, *Reviews of Modern Physics* **47**, 557.

[1] Saleh, B.E.A. (1974). Trade-off between resolution and noise in restoration by superposition of images, *Applied Optics* **13**, 1833.

[1] Sawada, N., Kidode, M., Shinoda, H., Asada, H., Iwanaga, M., Watanabe, S., Mori, K., and Akiyama, M. (1981). An analytic correction method for satellite MSS geometric distortions, *Photogrammetric Engineering & Remote Sensing* **47**, 1195.

[1] Sawchuk, A.A. (1973). Space-variant system analysis of image motion, *J. Opt. Soc. Am.* **63**, 1052.

[1] Saxton, W.D. (1974). A new computer language for electron image processing, *Computer Graphics & Image Processing* **3**, 266.

[1] Schiske, P. (1974). Ein und mehrdeutigeit der phasen bestimmung aus bild und beuguns figur, *Optik* **40**, 261.

[1] Shanks, J.L. (1970). The design of stable two-dimensional recursive filters, in *Proceedings of the Kelly Communications Conference*, University of Missouri, Rolla.

[1] Silverman, H.S. and Pearson, A.E. (1973). On deconvolution using the discrete Fourier transform, *IEEE Transactions on Audio & Electroacoustics* **AU-21**, 112.

[1] Slepian, D. (1967). Linear least-squares filtering of distorted images, *J. Opt. Soc. Am.* **57**, 918.

[2] Slepian, D. (1967). Restoration of photographs blurred by image motion, *Bell Systems Technical Journal* **46**, 2353.

[3] Slepian, D. and Pollak, H.O. (1961) Prolate spheroidal wave functions, Fourier analysis and uncertainty – I, *Bell Systems Technical Journal* **40**, 43.

[1] Smith, H.A. (1966). Improvement of the resolution of a linear scanning device, *SIAM Journal of Applied Mathematics* **14,** 231.

[2] Smith, P.R., Peters, T.M., and Bates, R.H.T. (1973). Image reconstruction from finite numbers of projections, *Journal of Physics* A **6,** 361.

[1] Sondhi, M.M. (1972). Image restoration: the removal of spatially invariant degradations, *Proc. IEEE* **60,** 842.

[1] Stark, H., Cahana, D., and Webb, H. (1981). Restoration of arbitrarily finite-energy optical objects from limited spatial and spectral information, *J. Opt. Soc. Am.* **71,** 635.

[1] Stockham, T.G. Jr. (1972). Image processing in the context of a visual model, *Proc. IEEE* **60,** 828.

[2] Stockham, T.G., Cannon, T.M., and Ingebretson, R.B. (1975). Blind deconvolution through digital signal processing, *Proc. IEEE* **63,** 678.

[1] Stroke, G.W. (1972). Optical Computing, *IEEE Spectrum* **9**(12), 23.

[2] Stroke, G.W. and Halioua, M. (1973). Image improvement in high-resolution electron microscopy with coherent illumination (low-contrast objects) using holographic image-deblurring deconvolution, III, part A, Theory, *Optik* **37,** 192.

[3] Stroke, G.W. and Halioua, M. (1973). Image improvement in high resolution electron microscopy with coherent illumination (low-contrast objects) using holographic image-deblurring deconvolution III B, *Optik* **37,** 249.

[1] Stuller, J.A. (1972). An algebraic approach to image restoration filter design, *Computer Graphics & Image Processing* **1,** 107.

[1] Taylor, L.S. (1981). The phase retrieval problem, *IEEE Transactions on Antennas & Propagation* **AP-29,** 386.

[1] Ter-Pogossian, M.M., Raichle, M.E., and Sobel, B.E. (1980). Positron emission tomography, *Scientific American* **243**(4), 171.

[1] Tescher, A.G. and Andrews, H.C. (1972). Data compression and enhancement off sampled images, *Applied Optics* **11,** 919.

[1] Tetelbaum, S.I. (1957). About a method of obtaining volume images with the help of X-rays, *Isvestiya Kievskoga Politekhnich, Instituta* **22,** 154.

[1] Thompson, A.R., Clark, B.G., Wade, C.M., and Napier, P.J. (1980). The very large array, *Astrophysical Journal* Supplement Series **44,** 151.

[2] Thompson, C.J. (1982). Data handling in positron emission tomography, MARIA Workshop on Positron Emission Tomography (22 Apr. 1982), Edmonton, Alberta, Canada.

[1] Trussel, H.J. (1981). Processing of X-ray images, *Proc. IEEE* **69,** 615.

[2] Trussel, H.J. (1978). Notes on linear image restoration by maximising the a posteriori probability, *IEEE Transactions on Acoustics, Speech & Signal Processing,* **ASSP-26,** 174.

[3] Trussel, H.J. and Hunt, B.R. (1978). Sectioned methods for image restoration, *IEEE Transactions on Acoustics, Speech & Signal Processing* **ASSP-26,** 157.

[4] Trussel, H.J. and Hunt, B.R. (1978). Image restoration of space-variant blurs by sectioned methods, *IEEE Transactions on Acoustics, Speech & Signal Processing* **ASSP-26,** 608.

[1] Twomey, S. (1963). On the numerical solution of Fredholm integral equations of the first kind by the inversion of the linear system produced by quadrature, *J. ACM* **10,** 97.

[1]   Van Wie, P. and Stein, M. (1976). A Landsat digital image rectification system, *Report X-931-76-101*, Goddard Space Flight Centre, Greenbelt, Maryland.

[1]   Vander Lugt, A. (1968). A review of optical data processing techniques, *Optica Acta* **15**, 1.

[1]   Weigelt, G. and Wirnitzer, B. (1983). Image reconstruction by the speckle-masking method, *Optics Letters* **8**, 389.

[1]   Woolf, N.J. (1982). High resolution imaging from the ground, *Annual Reviews of Astronomy & Astrophysics* **20**, 367.

[1]   Zimmermann, F.S. and Gupta, S.C. (1973). A state variable approach to digital image processing, *Computers & Electrical Engineering* **1**, 255.

(7) Theses

T1   Cole, E.R. (1973). The removal of unknown image blurs by homomorphic filtering, Ph.D. thesis, Dept of Electrical Engineering, University of Utah, Salt Lake City.

T1   Fright, W.R. (1984). The Fourier phase problem, Ph.D. thesis, Electrical & Electronic Engineering Dept., University of Canterbury, Christchurch, New Zealand.

T1   Garden, K.L. (1984). An overview of computed tomography, Ph.D. thesis, Electrical & Electronic Engineering Dept, University of Canterbury, Christchurch, New Zealand.

T1   Gullberg, G.T. (1979). The attenuation radon transform: theory and application in medicine and biology, Ph.D. thesis, Lawrence Berkeley and Donner Laboratories, University of California at Berkeley, USA.

T1   McDonnell, M.J. (1975). Nonrecursive digital image restoration, Ph.D. thesis, Dept of Electrical Engineering, University of Canterbury, Christchurch, New Zealand.

T1   Milner, M.O. (1979). Error correction in images and imaging instruments, Ph.D. thesis, Dept. of Electrical Engineering, University of Canterbury, Christchurch, New Zealand.

T1   Peters, T.M. (1973). Image reconstruction from projections, Ph.D. thesis, Dept of Electrical Engineering, University of Canterbury, Christchurch, New Zealand.

T1   Van Hove, P.L. (1982). Signal reconstruction from Fourier transform amplitude, M.Sc. thesis, M.I.T., Cambridge, Mass., USA.

GLOSSARY

Selected abbreviations and symbols are listed here. The reader is reminded that our basic functional, algebraic, and set-theoretic conventions and notations are set down in §1.

We point out that some symbols are given different meanings in several sections of this book. Once re-defined in any particular section, however, that meaning is retained for the rest of the section. All of the symbols itemized below are invoked repeatedly, and with the same meaning, throughout the book. Each symbol is listed under the chapter where it first appears, and its defining equation is quoted whenever appropriate. Each chapter's list is alphabetical, with the English alphabet preceding the Greek, and with English type-faces ordered as follows: Italic, Script, Gothic, Orator. Non-alphabetical symbols are collected at the end of each chapter's list. We note, finally, that those symbols which are first introduced in Chapters VII – IX appear in so few sections that it seems pointless to include them here.

Included among the following abbreviations are certain other items of stark terminology.

Abbreviations

ADA, ANSI, FORTRAN, EPIC, FORTRAN, HELP, PASCAL, PLOT79	computer languages, software, program packages, etc.
ASCII, EBCDIC	codes
BPI	bits per inch
CCD	charge coupled device
CPU	central processing unit (of a computer)
CT	comput(eriz)ed tomography
DEC, Intel, RSX, RT-11, UNIX, VAX, VMS	trademarks labelling computers, operating systems, etc.
D.F.T.	discrete Fourier transform
DSIR	Department of Scientific and Industrial Research

EROS, Landsat, Seasat, SPOT	Earth Resources Satellites and concomitants
F.F.T.	fast Fourier transform
GCP	ground control point
I/O	input/output
LHS	left hand side
OTF	optical transfer function
pdf	probability density function
PEL	Physics and Engineering Laboratory (DSIR)
pep	positron–electron pair
PET	positron emission tomography
psf	point spread function
psi	point spread invariant (isoplanatic)
psv	point spread variant (non-isoplanatic)
RHS	right-hand side
SPECT	single photon emission tomography
VDU	visual display unit
1-D	one dimension or one-dimensional
2-D	two dimensions or two-dimensional

Chapter I

$a(x)$	actually recorded image
$b(x)$	ideal blurred image, see §4, (3.4) and (14.1), and (3.2)
$c(x)$	contamination, see (4.1)
$d\sigma(x)$	element of area in image space, see (3.2)
D	aperture (pupil) diameter of imaging instrument
$\mathcal{D}(\zeta, z)$	distortion introduced by propagation medium
$f(\mathbf{x})$	true image, see §4 and (3.1)
$g(\mathbf{x}, t)$	instantaneous image

$G(\mathbf{z}, t)$	radiation field in aperture (pupil) of imaging instrument
$h(\mathbf{x})$	point spread function (psf), see §4 and (3.3)
$r(\mathbf{x})$	ideal recorded image, (3.7)
$\mathfrak{r}(\mathbf{x})$	recordable image, (4.1)
t	time
$\mathbf{x}$	position vector in image space
$\mathbf{z}$	position vector in aperture (pupil) plane
ζ	position vector in source distribution
λ	wavelength of radiation
$\odot$	convolution operator; see list for Chapter II

Chapter II

$d\Sigma(\mathbf{u})$	element of area in Fourier space
$f(x)$	1-D image, (6.1)
$f(\mathbf{x})$	2-D positive true image, (8.1), see (3.10) and (6.13)
$f_{l,m}$	point sample of $f(\mathbf{x})$, (12.1)
$F(u)$	1-D spectrum of $f(x)$, (6.1)
$F(\mathbf{u})$	2-D spectrum, (6.12); note that $F(\mathbf{u}) = \mathsf{F}\{f(\mathbf{x})\}$
$F_{l,m}$	Fourier coefficient, (10.3)
$\mathfrak{f}(\mathbf{x})$	not-necessarily-positive true image
$\mathsf{F}\{\cdot\}$	Fourier transform operator, (6.15), see (6.17)
$H(\mathbf{u})$	optical transfer function (OTF); note that $H(\mathbf{u}) = \mathsf{F}\{h(\mathbf{x})\}$, see (8.13)
$\mathfrak{K}(\mathbf{z})$	apodization function, see (8.4)
$L^{(g)}$	extent of g, (7.12)
$\mathrm{lagrange}_j(\cdot)$	jth Lagrange polynomial, (11.13)
$p(\xi; \phi)$	projection at angle ϕ, (9.4)
$\mathfrak{p}(\mathbf{x})$	periodic image, (10.7)

$p_b(\mathbf{x})$	periodic overlapped ideal blurred image, (12.6)
$\mathfrak{P}(\mathbf{u})$	spectrum of periodic image, (10.10)
$\text{rect}(\cdot)$	rectangular function, (11.4) and (11.5)
$(r; \theta)$	cylindrical polar components of $\mathbf{x}$; see list for Chapter I
$\text{sinc}(\cdot)$	sinc function, (6.4)
$\text{tri}(\cdot)$	triangular function, (11.4) and (11.5)
x	coordinate in image space (1-D or 2-D)
(x, y)	Cartesian components of x; see list for Chapter I
u	coordinate in Fourier space (1-D or 2-D)
(u, v)	Cartesian components of $\mathbf{u}$
$\mathbf{u}$	position vector in Fourier space
$\mathbf{u}_l$	position vector of lth sample point in Fourier space
w	complex variable, (13.4)
$\mathbf{Z}_g^t$	set (of particular kind specified by t) of zeros in complex Fourier plane corresponding to 1-D function $g(\cdot)$
$\delta(\cdot)$	delta function, (6.3), see (6.18) and (7.17)
$\delta_{..}$	Kronecker delta, (9.9)
(ξ, η)	Cartesian coordinates in image space, rotated by ϕ with respect to (x, y) coordinates
$(\rho; \phi)$	cylindrical polar components of $\mathbf{u}$
Ω_s	frame enclosing (in any space) the quantity $s(\cdot)$
$\odot$	convolution operator, (7.1) and (7.5)
$*$	correlation operator, (7.2) and (7.6)

Chapter III

$a(\mathbf{x})$	preprocessed recorded image, (14.9) and (15.3)

$A(\mathbf{u})$	spectrum of $a(\mathbf{x})$, (16.3)
$B(\mathbf{u})$	spectrum of ideal blurred image, (14.2)
$\hat{f}(\mathbf{x})$	recoverable true image
$\hat{h}(\mathbf{x})$	modified point spread function
$H_{l,m}$	point sample of $H(\mathbf{u})$, (14.7)
L_1, L_2	x, y extents of a rectangular Γ
pre$\{\cdot\}$	preprocessing operator, (15.3)
R	radial extent of a circular Γ
$W(\mathbf{u})$	Wiener filter, (16.5)
Γ	recording frame
Γ_0	frame within which $f(\mathbf{x})$ can be faithfully restored
$\Phi(\mathbf{u})$	measure of noise-to-signal ratio, see (16.5)
Φ	filter constant
Ω	frame containing all parts of $f(\mathbf{x})$ spread by $h(\mathbf{x})$ into Γ, (15.2)

Chapter IV

$e(\mathbf{x})$	true image for ancillary phase problem, (22.6)
$\mathfrak{E}(\mathbf{u})$	spectrum of $e(\mathbf{x})$, (22.15)
$\tilde{f}(\mathbf{x})$	estimate of the true image
$(1/2\mathfrak{L}_1), (1/2\mathfrak{L}_2)$	spacings in u, v directions of primitive sampling grid
$\Psi(\mathbf{u})$	phase function

Chapter V

M	symbolical physical model
$p(\cdot, \cdot; \cdot)$	PET-projection, (28.16) and (28.17)
$\tilde{p}(\cdot; \cdot)$	modified projection, (33.8)
$p_{\text{comp}}(\cdot; \cdot)$	projection computed for hollow image, see (31.15)

$p_{\text{diff}}(\cdot\,;\cdot)$	difference projection, (31.15)
$p_{\text{dist}}(\cdot\,;\cdot)$	distorted projection, (31.1) and (31.2)
$p_{\text{err}}(\cdot\,;\cdot)$	error projection, (31.4)
$p_{\text{inc}}(\cdot\,;\cdot)$	incomplete projection, (30.1)
$p_{\text{SPECT}}(\cdot\,;\cdot)$	SPECT-projection, (26.37)
q	radiation intensity, (26.3), see (26.1) and (26.7)
$\hat{r}$	radius of circle circumscribing cross-section, see (25.1)
$\mathfrak{R}$	distance from radiating element, see (26.1)
$\hat{\rho}$	cut-off in Fourier space, (29.1) and (31.16)
$\Phi(\cdot,\cdot)$	detector beam (simplified detector response), (26.5) and (28.2)
$\Phi(\cdot,\cdot\,;\cdot\,;\cdot)$	detector response, (26.2)

Chapter VI

$a_m(\mathbf{x}),\ \tilde{a}_m(\mathbf{x})$	mth speckle image of $f(\mathbf{x})$, $\tilde{f}(\mathbf{x})$; (34.1), (34.9)
$A_m(\mathbf{u}),\ \tilde{A}_m(\mathbf{u})$	spectrum of $a_m(\mathbf{x})$; (34.11)
$a_m(\mathbf{x})$	mth recorded image
$C_{\text{comp}}(\mathbf{u}),\ \tilde{C}_{\text{comp}}(\mathbf{u})$	composite spectral contamination; (36.5), see (36.7)
$c_m(\mathbf{x}),\ \tilde{c}_m(\mathbf{x})$	contamination associated with $a_m(\mathbf{x})$, $\tilde{a}_m(\mathbf{x})$; see (34.1), (34.9)
$C_m(\mathbf{u}),\ \tilde{C}_m(\mathbf{u})$	spectrum of $c_m(\mathbf{x})$, $\tilde{c}_m(\mathbf{x})$; see (34.11)
D_a	scale length characterizing aberrations of imaging instrument, see (34.7)
D_0	scale length characterizing distortion introduced by propagation medium, see (34.2) and (34.7)
$f_{asa}(\mathbf{x})$	adjusted shift-and-add image, (38.3)
$f_{sa}(\mathbf{x})$	shift-and-add image, (38.2)
$f_{\text{LWH}}(\mathbf{x})$	Lynds–Worden–Harvey image, (38.7)
$h_m(\mathbf{x}),\ \tilde{h}_m(\mathbf{x})$	point spread function associated with $a_m(\mathbf{x})$, $\tilde{a}_m(\mathbf{x})$; see (34.1), (34.9)

$M, \tilde{M}$	total number of the $a_m(\mathbf{x})$, $\tilde{a}_m(\mathbf{x})$
$W_L(\mathbf{u})$	Labeyrie–Wiener filter, (36.8)
$\hat{\rho}$	diffraction limit of imaging instrument in Fourier space; see also list for Chapter V
ρ_s	reciprocal radius of seeing disc, or diffraction limit in presence of seeing (uncorrected by speckle processing)
$\langle Q_m \rangle$	average of the members of the set $\{Q_m; m = 1, 2, \ldots, \mathfrak{M}\}$ where $\mathfrak{M}$ is as large as practicable or convenient

INDEX